Principles of Horticulture

Principles of Horticulture

Principles of Horticulture

Fifth edition

C.R. Adams, K.M. Bamford and M.P. Early

AMSTERDAM • BOSTON • HEIDELBERG • LONDON
NEW YORK • OXFORD • PARIS • SAN DIEGO
SAN FRANCISCO • SINGAPORE • SYDNEY • TOKYO
Butterworth-Heinemann is an imprint of Elsevier

Butterworth-Heinemann is an imprint of Elsevier
The Boulevard, Langford Lane, Kidlington, Oxford OX5 1GB, UK
30 Corporate Drive, Suite 400, Burlington, MA 01803, USA

First published 1984
Reprinted 1985, 1987, 1988, 1990, 1991, 1992
Second edition 1993
Third edition 1998
Reprinted 1999
Fourth edition 2004
Firth edition 2008
Reprinted 2009

Notice
No responsibility is assumed by the publisher for any injury and/or damage to persons
or property as a matter of products liability, negligence or otherwise, or from any use
or operation of any methods, products, instructions or ideas contained in the material
herein. Because of rapid advances in the medical sciences, in particular, independent
verification of diagnoses and drug dosages should be made

British Library Cataloguing in Publication Data
A catalogue record for this book is available from the British Library

Library of Congress Cataloging-in-Publication Data
A catalog record for this book is available from the Library of Congress

ISBN: 978-0-7506-8694-5

For information on all Butterworth-Heinemann publications
visit our website at www.elsevierdirect.com

Printed and bound in *Slovenia*

09 10 11 12   10 9 8 7 6 5 4 3 2

Working together to grow
libraries in developing countries

www.elsevier.com | www.bookaid.org | www.sabre.org

ELSEVIER   BOOK AID
International   Sabre Foundation

Contents

Preface

By studying the principles of horticulture, one is able to learn how and why plants grow and develop. In this way, horticulturists are better able to understand the responses of the plant to various conditions, and therefore to perform their function more efficiently. They are able to *manipulate* the plant so that they achieve their own particular requirements of maximum yield and/or quality at the correct time. The text therefore introduces **the plant** in its own right, and explains how a correct naming method is vital for distinguishing one plant from another. The internal structure of the plant is studied in relation to the functions performed in order that we can understand why the plant takes it particular form. The environment of a plant contains many variable factors, all of which have their effects, and some of which can dramatically modify growth and development. It is therefore important to distinguish the effects of these factors in order to have precise control of growth. The environment which surrounds the parts of the plant above the ground includes factors such as light, day-length, temperature, carbon dioxide and oxygen, and all of these must ideally be provided in the correct proportions to achieve the type of growth and development required. The growing medium is the means of providing nutrients, water, air and usually anchorage for the plants.

In the wild, a plant will interact with other plants, often to different species and other organisms to create a balanced community. Ecology is the study of this balance. In growing plants for our own ends we have created a new type of community which creates problems – problems of competition for the environmental factors between one plant and another of the same species, between the crop plant and a weed, or between the plant and a pest or disease organism. These latter two competitive aspects create the need for **crop protection**.

It is only by identification of these competitive organisms (weeds, pests and diseases) that the horticulturist may select the correct method of control. With the larger pests there is little problem of recognition, but the smaller insects, mites, nematodes, fungi and bacteria are invisible to the naked eye and, in this situation, the grower must rely on the **symptoms** produced (type of damage). For this reason, the pests are covered under major headings of the organism, whereas the diseases are described under symptoms.

Symptoms (other than those caused by an organism) such as frost damage, herbicide damage and mineral deficiencies may be confused with pest or disease damage, and reference is made in the text to this problem. Weeds are broadly identified as perennial or annual problems. References at the end of each chapter encourage students to expand their knowledge of symptoms. In an understanding of crop protection, the **structure** and **life cycle** of the organism must be emphasized in order that specific measures, e.g. chemical control, may be used at the correct time and place to avoid complications such as phytotoxicity, resistant pest production or death of beneficial organisms. For this reason, each weed, pest and disease is described in such a way that **control** measures follow logically from an understanding of its biology. More detailed explanations of **specific** types of control, such as biological control, are contained in a separate chapter where concepts such as economic damage are discussed.

This book is not intended to be a reference source of weeds, pests and diseases; its aim is to show the *range* of these organisms in horticulture. References are given to texts which cover symptoms and life cycle stages of a wider range of organisms. Latin names of species are included in order that confusion about the varied common names may be avoided.

Growing media include soils and soil substitutes such as composts, aggregate culture and nutrient film technique. Usually the plant's water and mineral requirements are taken up from the growing medium by roots. Active roots need a supply of oxygen, and therefore the root environment must be managed to include aeration as well as to supply water and minerals. The growing medium must also provide anchorage and stability, to avoid soils that 'blow', trees that uproot in shallow soils or tall pot plants that topple in lightweight composts.

The components of the soil are described to enable satisfactory root environments to be produced and maintained where practicable. Soil conditions are modified by cultivations, irrigation, drainage and liming, while fertilizers are used to adjust the nutrient status to achieve the type of growth required.

The use of soil substitutes, and the management of plants grown in pots, troughs, peat bags and other containers where there is a restricted rooting zone, are also discussed in the final chapter.

The importance of the plant's aerial environment is given due consideration as a background to growing all plants notably their **microclimate**, its measurement and methods of modifying it. This is put in context by the inclusion of a full discussion of the **climate**, the underlying factors that drive the weather systems and the nature of local climates in the British Isles.

There has been an expansion of the **genetics** section to accommodate the need for more details especially with regard to genetic modification (GM) to reflect the interest in this topic in the industry. The changes in the classification system have been accommodated and the plant divisions revised without losing the familiar names of plant groups, such as monocotyledon, in the text. Concerns about **biodiversity** and the interest in plant conservation are addressed along with more detail on ecology and companion planting. More examples of plant adaptions have been provided and more emphasis has been given to the practical application of plant form in the leisure use of plants. The use of pesticides has been revised in the light of continued regulations about their use. More details have been included on the use of inert growing media such as **rockwool**.

Essential definitions have been picked out in tinted boxes alongside appropriate points in the text. Further details of some of the science associated with the principles of growing have been included for those who require more backgound; these topics have been identified by boxing off and tinting in grey.

The fifth edition is in full colour and has been reorganized to align closely with the syllabus of the very popular **RHS Certificate of Horticulture**. To this end, the chapters have been linked directly to the learning outcomes of the modules that cover The Plant, Horticultural Plant Health Problems, the Root Environment and Plant Nutrition. Introductions to Outdoor Food Production, Protected Cultivation, Garden Planning, Horticultural Plant Selection, Establishment and Maintenance have been expanded and a new chapter on Plant Propagation has been added. The expansion of these areas has made the essential relationship between scientific principles and horticultural practice more comprehensive with the essential extensive to help relate topics across the text.

This edition of the book continues to support not only the RHS Certificate of Horticulture and other Level Two qualifications, such as the National Certificates in Horticulture, but also provides an introduction to Level Three qualifications including the RHS Advanced Certificate and Diploma in Horticulture, Advanced National Certificates in Horticulture, National Diplomas in Horticulture and the associated Technical Certificates. The book continues to be an instructive source of information for keen gardeners, especially those studying Certificate in Gardening modules and wish to learn more of the underlying principles. Each chapter is fully supported with 'Further Reading' and self-assessment ('Check your Learning') sections.

Charles R Adams

Katherine M Bamford

Micheal P Early

Answers to 'Check Your Learning' on free website

A selection of answers to the 'Check Your Learning' sections found at the end of each chapter is available as a free download at http:/elsevierdirect.com/companions/9780750686945.

Acknowledgements

We are indebted to the following people without whom the new edition would not have been possible:

The dahlia featured on the cover is 'Western Spanish Dancer' and is with the kind permission of Aylett Nurseries Ltd.

Nick Blakemore provided the microscope photographs used on the cover and through the plant section of the new edition.

Thanks are also due to the following individuals, firms and organizations that provided photographs and tables:

Access Irrigation Limited

Agricultural Lime Producers' Association

Alison Cox

Cooper Pegler for sprayer

Dr C.C. Doncaster, Rothamsted Experimental Station

Dr P.R. Ellis, National Vegetable Research Station

Dr P. Evans, Rothamsted Experimental Station

Dr D. Govier, Rothamsted Experimental Station

Dr M. Hollings, Glasshouse Crops Research Institute

Dr M.S. Ledieu, Glasshouse Crops Research Institute

Dr E. Thomas, Rothamsted Experimental Station

Kenwick Farmhouse Nursery, Louth

KRN Houseplants

Micropropagation Services (EM) Ltd.for tissue culture photographs

Shell Chemicals

Syngenta Bioline for biological control

Soil Survey of England and Wales

Two figures illustrating weed biology and chemical weed control are reproduced after modification with permission of Drs H.A. Roberts, R.J. Chancellor and J.M. Thurston. Those illustrating the carbon and nitrogen cycles are adapted from diagrams devised by Dr E.G. Coker who also provided the photograph of the apple tree root system that he had excavated to expose the root system.

Contributions to the fifth edition were made by Chris Bird, Sparsholt College; Drs S.R. Dowbiggin and Jane Brooke, Capel Manor College; Anna Dourado; Colin Stirling, HortiCS; with essential technician support from David Carmichael and Terry Laverack.

Chapter 1   Horticulture in context

Summary

This chapter includes the following topics:

- **The nature of horticulture**
- **Manipulating plants**
- **Outdoor food production**
- **Protected culture**
- **Service horticulture**
- **Organic growing**

Figure 1.1 **Horticultural produce**

1

The nature of horticulture

Horticulture may be described as the practice of growing plants in a relatively intensive manner. This contrasts with agriculture, which, in most Western European countries, relies on a high level of machinery use over an extensive area of land, consequently involving few people in the production process. The boundary between the two is far from clear, especially when considering large-scale **outdoor production**. When vegetables, fruit and flowers are grown on a smaller scale, especially in gardens or market gardens, the difference is clearer cut and is characterized by a large labour input and the grower's use of technical manipulation of plant material. **Protected culture** is the more extreme form of this where the plants are grown under protective materials or in glasshouses.

There is a fundamental difference between production horticulture and service horticulture which is the development and upkeep of gardens and landscape for their amenity, cultural and recreational values. Increasingly horticulture can be seen to be involved with social well-being and welfare through the impact of plants for human physical and mental health. It encompasses environmental protection and conservation through large- and small-scale landscape design and management. The horticulturists involved will be engaged in plant selection, establishment and maintenance; many will be involved in aspects of garden planning such as surveying and design.

There may be some dispute about whether **countryside management** belongs within horticulture, dealing as it does with the upkeep and ecology of large semi-wild habitats. In a different way, the use of alternative materials to turf as seen on all-weather sports surfaces tests what is meant by the term horticulture.

This book concerns itself with the principles underlying the growing of plants in the following sectors of horticulture:

- **Outdoor production** of vegetables, fruit and/or flowers (see p5).
- **Protected cropping**, which enables plant material to be supplied outside its normal season and to ensure high quality, e.g. chrysanthemums, all the year round, tomatoes to a high specification over an extended season, and cucumbers from an area where the climate is not otherwise suitable. Plant propagation, providing seedlings and cuttings, serves outdoor growing as well as the glasshouse industry. Protected culture using low or walk-in polythene covered tunnels is increasingly important in the production of vegetables, salads, bedding plants and flowers.
- **Nursery stock** is concerned with the production of soil- or container-grown shrubs and trees. Young stock of fruit may also be established by this sector for sale to fruit growers: **soft fruit** (strawberries, etc.), **cane fruit** (raspberries, etc.) and **top fruit** (apples, pears, etc.).
- **Landscaping**, **garden construction** and maintenance that involve the skills of construction together with the development of planted

areas (**soft landscaping**). Closely associated with this sector is **grounds maintenance**, the maintenance of trees and woodlands (**arboriculture** and **tree surgery**), specialist features within the garden such as walls and patios (**hard landscaping**) and the use of water (**aquatic gardening**).

- **Interior landscaping** is the provision of semi-permanent plant arrangements inside conservatories, offices and many public buildings, and involves the skills of careful plant selection and maintenance.
- **Turf culture** includes decorative lawns and sports surfaces for football, cricket, golf, etc.
- **Professional gardening** covers the growing of plants in gardens including both public and private gardens and may reflect many aspects of the areas of horticulture described. It often embraces both the decorative and productive aspects of horticulture.
- **Garden centres** provide plants for sale to the public, which involves handling plants, maintaining them and providing horticultural advice. A few have some production on site, but stock is usually bought in.

The plant

There is a feature common to all the above aspects of horticulture; the grower or gardener benefits from knowing about the factors that may increase or decrease the plant's growth and development. The main aim of this book is to provide an understanding of how these factors contribute to the ideal performance of the plant in particular circumstances. In most cases this will mean optimum growth, e.g. lettuce, where a fast turnover of the crop with once over harvesting that grades out well is required. However, the aim may equally be restricted growth, as in the production of dwarf chrysanthemum pot plants. The main factors to be considered are summarized in Figure 1.2, which shows where in this book each aspect is discussed.

In all growing it is essential to have a clear idea of what is required so that all factors can be addressed to achieve the aim. This is what makes **market research** so essential in commercial horticulture; once it is known what is required in the market place then the choice of crop, cultivar, fertilizer regime, etc., can be made to produce it accurately.

It must be stressed that the incorrect functioning of any one factor may result in undesirable plant performance. It should also be understood that factors such as the soil conditions, which affect the underground parts of the plant, are just as important as those such as light, which affect the aerial parts. The nature of soil is dealt with in Chapter 17. Increasingly, plants are grown in alternatives to soil such as peat, bark, composted waste and inert materials which are reviewed in Chapter 22.

To manage plants effectively it is important to have a clear idea of what a **healthy plant** is like at all stages of its life. The appearance of abnormalities can then be identified at the earliest opportunity and

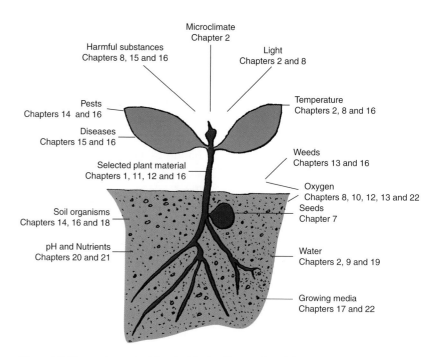

Figure 1.2 The requirements of the plant for healthy growth and development

appropriate action taken. This is straightforward for most plants, but it is essential to be aware of those which have peculiarities such as those whose healthy leaves are not normally green (variegated, purple, etc., see p82), dwarf forms, or those with contorted stems e.g. *Salix babylonica* var. *pekinensis* 'tortuosa'. The unhealthiness of plants is usually caused by pests (see Chapter 14) or disease (see Chapter 15). It should be noted that **physiological disorders** account for many of the symptoms of unhealthy growth which includes nutrient deficiencies or imbalance (see p127). Toxics in the growing medium (such as uncomposted bark, see p388) or excess of a nutrient (see p370) can present problems. Damage may also be attributable to environmental conditions such as frost, high and low temperatures, high wind (especially if laden with salt), a lack or excess of light (see p113) or water (see p122). Further details are given in Chapter 15.

Weather plays an important part in horticulture generally. It is not surprising that those involved in growing plants have such a keen interest in weather forecasting because of the direct effect of temperature, water and light on the growth of plants. Many growers will also wish to know whether the conditions are suitable for working in. Climate is dealt with in Chapter 2, which also pays particular attention to the microclimate (the environment the plant actually experiences).

A single plant growing in isolation with no competition is as unusual in horticulture as it is in nature. However, specimen plants such as leeks, marrows and potatoes, lovingly reared by enthusiasts looking for prizes in local shows, grow to enormous sizes when freed from competition. In landscaping, specimen plants are placed away from the influence of others, so that they not only stand out and act as a focal point, but also can attain perfection of form. A pot plant such as a fuchsia is isolated in its container, but the influence of other plants, and the consequent effect on its growth, depend on spacing. Generally, plants are to be found in groups, or communities (see Chapter 3).

Outdoor food production

Outdoor production of vegetables or fruit, whether on a commercial or garden scale, depends on many factors such as cultivation, propagation, timing, spacing, crop protection, harvesting and storage, but success is difficult unless the right site is selected in the first place.

Selecting a site

It is important that the plants have access to **light** to ensure good growth (see photosynthesis p113). This has a major effect on growth rate (see p110), but early harvesting of many crops is particularly desirable. This means there are advantages in growing on open sites with no overhanging trees and a southern rather than northern **aspect** (see p35).

A **free draining** soil is essential for most types of production (see drainage p343). This is not only because the plants grow better, but many of the cultural activities such as sowing, weeding and harvesting are easier to carry out at the right time (see soil consistency p342). **Earliness** and **timeliness** (p343) is also favoured by growing in light, well-drained soils which warm up quicker in the spring (see p29). Lighter soils are also easier to cultivate (see p307). For many crops, such as salads, where frequent cultivation is required the lighter soils are advantageous, but some crops such as cabbages benefit from the nature of heavier soils. In general, heavier soils are used to grow crops that do not need to be cultivated each year, such as soft fruit and top fruit in orchards, or are used for main crop production when the heavier soils are sufficiently dry to cultivate without structural damage. All horticultural soils should be well-drained unless deliberately growing 'boggy' plants.

Many tender crops, such as runner beans, tomatoes, sweet corn and the blossom of top fruit, are vulnerable to frost damage. This means the site should not be in a frost pocket (see p36). Slopes can be helpful in allowing cold air to drain off the growing area, but too steep slopes can become subject to **soil erosion** by water flow (see p298). Lighter soils, and seed, can be blown away on exposed sites (see p318).

Shelter is essential to diffuse the wind and reduce its detrimental effects. It plays an important part in extending the growing season. This can take the form of windbreaks, either natural ones such as trees or hedges or artificial ones such as webbing. Solid barriers like walls are not as effective as materials that diffuse the wind (see p37). Complete shelter is provided in the form of floating mulches, cloches, polytunnels and greenhouses (see protected culture p12).

Extending the season

Many fruits and vegetables are now regarded as commodity crops by the supermarkets and required year round. It is therefore necessary for British growers to extend the season of harvesting, within the bounds of our climate, to accommodate the market. Traditionally walled gardens provided a means to supply the 'big house' with out of season produce, but commercially this is now achieved with a range of techniques including various forms of protected cropping (see p12).

Cultural operations

Soil pH (acidity and alkalinity) levels are checked to ensure that the soil or substrate is suitable for the crop intended. If too low the appropriate amount of lime is added (see p361) or if too high sulphur can be used to acidify the soil (see p364).

Cultivations required in outdoor production depend on the plants, the site and the weather. Usually the soil is turned over, by digging or ploughing, to loosen it and to bury weeds and incorporate organic matter, then it is worked into a suitable tilth (with rakes or harrows) for seeds or to receive transplants (see p156). In many situations cultivation is supplemented or replaced by the use of rotavators (see p314). If there are layers in the soil that restrict water and root growth (see pans p312) these can be broken up with subsoilers (see p315).

Bed systems are used to avoid the problems associated with soil compaction by traffic (feet or machinery). On a garden scale, these are constructed so that all the growing area can be reached from a path so there is no need to step on it. These can be laid out in many ways, but should be no more than 1.2 metres across with the paths between minimized whilst allowing access for all activities through the growing season.

'No-dig' methods are particularly associated with organic growing (see p21). These include addition of large quantities of bulky organic matter applied to the surface to be incorporated by earthworms. This ensures the soil remains open (see p330) for good root growth as well as, usually, adding nutrients (see p376).

Freedom from weeds is fundamental to preparing land for the establishment of plants of all kinds. Whilst traditional methods involve turning over soil to bury the weeds several methods that use much less energy have become more common (see p314). Once planted the crop then has to be kept free of weeds by cultural methods or by using weed killers (see Chapter 16).

Propagation methods used for outdoor cropping include the use of seeds (p116), cuttings (p175) or grafting (p176).

Nutrient requirements are determined and are added in the form of fertilizers (see p373). They are usually applied as base dressings, top dressings, fertigation or a combination of methods (see p374).

Pest and disease control can be achieved by cultural, biological or chemical means (see Chapter 16) according to the production method adopted. This is helped by having knowledge and understanding of the causal organisms that affect the crop (Chapters 14 and 15).

Vegetable production

The choice of **cultivar** is an important decision that has to be made before growing starts. There are many possibilities for each crop, but a major consideration is the need for uniformity. Where this is important, e.g. for 'once over harvesting' or uniform size, then F1 hybrids are normally used even though they are more expensive (see p144). Required harvesting dates affect not only sowing dates but the selection of appropriate early, mid-season or late cultivars. Other factors for choice include size, shape, taste, cooking qualities, etc. Examples of carrot types to choose from are given in Table 1.1.

Table 1.1 Types of carrot shapes

Type	Features	Examples
Amsterdam	Small stumpy cylindrical roots	Amsterdam Forcing-3, Sweetheart
Autumn King	Large, late-maturing	Autumn King, 2 Vita Longa
Berlicum	Cylindrical, stumpy and late crop.	Camberly, Ingot
Chantenay	Stumpy and slightly tapered, for summer	Red Cored Supreme, Babycan
Nantes	Broader and longer	Nantes Express, Navarre, Newmarket
Paris Market	Small round or square roots, early harvest	Early French Frame, Little Finger

Most vegetables are grown in rows. This helps with many of the activities such as thinning and weed control (see p267). Seeds are often sown more thickly than is ideal for the full development of the plant; this ensures there are no gaps in the row and extra seedlings are removed before plant growth is affected. The final **plant density** depends on the crop concerned, but it is often adjusted to achieve specific market requirements, e.g. small carrots for canning require closer spacing than carrots grown for bunching. The arrangement of plants is also an important consideration in **spacing**; equidistant planting can be achieved by offsetting the rows (see Figure 1.3).

Seeds are often sown into a separate seedbed or into modular trays until they are big enough to be planted out, i.e. transplanted, into their final position. This enables the main cropped areas to be used with a minimum of wasted space. It is also a means of extending the season and speeding up plant growth by the use of greater protection and,

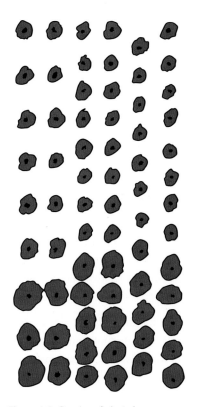

Figure 1.3 Spacing of plants in rows; offset rows to the right and mature plants to the bottom

where worthwhile, with extra heat. Larger plants are better able to overcome initial pest or disease attack in the field and also the risk of drying out.

Intercropping (the growing of one crop in between another) is uncommon in this country but worldwide is a commonly used technique for the following reasons:

- to encourage a quick growing plant in the space between slower ones in order to make best use of the space available;
- to enable one plant species to benefit from the presence of the others which provide extra nutrients e.g. legumes (see p366);
- to reduce pest and disease attacks (see also companion planting p54).

Successional cropping

Continuity of supply can be achieved by several means, most usually by the following:

- selecting cultivars with different development times (early to late cultivars);
- by using the same cultivar but planting on different dates.

These options can be combined to spread out the harvest and which can be achieved with some accuracy with knowledge of each cultivar and the use of accumulated temperature units (ATUs see p32).

Aftercare

After the crop is established, there are many activities to be undertaken according to the crop, the production method and the intended market. These operations include:

- feeding (see fertilizer, p373)
- weed control (see Chapter 13)
- irrigation (see p346)
- mulching (see p335)
- earthing up e.g. potatoes and leeks (see p46)
- pest and disease control. This is essential to ensure both the required yield and quality of produce. Examples of the important pests and diseases of vegetables are given in Chapters 14 and 15 and a survey of methods of control can be found in Chapter 16.

Harvesting

The stage of harvesting is critical depending upon the purpose of the crop. Recognizing the correct stage to sever a plant from its roots will affect its shelf life, storage or suitability for a particular market. Some vegetables which are harvested at a very immature stage are called 'baby' or 'mini'.

The method of harvesting will vary; wholesale packaging requires more protective leaf left on than a pre-packed product. Grading may take place at harvesting, e.g. lettuce, or in a packing shed after storage, e.g. onions.

Storage

An understanding of the physiology of the vegetable or plant material being stored is necessary to achieve the best possible results. Root vegetables are normally biennial and naturally prepared to be overwintered, whether in a store or outside (see p119). Annual vegetables are actively respiring at the time of picking (see p118), but with the correct temperature and humidity conditions the useful life can be extended considerably. Great care must be taken with all produce to be stored as any bruising or physical damage can become progressive in the store. Dormant vegetables can be cold stored, but care must be taken to prevent drying out. For this reason different types of store are used depending on the crop; ambient air cooling is used for most hard vegetables and refrigeration for perishable crops gives a fast pull-down of temperature and field heat (see p119).

Fruit production

Crops in the British Isles can be summarized as follows:

- **top (tree) fruit**; which in turn can be sub-divided into **pip fruit**, mainly apples and pears, and **stone fruit** (plums, cherries and peaches).
- **soft fruit** which in turn can be sub-divided into **bush fruit** (black, white and red currants; gooseberries, blueberries), **cane fruit** (raspberries, blackberries, loganberries and other hybrids; see p69) and **strawberries**.

There are many differences between vegetable and fruit growing, most of which are related to how long the crop is in the ground before replanting. Whereas most vegetables are in the soil for less than a year, fruit is in for much longer; typically strawberries last for two to three years, raspberries for eight to ten years and top fruit for some 15 to 20 years or more. Fruit plants should not be replanted in the same place (see p278).

The particular **site** requirements are as follows:

- **freedom from frost** is a major consideration (see p31) as most fruit species are vulnerable to low temperatures which damage blossom and reduce pollination (p134). Cold can also damage young tender growth which leads to less efficient leaves (p115) and russeting of fruit.
- **deep, well-drained loams** are ideal for most types of fruit growing. Unlike vegetable production, heavier soils are acceptable because the soil is not cultivated on a regular basis.
- **soil pH** should be adjusted before these long-term crops are established; most benefit from slightly acid soils (pH 6 to 6.5), but allowance should be made for the normal drop in pH over time (see p358). Blueberries and other Ericaceous fruits are the exception, requiring a pH of 4.5 to 5.5.

There are many production methods and the choice is mainly related to the space available, aftercare (such as pest and disease control) and the method of harvesting; taking fruits from large trees presents difficulties and making it easy for the public in 'pick your own' (PYO) situations is essential. Several methods lend themselves to smaller gardens, growing against walls or as hedges. These considerations greatly influence the selection of cultivar and rootstocks.

Top fruit can be grown in a natural or 'unrestricted' way in which case the size of the tree depends on the cultivar and whether it is grown as a standard, half standard or bush. Restricted forms include cordons, espalier, fan and columns (see Figure 1.4). **Rootstocks** play an important part in determining the size of top fruit trees, e.g. by grafting a cultivar with good fruiting qualities on to the roots of one with suitable dwarfing characteristics (see p177). Excess vigour, which can lead to vegetative growth (leafiness) at the expense of fruit, may be reduced by restricting nutrient and water uptake by growing in grass (see competition p46), ringing the bark (see p95) or, more rarely, root pruning. Soft and cane fruits are usually grown on their own unrestricted roots.

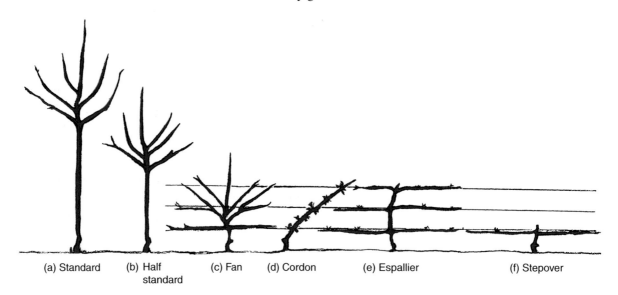

| (a) Standard | (b) Half standard | (c) Fan | (d) Cordon | (e) Espallier | (f) Stepover |

Figure 1.4 **Fruit tree forms**

Training and pruning plays an important part of the husbandry of fruit growing. The shape of trees and bushes is established in the early years ('formative pruning'). Suitable frameworks and wiring systems are set up for many of the growing systems (see Figure 1.4) and the new growth has to be tied in at appropriate times. Pruning plays a major part in maximizing flowering and fruiting, as does the bending down of branches (see p158). The shape created and maintained has a significant effect on pest and disease control; the aim is usually to have an open centre which reduces humidity around the foliage (see p159) and lets the sunlight into the centre of the tree to give a good fruit colour. Pruning is also undertaken to remove weak and diseased growth (see p159).

Fertilization of flowers is required before fruits are formed (see p137). In order for this to be successful **pollination** needs to take place (see p134).

Most top fruit is not self fertile. Therefore, another plant is needed to supply pollen and insects are required to carry it. Since successful pollination will only take place when both plants are in flower the choice of cultivars becomes limited; later flowering cultivars do not pollinate early flowering ones. Apple cultivars are placed in seven groups to help make this choice whereby selection is made from the same group (ideally) or an adjoining one. However, choice is further limited because some cultivars are incompatible with each other (p146). In particular, triploid cultivars, such as Bramley's Seedling, are unable to pollinate any other (see p146). Similar considerations apply to pears, but some plums, cherries and peaches are self fertile.

Propagation of top fruit is by grafting (see p176), raspberries by suckers (see p174), blackberries by tip layering and strawberries by runners.

Pest and disease control methods are discussed in Chapter 16. Note that **Certification Schemes** and **Plant Passports** are particularly important for plants that are propagated by vegetative means where viruses can be a significant problem. This is especially the case where they are grown for many years before renewal (see also p294).

Harvesting fruit for immediate sale or consumption must be undertaken at maturity to present the full flavour of the variety. Techniques involved in handling fruit to prevent bruising and subsequent rotting require an understanding of fruit physiology. Stone fruits, e.g. plums and cherries, are picked directly into the market container being graded at the same time because these fruits often have a very attractive bloom which is lost if handled too often. Soft fruits will not tolerate washing or excessive handling and grading is done at picking. With strawberries the stalk is not left attached, only the calyx, to prevent it sticking into an adjoining fruit and causing a rot. Machine harvesting of raspberries for the processing industry is less important now as most fruit is grown for the dessert market and is often protected during harvest by temporary, polythene covered structures known as 'Spanish Tunnels' or 'Rain Sheds'.

Storage of fruit crops requires considerable skill and technique. Pip fruits, e.g. apples and pears, must be at an exact stage of maturity for satisfactory storage. If storage is to be for a long time, e.g. the following spring, then controlled atmosphere storage is used, where the levels of CO_2 and O_2 are controlled as well as temperature and humidity.

Soft fruit crops are harvested during the summer when the ambient air temperature is high and the fruit will continue to ripen after it has been picked. It is therefore essential to lower the temperature of the fruit quickly, known as removing 'field heat'. Refrigerated storage is used, but excessively low temperatures will cause the fruit to respire even more quickly when removed from store (see p119). This causes punnets (fruit containers) to mist up and the fruit to rot more quickly. The maintenance of the fruit at a cool temperature from grower to consumer is referred to as 'cool chain marketing'.

Protected culture

Protection for plants can be in the form of simple coverings such as floating mulches, cloches or cold frames and more complex structures such as polytunnels or glasshouses.

The advantage of protection by these various methods is that to a greater or lesser extent they **modify weather** conditions, particularly wind, and so keep the environment around the plants warmer. This factor enables plants to be grown over a longer season, which is advantageous where continuity of supply, or earlier or later produce commands a premium. In leisure horticulture, the protection offered enables a wider range of plants to be kept, propagated and displayed.

The changed environment in protected cropping necessitates a careful management approach to watering (p350) and ventilation. Any plants requiring insect pollination have to be catered for (p137). Pests, diseases and weeds can also benefit from the warmer conditions and tropical species assume more importance.

Glasshouses, or conservatories, enable tender plants (see p156) to be grown all year round, especially if a source of **heat** is also available. Half hardy plants can be 'brought on' earlier and similarly plants can be grown from seed and planted out when conditions are suitable after a period of 'hardening off' (p156).

The closed environment makes it possible to maximize crop growth by using supplementary lighting, shade, and raising carbon dioxide levels (see p113).

Day length can be modified by the use of night lighting and blackouts to encourage flowering out of season (see p161). A wider range of biological control is possible within an enclosed zone (see p271). Greenhouses also allow work to continue even when the weather is unsuitable outside.

There are many designs of greenhouses, some of which are illustrated in Figure 1.5. Others are much more ornamental rather than purely functional. They range from the grand, as seen in the Botanic Gardens, to the modest in the smaller garden. Although the structures can be clear glass to the ground, there are many situations where brick is used up to bench level e.g. Alpine Houses. Many older 'vinery' style houses were substantially underground to conserve heat.

Structural materials used for glasshouses depend again on their intended purpose, but most are either aluminium and steel construction or wood (usually Western Red Cedar). Those which are for commercial production tend to be made of aluminium and steel with an emphasis on maximizing light (see p113) by increasing the height of the gutter and using larger panes of glass. Aluminium is lightweight and very suitable as glazing bars for glasshouse roofs, it is also virtually maintenance free, but does transmit heat away more than alternatives such as wood. Where more attractive structures are preferred, wood is often chosen although

Figure 1.5 **Glasshouses**

such structures are less efficient in light transmission and require more maintenance.

Cladding materials are usually glass or plastic although there are many types of plastic available. Glass has superior light transmission and heat retention. Plastics tend to be cheaper but are less durable. They have poorer light transmission when new and most deteriorate more rapidly than glass. Polycarbonate is often used in garden centres where the danger of glass overhead is considered to be too great in public areas. The biodomes at the Eden Project in Cornwall are made up of hexagonal panels made of thermoplastic ETFE cushions (see Figure 1.6).

Orientation of the glasshouse depends on the intended purpose. For many commercial glasshouses the need for winter light is the most

Figure 1.6 **Geodesic biome domes** at the Eden Project

significant consideration, this is achieved with an east–west orientation. However, the most even light distribution occurs when the house is orientated north–south which may also be the choice if several houses are in a block. For many decorative structures the orientation is subservient to other considerations.

The **siting** should ensure an open position to maximize light, but with shelter from wind. Frost pockets need to be avoided (see p36) and there should be good access which meets the needs of the intended use. Water is needed for irrigation and normally an electricity supply needs to be available.

Light availability is emphasized in the selection of structure, cladding and siting, as this is fundamental to the growth of plants (see photosynthesis p110). **Supplementary lighting** in the greenhouse is advantageous in order to add to incoming light when this is too low (see p114). More rarely, **total lighting** can be used when plants are grown with no natural light such as in growth cabinets for experimental purposes. Low level lighting to adjust day length is used to initiate flowering out of season, e.g. year round chrysanthemums, poinsettia for the Christmas market (see **photoperiodism** p160).

Careful **water management** is essential in the glasshouse where plants are excluded from rainfall. A suitable supply of water, free from toxins and pathogens (see p351), is a major consideration especially with increasing emphasis on water conservation (see p351). For many, water is supplied by hoses or watering cans with spray controlled with the use of a lance or rose. There are many systems which lend themselves to reduced manual input, and on both small and large scale automatic watering is preferred, using one or other of the following:

- overhead spraying
- low level spraying
- seep hose
- trickle or drip lines
- ebb and flow
- capillary matting or sand beds.

Water is not only used to supply plant needs directly, but also to help cool greenhouses. '**Damping down**' is the practice of hosing water on to the floor, usually in the morning, so that the evaporation that follows takes heat out of the air (see p37). This increases the humidity in the environment (see p39) which can advantageously create a good environment for plant growth. On the other hand, if done at the wrong time it can encourage some pests and diseases (see p267). Water can also be used to apply nutrients through a dilutor, either as a one-off event or at each watering occasion; this is known as 'fertigation' and enables the grower to provide the exact nutritional requirement for the plant at particular stages of its development.

Heating can be supplied by a variety of methods including paraffin, electricity, methane (mains gas), propane (bottled gas) and, less commonly now, solid fuel. Some commercial growers are installing biomass boilers and some are in a position to use waste heat from other processes. Fuel costs and environmental considerations have put increasing emphasis on reducing the need for heat (choice of plants, use of thermal screens, etc.) and reducing heat losses with **insulation** materials such as bubble wrap (with consequent reduction in light transmission).

Ventilation is essential in order to help control temperature and humidity (see p39). Air is effectively circulated by having hinged panes set in the roof and the sides (these are often louvre panes). The movement of air is often further enhanced by the use of fans.

Shading is used to reduce the incoming radiation (see p113). Although much emphasis is put on ensuring good light transmission, particularly for winter production, the high radiation levels in summer can lead to temperatures which are too high even with efficient ventilation. Traditionally, shading was achieved by applying a lime wash. This has been superseded by modern materials which are easier to remove and some even become less opaque when wet to maintain good light levels when it is raining. Most modern production units have mechanized blinds which can also help retain heat overnight. Many ornamental houses will have attractive alternatives such as external shades in natural materials.

Growing media options in protected culture are very extensive, but the choice depends on whether the plants are grown in soil, in containers on the ground or in containers on benching. **Border soils** have been used over the years, but they have many disadvantages, especially with regard to pest and disease problems and the expense of controlling this (see soil sterilization p265). A range of composts is available for those who choose to grow in containers (see p390). However, a significant proportion of commercial glasshouse production uses one of the hydroponics systems (see p394).

Pest and disease control has special considerations because the improved conditions for plants can also lead to major pest and disease outbreaks which develop quickly. If the atmosphere becomes wet, too humid or too dry even more problems can be expected. Furthermore this environment supports organisms not commonly found outdoors such as two-spotted red spider mites (see p224). Besides a range of cultural and chemical methods, the enclosed space makes it possible to use a wider range of biological controls than is possible outside (see p275).

Automatic systems to control temperature, ventilation and lighting have developed over the years to reduce the manual input (and the unsociable hours) required to manage conditions through the growing season. Some of the most exciting developments have occurred as computerized **systems** have been introduced to integrate the control of light, temperature and humidity. In order to control the conditions indoors the systems are usually linked to weather stations (see p39) to provide the required information about the current wind strength and direction, rain and light levels (see Figure 1.7). The use of the computer has made it possible for the whole environment of the glasshouse and the ancillary equipment to be fully integrated and controlled to provide the optimum growing conditions in the most efficient manner. It has also enabled more sophisticated growing regimes to be introduced.

Polytunnels provide a cheaper means of providing an enclosed protected area. They are usually constructed of steel hoops set in the ground and clad with polythene, but in some cases, such as for nursery stock, a net cover is more appropriate (see Figure 1.8). They are not usually considered to be attractive enough for consideration outside commercial production although they are often seen in garden centres.

Figure 1.7 **Glasshouse weather station**

Figure 1.8 **Net tunnel**

Walk-in tunnels offer many of the features of a greenhouse, but there are considerable drawbacks besides looks; they tend to have limited ventilation and, despite use of ultra violet inhibitors, the cladding is short lived (3–6 years). Nevertheless there have been steady improvements in design and there are many hybrids available between the basic polytunnel and the true traditional greenhouse, utilizing polycarbonate either as double or triple glazing.

Low tunnels (with wire hoops 30 to 50 cm high) are commonly used to protect rows of vegetables. These are put in place after sowing or planting; access and ventilation is gained thereafter by pulling up the sides.

Cold frames are mainly used to raise plants from seed and to harden off plants from the greenhouse ready to be planted outdoors. The simple 'light' (a pane of glass or plastic in a frame) is hinged on the base of wood or brick and propped up to provide ventilation and exposure to outdoor temperatures. The degree to which plants are exposed to the outdoor conditions is steadily increased as the time for planting out approaches. A **frameyard** is a collection of cold frames.

Cloches were originally glass cases put over individual plants for protection (cloche comes from the name of the cover used in old clocks). They are now more usually sheets of glass or plastic clipped together over individual plants, or rows of them can cover a line of vegetables (mostly superseded today by low tunnels in commercial production).

Floating mulches are lightweight coverings laid loosely over a row or bed of plants (see Figure 1.9) and held in place by stones or earth at intervals. They provide some protection against frost, speed up germination and early growth and provide a barrier against some pests.

Figure 1.9 **Fleece**; an example of a floating mulch

They take three main forms:

- **fleece**, which is a light, non-woven material (polypropylene fibre) permeable across its entire surface allowing light, air and water to penetrate freely. Humidity can be a problem as the temperatures rise.
- **perforated plastic film** is a thin gauge plastic film perforated with holes which allow it to stretch as the plants grow. High humidity is less of a problem because of the holes. Films are made with varying concentration of holes which allow for the requirements of different crops. The greater the number of holes the less the harvest date is advanced but the longer the cover can stay on the plants.
- **fine netting** does not offer the same protection from the elements, but does help keep off pest attacks.

Service horticulture

In contrast to the production of plants for food and flowers, those in service horticulture (embracing the many facets of landscaping, professional gardening and turf culture) are engaged in **plant selection**, **establishment** and **maintenance**. This will mainly involve:

- trees and shrubs;
- hedges, windbreaks and shelter belts;
- climbing plants;
- decorative annuals, biennials, perennial plants;
- ground cover;
- alpines;
- ornamental grasses and turf for lawns or sports surfaces.

Many will be involved in aspects of garden planning such as surveying and design.

Site requirements

For many aspects of this part of horticulture these will be similar to that for the production of plants, but it is much more common to find that the choice of plants is made to fit in with the site characteristics, i.e. 'go with the flow'. This is because the site (the garden, the park, the recreational area) already exists and it is often too expensive to change except on a small scale, e.g. for acid loving plants Rhododendron and Ericaceous species (see p364). The characteristics of the site need to be determined when planning their use and (as for outdoor production) this will include climate, topography, aspect, soil(s), drainage, shade, access, etc. However, there will often be more consideration given to view lines, incorporating existing features of value and accommodating utilities such as sheds, storage, maintenance and composting areas.

Design

Substantial plant knowledge is needed to help fulfil the principles of design which encompass:

- **unity** (or harmony); this is ensuring that there are strong links between the components, i.e. the individual parts of the design relating to each other. This encompasses all aspects such as continuity of materials, style or ideas (e.g. 'Japanese', 'chic' or 'rural');
- **simplicity**; to bring a sense of serenity, avoiding clutter by limiting the number of different materials used and repeating plants, colours and materials around the garden;
- **repetition** of shapes, materials, patches of colour to ensure unity, but also in order to introduce rhythm by the spacing and regularity of the repetition (see Figure 1.10);
- **focal points** are features of the garden that draw the eye, such as statues, furniture and individual plants, only one of which should be

Figure 1.10 **Show garden** illustrating unity, simplicity and repetition

noticeable at a time. These are used to create a series of set pieces for
viewing and to move the viewer through the garden;

- **scale**; plantings, materials, features, patio and path sizes should be
  in proportion with each other, e.g. only small trees are likely to look
  right in small gardens;
- **balance** can be achieved most easily by developing a symmetrical
  garden, but success with other approaches is possible by considering
  less formal ways of balancing visual components, e.g. groups of
  evergreens with deciduous trees; ponds with lawns; several small
  plants with a single shrub; open area with planted areas;
- **interest**; much of the interest is related to the selection and grouping
  of plants based on their form, colours and textures.

Decisions need to be made with regard to the overall **style** to be
achieved. The need for unity suggests that mixing styles is to be avoided
or handled with care. This is particularly true for the choice between
formal and informal approaches to the garden or landscape.

Propagation

Nursery stock growers specialize in propagating plants which are sold
on to other parts of the industry. Other parts of the industry may also
propagate their own plants. Plants can be grown from seed (see p166),
from division, layering, cuttings, micro-propagation, grafting or budding
(see vegetative propagation p172).

Sources of plants

The source depends on the type and quantity, but is usually from
specialist nurseries, garden centres or mail order, including the Internet.
Plants are supplied in the following ways:

- **Bare rooted plants** are taken from open ground in the dormant period
  (p115). Whilst cheaper, these are only available for a limited period

and need to be planted out in the autumn or spring when conditions are suitable; in practice this is mainly October and March. Roots should be kept moist until planted and covered with wet sacking while waiting. Plants received well before the time for permanent planting out should be 'heeled in' (i.e. temporary planting in a trench to cover the roots).

- **Root balled plants** are grown in open ground, but removed with soil, and the rootball is secured until used by sacking (hessian). This natural material does not need to be removed at planting and will break down in the soil. This reduces the problems associated with transplanting larger plants.
- **Containerized** plants are also grown in open ground, but transferred to containers. Care needs to be taken to ensure that the root system has established before planting out unless treated as a bare-rooted stock.
- **Container-grown** plants, in contrast, are grown in containers from the time they are young plants (rather than transferred to containers from open ground). This makes it possible to plant any time of the year when conditions are suitable. Most plants supplied in garden centres are available in this form.

It is essential that care is taken when buying plants. Besides ensuring that the best form of the plants are being purchased and correctly labelled, the plants must be healthy and 'well grown'; the plants should be compact and bushy (see etiolated p153), free from pest or disease and with appropriately coloured leaves (no signs of mineral deficiency; see p127). The roots of container plants should be examined to ensure that they are visible and white rather than brown. The contents of the container should not be rootbound and the growing medium not too wet or dry.

Establishment

The site needs to be prepared to receive the plant at the right time of the year. The soil should be cultivated to produce the appropriate structure and tilth (see p313) and base dressings of fertilizers applied. Plants should not go into the ground when it is dry, waterlogged or frozen. After sowing or planting out, care has to be taken particularly with regard to watering and weed control, also with protection from pests and diseases.

Maintenance activity is ongoing (as anyone who looks after a garden will know). There are many things to do almost every month of the year to keep the planting in good order, including:

- mowing turf
- irrigation/watering
- feeding
- hedge cutting, clipping topiary
- pruning trees and shrubs
- weeds, pest and disease control
- staking

- dead heading
- dividing perennials.

Interior plant care

Interior spaces in offices, shops, schools, etc., can be decorated and benefit from an enhanced atmosphere using mobile containers. Often carried out on contract, this work requires all the care of protected cropping with particular attention being paid to watering (often spaces are centrally heated) and lighting (plants are often pushed into an otherwise little-used dark corner). The problems of transport and associated variation in environmental conditions must also be considered.

Organic growing

Organic, or **ecological**, **growers** view their activities as an integrated whole and try to establish a sustainable way forward by conserving non-renewable resources and eliminating reliance on external inputs. Where their growing depends directly, or indirectly (e.g. the use of straw or farmyard manure), on the use of animals due consideration is given to their welfare and at all times the impact of their activities on the wider environment is given careful consideration.

The soil is managed with as little disturbance as possible to the balance of organisms present. Organic growers maintain **soil fertility** by the incorporation of animal manures (see p330), composted material (see p333), green manure or grass–clover leys (p332). The intention is to ensure plants receive a steady, balanced release of nutrients through their roots; 'feed the soil, not the plant'. Besides the release of nutrients by decomposition (see p324), the stimulated earthworm activity incorporates organic matter deep down the soil profile, improving soil structure which can eliminate the need for cultivation (see earthworms, p321).

The main cause of species imbalance is considered to be the use of many **pesticides** and **quick-release fertilizers**. Control of pests and diseases is primarily achieved by a combination of resistant cultivars (p290) and 'safe' pesticides derived from plant extracts (p282), by careful rotation of plant species (p267) and by the use of naturally occurring predators and parasites (p271). Weeds are controlled by using a range of cultural methods including mechanical and heat-producing weed control equipment (p264). The balanced nutrition of the crop is thought to induce greater resistance to pests and diseases (p60). The European Union Regulations (1991) on the 'organic production of agricultural products' specify the substances that may be used as 'plant-protection products (see Table 16.4), detergents, fertilizers, or soil conditioners' (see Table 21.3).

Those intending to sell produce with an organic label need to comply with the standards originally set by the International Federation of

Organic Agricultural Movement (IFOAM). These standards set out the principles and practices of organic systems that, within the economic constraints and technology of a particular time, promote:

- the use of management practices which sustain soil health and fertility;
- the production of high levels of nutritious food;
- minimal dependence on non-renewable forms of energy and burning of fossil food;
- the lowest practical levels of environmental pollution;
- enhancement of the landscape and wild life habitat;
- high standards of animal welfare and contentment.

Certification is organized nationally with a symbol available to those who meet and continue to meet the requirements. In the UK, the Soil Association is licensed for this purpose.

Check your learning

1. State what is meant by nursery stock production.

2. Explain why market research is advisable before starting to grow a crop.

3. Explain what is meant by a healthy plant.

4. Explain why most crops are grown in rows.

5. State the different methods of growing plants earlier in the year.

6. State the advantages a wooden structure for a glasshouse in a garden situation.

7. Explain what is meant by 'hardening off' plants and why it is necessary.

8. Explain how organic growers can maintain the fertility of their soils.

Further reading

Armitage, A.M. (1994). *Ornamental Bedding Plants*. CABI.

Baker, H. (1992). *RHS Fruit*. Mitchell Beazley.

Baker, H. (1998). *The Fruit Garden Displayed*. 9th edn. Cassell.

Beckett, K.A. (1999). *RHS Growing Under Glass*. Mitchell Beazley.

Beytes, C. (ed.) (2003). *Ball Redbook: Greenhouses and Equipment*. Vol. 1. 17th edn. Ball Publishing.

Brickell, C. (ed.) (2006). *RHS Encyclopedia of Plants and Flowers*. Dorling Kindersley.

Brickell, C. (ed.) (2003). *RHS A–Z Encyclopedia of Garden Plants*. 2 vols. 3rd edn. Dorling Kindersley.

Brookes, J. (2001). *Garden Design*. Dorling Kindersley.

Brookes, J. (2002). *Garden Masterclass*. Dorling Kindersley.

Brown, S. (2005). *Sports Turf and Amenity Grassland Management*. Crowood Press.

Caplan, B. (1992). *Complete Manual of Organic Gardening*. Headline Publishing.

Eames, A. (1994). *Commercial Bedding Plant Production*. Grower Books.

Edmonds, J. (2000). *Container Plant Manual*. Grower Books.

Fedor, J. (2001). *Organic Gardening for the 21st Century*. Frances Lincoln.

Hamrick, D. (ed.) (2003). *Ball Redbook: Crop Production.* Vol. 2. 17th edn. Ball Publishing.

Hessayon, D.G. (1993). *The Garden Expert.* Expert Publications.

Lamb, K. *et al.* (1995). *Nursery Stock Manual.* Revised edn. Grower Books.

Lampkin, N. (1990). *Organic Farming.* Farming Press.

Larcom, J. (1994). *The Vegetable Garden Displayed.* Revised edn. BT Batsford.

Larcom, J. (2002). *Grow Your Own Vegetables.* Frances Lincoln.

Mannion, A.M. and Bowlby, S.R. (eds) (1992). *Environmental Issues in the 1990s.* John Wiley & Sons.

Pears, P. and Strickland, S. (1999). *Organic Gardening.* RHS. Mitchell Beazley.

Pollock, M. (ed.) (2002). *Fruit and Vegetable Gardening.* MacMillan.

Power, P. (2007). *How to Start Your Own Gardening Business: An Insider Guide to Setting Yourself Up as a Professional Gardener.* 2nd edn. How to Books.

Staines, R. (1992). *Market Gardening.* Fulcrum Publishing.

Swithenbank, A. (2006). *The Greenhouse Gardener.* Frances Lincoln.

Thomas, H. and Wooster, S. (2008). *The Complete Planting Design Course: Plans and Styles for Every Garden.* Mitchell Beazley.

Toogood, A. (2003). *Flowers.* Harper Collins.

Williams, R. (1995). *The Garden Designer.* Frances Lincoln.

Wilson, A. (ed.) (2007). *Garden Plans.* Mitchell Beazley.

Chapter 2   Climate and microclimate

Summary

This chapter includes the following topics:

- **The sun's energy**
- **Effect of latitude**
- **Weather systems**
- **Weather and climate**
- **Climate of British Isles**
- **Growing seasons**
- **World climate**
- **Local climate**
- **Microclimate**
- **Weather instruments**

Figure 2.1 **Agrometerological station** showing rain gauge (left), Stevenson's screen (centre) and anemometer (right).

The Sun's energy

The energy that drives our weather systems comes from the sun in the form of solar radiation. The sun radiates waves of **electro-magnetic energy** and high-energy particles into space. This type of energy can pass through a vacuum and through gases. The **Earth intercepts the radiation energy** and, as these energy waves pass through the atmosphere, they are absorbed, scattered and reflected by gases, air molecules, small particles and cloud masses (see Figure 2.2).

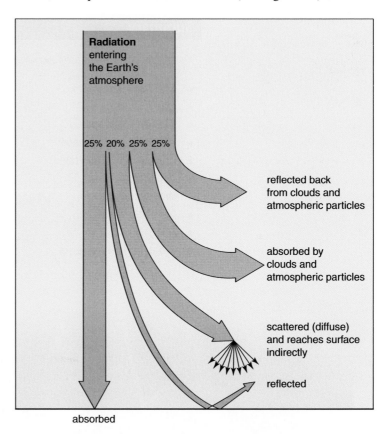

Figure 2.2 **Radiation energy** reaching the Earth's surface showing the proportions that are reflected back and absorbed as it passes through the atmosphere and that which reaches plants indirectly. About 5 per cent of the radiation strikes the Earth's surface but is reflected back (this is considerably more if the surface is light coloured, e.g. snow, and as the angle of incidence is increased).

About a quarter of the total radiation entering the atmosphere reaches the Earth's surface directly. Another 18 per cent arrives indirectly after being scattered (diffused). The surface is warmed as the molecules of rock, soil, and water at the surface become excited by the incoming radiation; the energy in the electro-magnetic waves is converted to heat energy as the surface material absorbs the radiation. A reasonable estimate of energy can be calculated from the relationship between radiation and sunshine levels. The amounts received in the British Isles are shown in Figure 2.3 where the differences between winter and summer are illustrated.

However, the nature of the surface has a significant effect on the proportion of the incoming radiation that is absorbed. The sea can

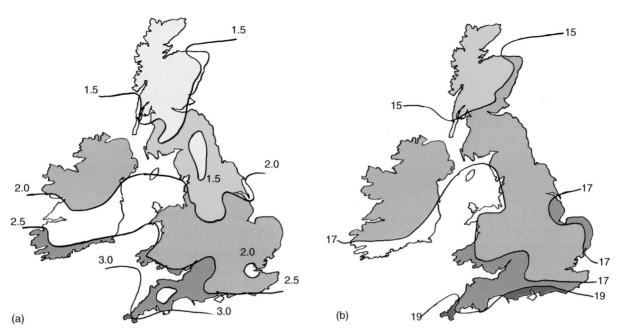

Figure 2.3 **Radiation received in the British Isles**; mean daily radiation given in megajoules per metre square. (a) January (b) July.

absorb over 90 per cent of radiation when the sun is overhead, whereas for land it is generally between 60 and 90 per cent. Across the Earth darker areas tend to absorb more energy than lighter ones; dark soils warm up more quickly than light ones; afforested areas more than lighter, bare areas with grass are between these values. Where the surface is white (ice or snow) nearly all the radiation is reflected.

Effect of latitude

Over the Earth's surface some areas become warmed more than others because of the differences in the quantity of radiation absorbed. Most energy is received around the Equator where the sun is directly overhead and the radiation hits the surface at a right angle. In higher latitudes such as the British Isles more of the radiation is lost as it travels further through the atmosphere. Furthermore, the energy waves strikes the ground at an acute angle, leading to a high proportion being reflected before affecting the molecules at the surface (see Figure 2.4).

As a consequence of the above, more energy is received than lost over the span of a year in the region either side of the Equator between the Tropic of Capricorn and Tropic of Cancer. In contrast, to the north and south of these areas more energy radiates out into space, which would lead to all parts of this region becoming very cold. However, air and water (making up the Earth's atmosphere and oceans) are able to redistribute the heat.

Movement of heat and weather systems

Heat energy moves from warmer areas (i.e. those at a higher temperature) into cooler areas (i.e. those at a lower temperature) and there are three types of energy movement involved. **Radiation** energy moves efficiently

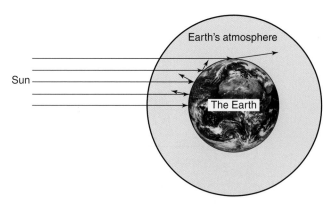

Figure 2.4 Effect of angle of incidence on heating at the Earth's surface. A higher proportion of the incoming radiation is reflected as the angle of incidence increases. Note also that a higher proportion of the incoming radiation is absorbed or reflected back as it travels longer through the atmosphere in the higher latitudes.

through air (or a vacuum), but not through water or solids. Heat is transferred from the Earth's surface to the lower layers by **conduction**. As soil surfaces warm up in the spring, temperatures in the lower layers lag behind, but this is reversed in the autumn as the surface cools and heat is conducted upwards from the warmer lower layers. At about one metre down the soil temperature tends to be the same all the year round (about 10°C in lowland Britain).

Heat generated at the Earth's surface is also available for redistribution into the atmosphere. However, air is a poor conductor of heat (which explains its usefulness in materials used for insulation such as polystyrene foam, glass fibre and wool). It means that, initially, only the air immediately in contact with the warmed surface gains energy. Although the warming of the air layers above would occur only very slowly by conduction, it is the process of **convection** that warms the atmosphere above. As fluids are warmed they expand, take up more room and become lighter. Warmed air at the surface becomes less dense than that above, so air begins to circulate with the lighter air rising, and the cooler denser air falling to take its place; just as with a convector heater warming up a room. This circulation of air is referred to as **wind**.

In contrast, the water in seas and lakes is warmed at the surface making it less dense which tends to keep it near the surface. The lower layers gain heat very slowly by conduction and generally depend on gaining heat from the surface by turbulence. Large-scale water **currents** are created by the effect of tides and the winds blowing over them.

On a global scale, the differences in temperature at the Earth's surface lead to our major **weather systems**. Convection currents occur across the world in response to the position of the hotter and colder areas and the influence of the Earth's spin (the Coriolis Effect). These global air movements, known as the trade winds, set in motion the sea currents, follow the same path but are modified as they are deflected by the continental land masses (see Figure 2.5).

Weather and climate

Weather is the manifestation of the state of the atmosphere. Plant growth and horticultural operations are affected by weather; the influence of rain and sunshine are very familiar, but other factors such as frost, wind, and humidity have important effects. It is not surprising that growers usually have a keen interest in the weather and often seek to modify its effect on their plants. Whilst most people depend on public weather forecasting, some growers are prepared to pay for extra information and others believe in making their own forecasts, especially if their locality tends

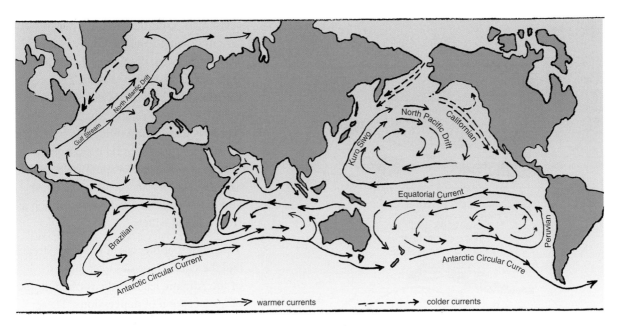

Figure 2.5 Global sea and wind movements. Warmer and colder water currents set in motion by the wind circulation around the world.

to have different weather from the rest of the forecast area. Weather forecasting is well covered in the literature and only its component parts are considered here.

Climate can be thought of as a description of the weather experienced by an area over a long period of time. More accurately, it is the long-term state of the atmosphere. Usually the descriptions apply to large areas dominated by atmosphere systems (global, countrywide or regional), but local climate reflects the influence of the topography (hills and valleys), altitude and large bodies of water (lakes and seas).

Climate of the British Isles

The British Isles has a **maritime** climate, characterized by mild winters and relatively cool summers, which is a consequence of its proximity to the sea. This is because water has a much larger **heat capacity** than materials making up the land. As a consequence, it takes more heat energy to raise the temperature of water one degree, and there is more heat energy to give up when the water cools by one degree, when compared with rock and soil. Consequently bodies of water warm up and cool down more slowly than adjoining land. The nearby sea thus prevents coastal areas becoming as cold in the winter as inland areas and also helps maintain temperatures well into the autumn.

In contrast, inland areas on the great landmasses at the same latitude have a more extreme climate, with very cold winters and hot summers; the features of a **continental climate**. Whereas most of the British Isles lowland is normally above freezing for most of the winter, average mid-winter temperatures for Moscow and Hudson Bay (both continental climate situations) are nearer −15°C.

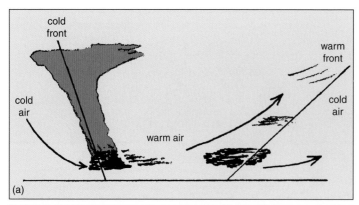

Figure 2.6 **Cloud formation and rainfall** caused by (a) fronts and (b) higher ground (orographic rain). Note: warm air caused to rise over cold air or higher ground forms cloud when the air reaches the dew point of the air mass.

The **North Atlantic Drift**, the ocean current flowing from the Gulf of Mexico towards Norway, dominates the climate of the British Isles (see Figure 2.5). The effect of the warm water, and the prevailing southwesterly winds blowing over it, is particularly influential in the winter. It creates mild conditions compared with places in similar latitudes, such as Labrador and the Russian coast well to the north of Vladivostok, which are frozen in the winter.

The mixing of this warm moist air stream and the cold air masses over the rest of the Atlantic leads to the formation of a succession of **depressions**. These regularly pass over the British Isles bringing the characteristic unsettled weather; with clouds and rain where cold air meets the moist warm air in the slowly swirling air mass. Furthermore, the moist air is also cooled as it is forced to rise over the hills to the west of the islands giving rise to **orographic rain**.

In both instances clouds form when the dew point is reached (see p41). This leads to much higher rainfall levels in the west and north compared with the south and east of the British Isles. In contrast, a **rain shadow** is created on the opposite side of the hills because, once the air has lost water vapour and falls to lower warmer levels, there is less likelihood of the dew point being reached again (see Figure 2.6). Depressions are also associated with windier weather.

The sequence of depressions (low-pressure areas) is displaced from time to time by the development of **high-pressure areas** (anti-cyclones). These usually bring periods of settled drier weather. In the summer these are associated with hotter weather with air drawn in from the hot European land mass or North Africa. In the winter, clear cold weather occurs as air is drawn in from the very cold, dry continental landmass. In the spring, these anti-cyclones often lead to radiation **frosts**, which are damaging to young plants and top fruit blossom.

The growing season

The outdoor growing season is considered to be the time when temperatures are high enough for plant growth. Temperate plants usually start growing when the daily mean temperatures are above 6°C. Spring in the southwest of the British Isles usually begins in March, but there is nearly a two-month difference between its start in this area and the northeast (see Figure 2.7).

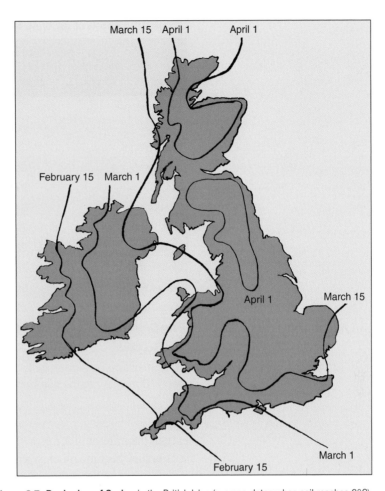

Figure 2.7 **Beginning of Spring** in the British Isles (average dates when soil reaches 6°C).

In contrast, as temperatures drop below 6°C the growing season draws to a close. This occurs in the autumn, but in the southwest of England and the west of Ireland this does not occur until December, and on the coast in those areas there can be 365 growing days per year. Within the general picture there are variations of growth periods related to altitude, aspect, frost pockets, proximity of heat stores, shelter and shade: the so-called local climates and microclimates (see p37). However, for most of mainland UK, the potential growing season spans between eight and nine months. Examples are given in Table 2.1.

Although this length of growing period will be a straightforward guide to grass growing days and the corresponding need for mowing, many other plants will stop growing as they complete their life cycle well before low temperatures affect them. Furthermore, there are plants whose growing season is defined differently. For example, most plants introduced from tropical or sub-tropical areas do not start growing until a mean daily temperature of 10°C is experienced. More significantly, they are restricted by their intolerance to cold so for many their outdoor season runs from the **last frost** of spring to the **first frost** of autumn.

Proximity to the sea not only increases the length of the growing season, but also reduces its intensity, i.e. the extent to which temperatures

Table 2.1 **Length of growing season** in the British Isles

Area	Length of growing season in days*	Time of year	
		start	finish
S-W Ireland	320	Feb 15	Jan 7
Cornwall	320	Feb 15	Jan 7
Isle of Wight	300	March 1	Jan 1
Anglesey	275	March 1	Dec 15
South Wales	270	March 15	Dec 15
East Lancs	270	March 15	Dec 1
East Kent	265	March 15	Nov 28
N. Ireland	265	March 15	Nov 28
Lincolnshire	255	March 21	Nov 25
Warwickshire	250	March 21	Nov 22
West Scotland	250	March 21	Nov 20
East Scotland	240	March 28	Nov 15
N-E Scotland	235	April 1	Nov 10

*Length of season is given for lower land in the area; reduce by 15 days for each 100 m rise into the hills (approximately 5 days per 100 feet).

exceed the minimum for growth. Although inland areas have a shorter season they become much warmer more quickly before cooling down more rapidly in the autumn. The differences in intensity can be expressed in terms of accumulated temperature units.

Accumulated temperature units (ATUs) are an attempt to relate plant growth and development to temperature and to the duration of each temperature. There is an assumption that the rate of plant growth and development increases with temperature. This is successful over the normal range of temperatures that affect most crops. On the basis that most temperate plants begin to grow at temperatures above 6°C the simplest method accredits each day with the number of degrees above the base line of 6°C and accumulates them (note that negative values are not included). A second method calculates ATUs from weather records on a monthly rather than daily basis. Examples are given in Table 2.2.

Methods such as these can be used to predict likely harvest dates from different sowing dates. Growers may also use such information to calculate the sowing date required to achieve a desired harvest date. In the production of crops for the freezing industry, it has been possible to smooth out the supply to the factory by this method. For example, a steady supply of peas over six weeks can be organized by using the local weather statistics to calculate when a range of early to late varieties of peas (i.e. with different harvest ATUs) should be sown.

More accurate methods, such as the **Ontario Units**, use day and night temperatures in the calculation. These have been used to study the growth

Table 2.2 **Examples of Accumulated Temperature Units** (ATUs) calculated on (a) a daily basis and (b) monthly basis

a) Accumulated Temperature Units (ATUs) calculated on a daily basis with a base line of 6°C

Date March	Average temp. (°C)		Temperature units in day-degrees	ATUs in day-degrees
1	6	(6 − 6 = 0)	1 × 0 = 0	0
2	7	(7 − 6 = 1)	1 × 1 = 1	0 + 1 = **1**
3	7	(7 − 6 = 1)	1 × 1 = 1	1 + 1 = **2**
4	5	(5 − 6 = '0')	1 × (0) = 0	2 + 0 = **2**
5	8	(8 − 6 = 2)	1 × 2 = 2	2 + 2 = **4**
6	7	(7 − 1 = 1)	1 × 1 = 1	4 + 1 = **5**
7	8	(8 − 6 = 2)	1 × 2 = 2	5 + 2 = **7**

b) Accumulated Temperature Units (ATUs) calculated on a monthly basis

Month	Average temperature		Temperature units in day degrees	ATUs in day-degrees
February	5	(5 − 6 = '0')	28 × 0 = 0	0
March	7	(7 − 6 = 1)	31 × 1 = 31	31
April	8	(8 − 6 = 2)	30 × 2 = 60	91
May	11	(11 − 6 = 5)	31 × 5 = 155	246
June	13	(13 − 6 = 7)	30 × 7 = 210	456
July	14	(14 − 6 = 8)	31 × 8 = 248	704

This method provides a basis for comparing the growing potential of different areas (see Table 2.3 and Figure 2.7).

Table 2.3 **Accumulated Heat Units** for different places in Europe

	Accumulated Heat Units (AHUs)		
Location	May to June	July to Sept	Total
Edinburgh	300	700	1000
Glasgow	250	650	900
Belfast	300	700	1000
Manchester	425	875	1300
Norwich	430	950	1380
Birmingham	450	900	1350
Amsterdam	480	980	1460
Swansea	450	900	1350
London	470	950	1420
Littlehampton	450	950	1400
Channel Isles	480	970	1450
Paris	550	1100	1650
Bordeaux	600	1200	1800
Marseilles	800	1500	2300

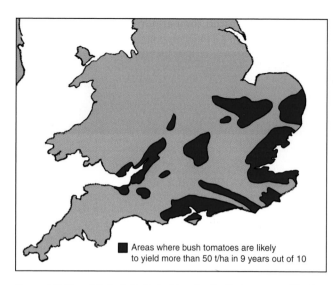

Figure 2.8 **Use of Ontario Units** to determine the likely success of bush tomato crops.

of tropical crops such as sunflowers, tomatoes and sweet corn grown in a temperate area. Using this approach the extent to which bush tomatoes could be grown in Southern England for an expected yield of 50 tonnes per hectare in nine years out of ten could be mapped (see Figure 2.8).

The accumulated heat unit concept can also be used to estimate greenhouse heating requirements by measuring the extent to which the outside temperature falls below a base or control temperature, called 'degrees of cold'. In January, a greenhouse maintained at 18°C at Littlehampton on the coast in Sussex accumulates, on average, 420 cold-degrees compared with 430 for the same structure inland at Kew, near London. For a hectare of glass this difference of 10 cold-degrees C is the equivalent of burning an extra 5000 litres of oil. This provides a useful means of assessing possible horticultural sites when other data such as solar heating and wind speed are all brought together. Other methods based on this concept enable growers to calculate when different varieties of rhubarb will start growing and the energy requirements for chill stores and refrigeration units.

World climates

In addition to **maritime** and **continental** climates already mentioned there are many others, including the **Mediterranean** climate (as found in southern parts of Europe, California, South Africa, Australia and central Chile) that is typified by hot dry summers and mild winters. The characteristics of a range of the world climate types are given in Table 2.4. Plants native to these other areas can present a challenge for those wishing to grow them in the British Isles. To some extent plants are tolerant, but care must be taken when dealing with the plant's degree of hardiness (its

Table 2.4 A summary of some of the **world climates**

| Climatic region | Temperature | | | Rainfall | |
	Range	Winter	Summer	Distribution	Total
Temperate					
maritime	narrow	mild	warm	even	moderate
continental	wide	very cold	very warm	summer max	moderate
Mediterranean	moderate	mild	hot	winter max	moderate
Sub-tropical	moderate	mild*	hot	summer max	high
Tropical maritime	narrow	warm**	hot	even	moderate
Equatorial	narrow	hot**	hot	even	high

*frosts uncommon.

**no frosts.

ability to withstand all the features in the climate to which it is exposed). Plant species that originated in **sub-tropical** areas (such as south-east China and USA) tend to be vulnerable to frosts and those from **tropical** and **equatorial** regions are most commonly associated with growing under complete protection such as in conservatories and hothouses.

Local climate

Most people will be aware that even their regional weather forecast does not do justice to the whole of the area. The local climate reflects the influence of the topography (hills and valleys), altitude and lakes and seas that modifies the general influence of the atmospheric conditions.

Coastal areas

These are subject to the moderating influence of the body of water (see p29). Water has a large heat capacity compared with other materials and this modifies the temperature of the surroundings.

Altitude

The climate of the area is affected by altitude; there is a fall in temperature with increase in height above sea level of nearly 1°C for each 100 m. The frequency of snow is an obvious manifestation of the effect. In the southwest of England there are typically only 5 days of snow falling at sea level each year whereas there are 8 days at 300 m. At higher altitudes the effect is more dramatic; in Scotland there are nearer 35 days per year at sea level, 38 days at 300 m, but 60 days at 600 m.

The colder conditions at higher altitudes have a direct effect on the growing season. On the southwest coast of England there are nearly 365 growing days per year, but this decreases by 9 days for each 30 m above sea level. In northern England and Scotland there are only about 250 growing days which are reduced by 5 days per 30 m rise, i.e. to just 200 days at 300 m (1000 feet) above sea level in northern England.

Topography

The presence of slopes modifies climate by its aspect and its effect on air drainage. **Aspect** is the combination of the slope and the direction that it faces. North-facing slopes offer plants less sunlight than a south-facing one. This is dramatically illustrated when observing the snow on opposite sides of an east–west valley (or roofs in a street), when the north facing sides are left white long after the snow has melted on the other side (see Figure 2.9); much

Figure 2.9 **Effect of aspect**; note the difference between the north-facing slope (right) with snow still lying after it has melted on the south-facing slope (left).

more radiation is intercepted by the surface on the south facing slope. Closer examination reveals considerable differences in the growth of the plants in these situations and it is quite likely that different species grow better in one situation compared with the other. Plants on such slopes experience not only different levels of light and heat, but also different water regimes; south-facing slopes can be less favourable for some plants because they are too dry.

Air drainage

Cold air tends to fall, because it is denser than warm air, and collects at the bottom of slopes such as in valleys. **Frost pockets** occur where cold air collects; plants in such areas are more likely to experience frosts than those on similar land around them. This is why orchards, where blossom is vulnerable to frost damage, are established on the slopes away from the valley floor. Cold air can also collect in hollows on the way down slopes. It can also develop as a result of barriers, such as walls and solid fences, placed across the slope (see Figure 2.10).

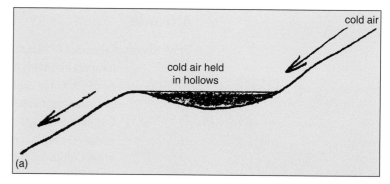

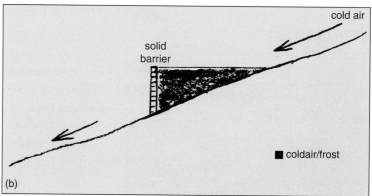

Figure 2.10 **The creation of frost pockets**: (a) natural hollows on the sides of valleys. (b) effect of solid barriers preventing the drainage of cold air.

Permeable barriers, such as trees making up shelterbelts, are less of a problem as the cold air is able to leak through. Frost susceptible plants grown where there is good **air drainage** may well experience a considerably longer growing season. Gardens on slopes can be modified to advantage by having a low-permeable hedge above (a woodland is even better) and a very permeable one on the lower boundary.

Microclimate

The features of the immediate surroundings of the plant can further modify the local climate to create the precise conditions experienced by the plant. This is known as its **microclimate**. The significant factors that affect plants include proximity to a body of water or other heat stores, shelter or exposure, shade, altitude, aspect and air drainage. The modifications for improvement, such as barriers reducing the effect of wind, or making worse, such as barriers causing frost pockets, can be natural or artificial. The microclimate can vary over very small distances. Gardeners will be familiar with the differences across their garden from the cool, shady areas to the hot, sunny positions and the implications this has in terms of the choice of plants and their management.

Growers improve the microclimate of plants when they establish windbreaks, darken the soil, wrap tender plants in straw, etc. More elaborate attempts involve the use of fleece, cold frames, cloches, polytunnels, glasshouses and conservatories. Automatically controlled, fully equipped greenhouses with irrigation, heating, ventilation fans, supplementary lighting and carbon dioxide are extreme examples of an attempt to create the ideal microclimate for plants.

Heat stores are materials such as water and brickwork, which collect heat energy and then release it to the immediate environment that would otherwise experience more severe drops in temperature. Gardeners can make good use of brick walls to extend the growing season and to grow plants that would otherwise be vulnerable to low temperatures. Water can also be used to prevent frost damage when sprayed on to fruit trees. It protects the blossom because of its **latent heat**; the energy that has to be removed from the water at 0°C to turn it to ice. This effect is considerable, and until the water on the surface has frozen, the plant tissues below are protected from freezing.

Shelter that reduces the effect of wind comes in many different forms. Plants that are grown in groups, or stands, experience different conditions from those that stand alone. As well as the self-sheltering from the effect of wind, the grouped plants also tend to retain a moister atmosphere, which can be an advantage but can also create conditions conducive to pest and disease attack. Walls, fences, hedges and the introduction of shelterbelts also moderate winds, but there are some important differences in the effect they have. The reduction in flow downwind depends on the height of the barrier although there is a smaller but significant effect on the windward side (see Figure 2.11).

The diagram also shows how turbulence can be created in the lee of the barrier, which can lead to plants being damaged by down forces. Solid materials such as brick and wooden fences create the most turbulence. In contrast, hedges and meshes filter the wind; the best effect comes from those with equal gap to solid presented to the wind. However, the introduction of a shelterbelt can bring problems if it holds back cold air to create a frost pocket (see p36).

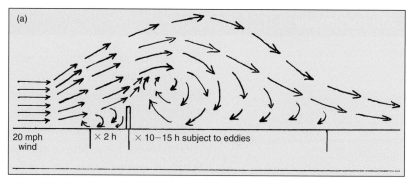

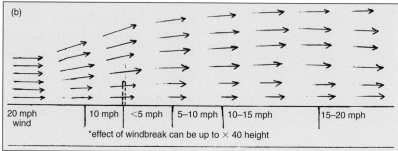

Figure 2.11 **The effect of windbreaks**: (a) solid barriers tend to create eddies to windward and, more extensively, to leeward, (b) a permeable barrier tends to filter the air and reduce its speed without setting up eddies.

Shade reduces the radiation that the plant and its surroundings receive. This tends to produce a cooler, moister environment in which some species thrive (see ecology p48). This should be taken into account when selecting plants for different positions outside in gardens. The grower will deliberately introduce shading on propagation units or on greenhouses in summer to prevent plants being exposed to high temperatures and to reduce water losses.

Plant selection

The horticulturist is always confronted with choices; plants can be selected to fit the microclimate or attempts can be made to change the microclimate to suit the plant that is desired.

Forecasting

Not only is there an interest in weather forecasting in order to plan operations such as cultivations, planting, frost protection, etc., but also for predicting pest and disease attacks, many of which are linked to factors such as temperature and humidity. Examples of outbreaks and methods of predicting them, such as **critical periods** that are used to predict potato blight.

Measurement

A range of instruments in an agro-meteorological weather station are used to measure precipitation, temperature, wind and humidity. The measurements are normally made at 09.00 Greenwich Mean Time (GMT) each day. Most of the instruments are housed in a characteristic Stevenson's Screen although there are usually other instruments on the designated ground or mounted on poles nearby (see Figure 2.1).

Temperature

The normal method of measuring the **air temperature** uses a vertically mounted mercury-in-glass thermometer that is able to read to the nearest 0.1°C. In order to obtain an accurate result, thermometers used must be protected from direct radiation i.e. the readings must be made 'in the shade'.

In meteorological stations, they are held in the **Stevenson's Screen** (see Figures 2.1 and 2.12), which is designed to ensure that accurate results are obtained at a standard distance from the ground. The screen's most obvious feature is the slatted sides, which ensure that the sun does not shine directly on to the instruments (see radiation p26) whilst allowing the free flow of air around the instruments. The whole screen is painted white to reflect radiation that, along with its insulated top and base, keeps the conditions inside similar to that of the surrounding air. In controlled environments such as glasshouses the environment is monitored by instruments held in an **aspirated screen**, which draws air across the instruments to give a more accurate indication of the surrounding conditions (see p116).

The dry bulb thermometer is paired with a wet bulb thermometer that has, around its bulb, a muslin bag kept wet with distilled water. In combination, they are used to determine the **humidity** (see below). Robust mercury-in-glass thermometers set in sleeves are also used to determine soil temperatures; temperatures both at the soil surface and at 300 mm depth are usually recorded in agro-meteorological stations.

The highest and lowest temperatures over the day (and night) are recorded on the **Max-Min (maximum and minimum) thermometers** (see Figure 2.12) mounted horizontally on the floor of the screen. The maximum thermometer is a mercury-in-glass design, but with a constriction in the narrow tube near the bulb that contains the mercury. This allows the mercury to expand as it warms up, but when temperatures fall the mercury cannot pass back into the bulb and so the highest temperature achieved can be read off ('today's high'). Shaking the contents back into the bulb resets it. The minimum thermometer contains alcohol. This expands as it warms but as it contracts to the lowest temperature ('tonight's low') a thin marker is pulled down by the retreating liquid. Because it is lightly sprung, the marker

Figure 2.12 **Stevenson's Screen** showing Max-Min Thermometer (horizontal) and Wet and Dry Thermometer (vertical).

is left behind whenever the temperature rises. Using a magnet, or tilting, to bring the marker back to the surface of the liquid in the tube, can reset the thermometer. In addition to the screen reading, there are other lowest-temperature thermometers placed at ground level giving 'over bare soil' and 'grass' temperatures (see Figure 2.1).

Precipitation

The term precipitation covers all the ways in which water reaches the ground as rain, snow and hail. It is usually measured with a **rain gauge** (see Figure 2.13).

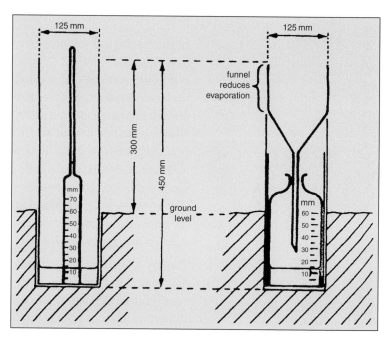

Figure 2.13 **Rain gauges**. A simple rain gauge consists of a straight-sided can in which the depth of water accumulated each day can be measured with a dipstick. An improved design incorporates a funnel, to reduce evaporation, and a calibrated collection bottle. A rain gauge should be set firmly in the soil away from overhanging trees etc. and the rim should be 300 mm above ground to prevent water flowing or bouncing in from surrounding ground.

Simple rain gauges are based on straight-sided cans set in the ground with a dipstick used to determine the depth of water collected. Accurate readings to provide daily totals are achieved with a design that maximizes collection, but minimizes evaporation losses by intercepting the precipitation water in a funnel. This leads to a tapered measuring glass calibrated to 0.1 mm. These gauges are positioned away from anything that affects the local airflow e.g. buildings, trees and shrubs. They are set in the ground but with the rim above it to prevent water running in from the surroundings.

Recording rain gauges are available which also give more details of the pattern of rainfall within twenty-four hour periods. The 'tipping bucket' type has two open containers on a see-saw mechanism so arranged that as one bucket is filled, it tips and this is recorded on a continuous chart; meanwhile the other bucket is moved into position to continue collection.

Humidity

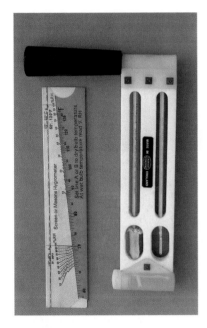

Figure 2.14 **Whirling hygrometer with calculator.**

Humidity is the amount of water vapour held in the atmosphere. In everyday language it is something described as 'close' (warm and sticky), 'dry' (little water in the air), 'damp' (cool and moist) and 'buoyant' (comfortable atmosphere). It is usually expressed more accurately in terms of relative humidity (see below).

The **whirling hygrometer** (psychrometer) is the most accurate portable instrument used for taking air measurements (see Figure 2.14). The wet and dry bulb thermometers are mounted such that they can be rotated around a shaft held in the hand, rather like a football rattle. Whilst the dry bulb gives the actual temperature of the surroundings, the wet bulb temperature is depressed by the evaporation of the water on its surface (the same cooling effect you feel when you have a wet skin). The drier the air, the greater the cooling effect, i.e. a greater wet bulb depression.

The relative humidity is calculated using hygrometric tables after the full depression of the wet bulb temperature has been found (see example below).

Hygrometers made out of hair, which lengthens as the humidity increases, are also used to indicate humidity levels. These can be connected to a pen that traces the changes on a revolving drum carrying a hygrogram chart. Other hygrometers are based on the moisture absorbing properties of different materials including the low technology 'bunch of seaweed'.

Relative Humidity

The quantity of water vapour held in the air depends on temperature, as shown below:

0°C	3 g of water per kilogram of air
10°C	7 g
20°C	14 g
30°C	26 g

The maximum figure for each temperature is known as its saturation point or **dew point**, and if such air is cooled further, then water vapour condenses into liquid water. One kilogram of saturated air at 20°C would give up 7 g of water as its temperature fell to 10°C. Indoors this is seen as 'condensation' on the coolest surfaces in the vicinity. Outdoors it happens when warm air mixes with cold air. Droplets of water form as clouds, fog and mist; dew forms on cool surfaces near the ground. If the air is holding less than the maximum amount of water it has **drying capacity** i.e. it can take up water from its surroundings.

One of the most commonly used measurements of humidity is relative humidity (RH) which is the ratio, expressed as a percentage, of the actual quantity of water vapour contained in a sample of air to the amount it could contain if saturated at the dry bulb temperature. This is usually estimated by using the wet and dry bulb temperatures in conjunction with hygrometric tables.

1. If the absolute humidity for air at 20°C (on the dry bulb) is found to be 14 g/kg this compares with the maximum of 14 g that can be held when such air is saturated. Therefore, the RH is 100 per cent.
2. It can be seen that RH falls to 25 per cent when the wet bulb depression shows only 3.5 g of water are present. This means that its drying capacity has increased (it can now take up 10.5 g of water before it becomes saturated).
3. An example of working out the relative humidity from wet and dry bulb measurements is given below in Tables 2.5 and 2.6. In example 4, when the dry bulb reading is 20°C and the depression of the wet bulb is 2.5°C then the relative humidity is 78 per cent.

Table 2.5 Calculation of relative humidity from wet and dry bulb measurements

Example	1	2	3	4	5
Dry bulb reading °C (A)	25	25	25	20	10
Wet bulb reading 26°C	21	19.5	18	22.5	6.5
Depression of wet bulb °C (B)	4	5.5	7	2.5	2.5
Relative humidity (%)*	70	60	50	78	71

*found from tables supplied with the hygrometer by reading along to the dry bulb reading row (A) then find where the column intersects the depression of the wet bulb figure (B) as shown in Table 2.6 below.

Table 2.6 Determination of relative humidity from the wet bulb depression

Depression of wet bulb from Table 2.5 in °C	Dry bulb reading (in °C)															
	10	11	12	13	14	15	16	17	18	19	20	21	22	23	24	25
0.5	94	94	94	95	95	95	95	95	95	95	96	96	96	96	96	96
1.0	88	88	89	89	90	90	90	90	91	91	91	91	92	92	92	92
1.5	82	83	83	84	84	85	85	86	86	86	87	87	88	88	88	88
2.0	76	77	78	79	79	80	81	82	82	83	83	83	83	84	84	84
2.5	71	72	73	74	74	75	76	77	77	78	78	79	80	80	80	81
3.0	65	66	68	69	70	71	71	72	72	74	74	75	76	76	77	77
3.5	60	61	62	64	65	66	67	68	69	70	70	71	72	72	73	74
4.0	54	56	57	59	60	61	63	64	65	65	66	67	68	69	69	69

Note: Example 4 from Table 2.5 illustrated in grey to show intersection of the dry bulb temperature of 20°C with the depression of the wet bulb by 2.5°C.

Wind

Wind speed is measured with an **anemometer**, which is made up of three hemispherical cups on a vertical shaft ideally set 10 metres above the ground (see Figures 1.7 and 2.1). The wind puts a greater pressure on the inside of the concave surface than on the convex one so that the shaft is spun round; the rotation is displayed on a dial usually calibrated in knots (nautical miles per hour) or metres per second. An older and still much used visual method is the **Beaufort Scale**; originally based on observations made at sea, it is used to indicate the wind forces at sea or on land (see Table 2.7).

Wind direction is indicated with a wind vane, which is often combined with an anemometer. Decorative wind vanes are a familiar sight, but the standard meteorological design comprises a pointer with a streamlined vertical plate on one end mounted so that it can rotate freely. The arrow shape points into the wind and the movements over a minute or so are averaged. The direction the wind is **coming from** is recorded as the number of degrees read clockwise from true north, i.e. a westerly wind

Table 2.7 The Beaufort Scale

Force	Description for use on land	Equivalent wind speed	
		m/sec	approx miles/hour
0	**Calm**; smoke rises vertically	0	0–1
1	**Light air**; wind direction seen by smoke drift rather than by wind vanes	2	1–3
2	**Light breeze**; wind felt on face, vane moves, leaves rustle	5	4–7
3	**Gentle breeze**; light flags lift, leaves and small twigs move	9	8–12
4	**Moderate breeze**; small branches move, dust and loose paper move	13	13–18
5	**Fresh breeze**; small leafy trees sway, crested wavelets on lakes	19	19–24
6	**Strong breeze**; large branches sway, umbrellas difficult to use	24	25–31
7	**Near gale**; whole trees move, difficult to walk against	30	32–38
8	**Gale**; small twigs break off, impedes all walking	37	39–46
9	**Strong gale**; slight structural damage	44	47–55
10	**Storm**; trees uprooted, considerable structural damage	52	55–63

is given as 270, south-easterly as 135 and a northerly one as 360 (000 is used for recording no wind).

Light

The units used when measuring the intensity of all wavelengths are watts per square metre (W/sq m) whereas lux (lumens/sq m) are used when only light in the photosynthetic range is being measured. More usually in horticulture, the **light integral** is used. The light sensors used for this measure the light received over a period of time and expressed as gram calories per square centimetre (gcals/sq m) or megajoules per square metre (MJ/sq m). These are used to calculate the irrigation need of plants in protected culture.

The usual method of measuring **sunlight** at a meteorological station is the Campbell-Stokes Sunshine Recorder (Figure 2.15), a glass sphere that focuses the sun's rays on to a sensitive card; the burnt trail indicates the periods of bright sunshine (see p26). Another approach is to use a solarimeter, which converts the incoming solar radiation to heat and then to electrical energy that can be displayed on a dial.

Automation

Increasingly instrumentation is automated and, in protected culture, linked to computers programmed to maintain the desired environment by adjusting the ventilation and boiler settings. To achieve this, the computer is informed by external instruments measuring wind speed, air temperature and humidity, and internally by those measuring CO_2 levels, ventilation settings, heating pipe temperature, air temperature and humidity.

Figure 2.15 **Campbell-Stokes Sunshine Recorder**

Check your learning

1. State the methods by which heat energy moves.

2. Explain why the British Isles is said to have a maritime climate.

3. Calculate, using the information in Table 2.2, the date of the second sowing of a variety of peas to be harvested in June, 2 days after the first sown on March 1.

4. Explain what is meant by a microclimate.

5. Determine the relative humidity of a site which has been tested with a whirling hygrometer and the dry bulb reading is 15°C and the wet bulb reading is 13°C.

Further reading

Bakker, J.C. (1995). *Greenhouse Climate Control – an Integrated Approach.* Wageningen Press.

Barry, R. *et al.* (1992). *Atmosphere, Weather and Climate.* 6th edn. Routledge.

Kamp, P.G.H. and Timmerman, G.J. (1996). *Computerised Environmental Control in Greenhouses.* IPC Plant.

Reynolds, R. (2000). *Guide to Weather.* Philips.

Taylor, J. (1996). *Weather in the Garden.* John Murray.

Chapter 3   Environment and ecology

Summary

This chapter includes the following topics:

- **The role of plants within plant communities**
- **Ecosystems**
- **Environmental factors**
- **Conservation**

Figure 3.1 A limestone valley

Plant communities

Plant communities can be viewed from the natural wild habitat to the more ordered situation in horticulture. Neighbouring plants can have a significant effect on each other, since there is competition for factors such as root space, nutrient supply and light. In the natural wild habitat, competition is usually between different species. In subsistence horticulture in the tropics, different crops are often inter-planted (see also companion planting, p54). In Western Europe, crops are usually planted as single species communities.

Single species communities

When a plant community is made up of one species it is referred to as a **monoculture**. Most fields of vegetables such as carrots have a single species in them. On a football field there may be only ryegrass (*Lolium*) with all plants a few millimetres apart. Each plant species, whether growing in the wild or in the garden, may be considered in terms of its own characteristic spacing distance (or plant density). For example, in a decorative border, the bedding plant *Alyssum* will be planted at 15 cm intervals while the larger *Pelargonium* will require 45 cm between plants. For decorative effect, larger plants are normally placed towards the back of the border and at a wider spacing.

In a field of potatoes, the plant spacing will be closer within the row (40 cm) than between the rows (70 cm) so that suitable soil ridges can be produced to encourage tuber production, and machinery can pass unhindered along the row. In nursery stock production, small trees are often planted in a square formation with a spacing ideal for the plant species, e.g. the conifer *Chamaecyparis* at 1.5 metres. The recent trend in producing commercial top fruit, e.g. apples, is towards small trees (using dwarf rootstocks) in order to produce manageable plants with easily harvested fruit. This has resulted in spacing reduced from 6 to 4 metres.

Too much competition for soil space by the roots of adjacent plants, or for light by their leaves, would quickly lead to reduced growth. Three ways of overcoming this problem may be seen in the horticulturist's activities of transplanting seedlings from trays into pots, increasing the spacing of pot plants in greenhouses, and hoeing out a proportion of young vegetable seedlings from a densely sown row. An interesting horticultural practice, which reduces root competition, is the **deep-bed** system, in which a one metre

Figure 3.2 Regular spacing in pot plants

depth of well-structured and fertilized soil enables deep root penetration. However, growers often deliberately grow plants closer to restrict growth in order to produce the correct size and the desired uniformity, as in the growing of carrots for the processing companies.

Whilst spacing is a vital aspect of plant growth, it should be realized that the grower might need to adjust the physical environment in one of many other specific ways in order to favour a chosen plant species. This may involve the selection of the correct light intensity; a rose, for example, whether in the garden, greenhouse or conservatory, will respond best to high light levels, whereas a fern will grow better in low light.

Another factor may be the artificial alteration of day length, as in the use of 'black-outs' and cyclic lighting in the commercial production of chrysanthemums to induce flowering. Correct soil acidity (pH) is a vital aspect of good growing; heathers prefer high acidity, whilst saxifrages grow more actively in non-acid (alkaline) soils. Soil texture, e.g. on golf greens, may need to be adjusted to a loamy sand type at the time of green preparation in order to reduce compaction and maintain drainage.

Each species of plant has particular requirements, and it requires the skill of the horticulturist to bring all these together. In greenhouse production, sophisticated control equipment may monitor air and root-medium conditions every few minutes, in order to provide the ideal day and night requirements.

This aspect of single species communities emphasizes the great contrast between production horticulture and the mixed plantings in ornamental horticulture. This inter-species competition is even more marked in the natural habitat of a broad-leaved temperate woodland habitat and reaches its greatest diversity in tropical lowland forests where as many as 200 tree species may be found in one hectare.

Plant species as plant communities

The subject of ecology deals with the inter-relationship of plant (and animal) species and their environment. Below are described some of the ecological concepts which most commonly apply to the natural environment, where human interference is minimal. It will be seen, however, that such concepts also have relevance to horticulture in spite of its more controlled environment.

Firstly, the structure, physiology and life cycle properties of a plant species should be seen as closely related to its position within a habitat, giving it a competitive advantage. Small short-lived ephemerals such as groundsel (*Senecio*) with its rapid seed production and low dormancy are able to achieve a speedy colonization of bare ground. The spreading perennial, bramble (*Rubus*) has thorns that ease its climbing habit over other plant species and a tolerance of low light conditions that assumes greater importance as tree species grow above it. Woody species such as oak (*Quercus*) quickly create a well-developed root system that supplies the water and minerals for their dominance of the habitat.

Aquatic species such as pondweed (*Potamogeton* spp) often have air spaces in their roots to aid oxygen and carbon dioxide diffusion.

Ecology terms

For a marsh willow-herb (*Epilobium palustre*), its only habitat is in slightly acidic ponds. In contrast, a species such as a blackberry (*Rubus fruticosus*) may be found in more than one habitat, e.g. heath land, woodland and in hedges. The common rat (*Rattus norvegicus*), often associated with humans, is also seen in various habitats (e.g. farms, sewers, hedgerows and food stores).

Within the term 'habitat', distinction can be drawn between closed plant communities and open plant communities.

These two terms can only be used in a relative way because the radiation from the sun, the gases in the atmosphere, and migrant species prevent a true closed system being established within the natural environment.

A couple of general points may be added about the vegetation of the British Isles. The British Isles at present has about 1700 plant species. Fossil and pollen evidence suggests that before the Ice Age there was a much larger number of plant species, possibly comparable to the 4000 species now seen in Italy (a country with a similar land area to that of Britain). In Neolithic times, when humans began occupying this area, most of Britain was covered by mixed oak forest. Since that time progressive clearing of most of the land has occurred, especially below the altitude limits for cattle (450 m) and for crops (250 m).

Plant associations

In natural habitats, it is seen that a number of plant species (and associated animals) are grouped together, and that away from this habitat they are not commonly found. Two habitat examples can be given. In south-east Britain, in a low rainfall, chalk grassland habitat there will often be greater knapweed (*Centaurea scabiosa*), salad burnet (*Poterium sanguisorba*) and bee orchid (*Ophrys apifera*). In the very different high rainfall, acid bogs of northern Britain, cotton grass (*Eriophorum vaginatum*), bog myrtle (*Myrtus gale*) and sundew (*Drosera anglica*) are commonly found together. Other habitat species such as bluebell (in dense broadleaved woodland), bilberry (in dry acid moor), mossy saxifrage (in wet north-facing cliffs), broom (in dry acid soils) and water violet (in wet calcareous soils) can be mentioned. It should be noted that successful weeds such as chickweed are not habitat-restricted (see Chapter 13) in this way.

Niche

The role of a species within its habitat.

For a *Sphagnum* moss, its niche would be as a dominant species within an acid bog. The term 'niche' carries with it an idea of the specialization that a species may exhibit within a community of other plants and animals. A niche involves, for plants, such factors as temperature, light intensity, humidity, pH, nutrient levels, etc. For animals such as pests

Habitat

The area or environment where an organism or community normally lives or occurs.

A **closed plant community** is one that receives only minimal contact with outside organisms (and also materials). A small isolated island community in the middle of a large lake would be one example.

An **open plant community** is one that receives continuous exposure to other organisms (and materials) from outside. A marine shoreline community would be one example.

'Semi natural vegetation' is vegetation not planted by humans, but influenced by human actions such as grazing, logging and other types of disruption of the plant communities. The vegetation described as 'semi natural' has been able to recover to such an extent that wild species composition and ecological processes are similar to those found in an undisturbed state.

and their predators, there are also factors such as preferred food and chosen time of activity determining the niche. The niche of an aphid is as a remover of phloem sugars from its host plant.

The term is sometimes hard to apply in an exact way, since each species shows a certain tolerance of the factors mentioned above, but it is useful in emphasizing specialization within a habitat. The biologist, Gause, showed that no two species can exist together if they occupy the same niche. One species will, sooner or later, start to dominate.

For the horticulturalist, the important concept here is that for each species planted in the ground, there is an ideal combination of factors to be considered if the plant is to grow well. Although this concept is an important one, it cannot be taken to an extreme. Most plants tolerate a range of conditions, but the closer the grower gets to the ideal, the more likely they are to establish a healthy plant.

Biome

A major regional or global community of organisms, such as a grassland or desert, characterized by the dominant forms of plant life and the prevailing climate.

This term refers to a wider grouping of organisms than that of a habitat. As with the term habitat, the term 'biome' is biological in emphasis, concentrating on the species present. This is in contrast to the broader ecosystem concept described below. Commonly recognized biomes would be 'temperate woodland', 'tropical rainforest', 'desert', 'alpine' and 'steppe'. About 35 types of biome are recognized worldwide, the classification being based largely on climate, on whether they are land- or water-based, on geology and soil, and on altitude above sea level. Each example of a biome will have within it many habitats. Different biomes may be characterized by markedly different potential for annual growth. For example, a square metre of temperate forest biome may produce ten times the growth of an alpine biome.

Ecosystems

This term brings emphasis to both the community of living organisms and to their non-living environment. Examples of an ecosystem are a wood, a meadow, a chalk hillside, a shoreline and a pond. Implicit within this term (unlike the terms habitat, niche, and biome) is the idea of a whole integrated system, involving both the living (**biotic**) plant and animal species, and the non-living (**abiotic**) units such as soil and climate, all reacting together within the ecosystem.

Ecosystems can be described in terms of their energy flow, showing how much light is stored (or lost) within the system as plant products such as starch (in the plant) or as organic matter (in the soil). Several other systems such as carbon, nitrogen, and sulphur cycles and water conservation may also be presented as features of the ecosystem in question (see page 324).

A **dominant species** is a species within a given habitat that exerts its influence on other species to the greatest extent and is usually the largest species member. In mixed oak woodland, the oak is the dominant plant species.

Ecosystem

An ecological community, together with its environment, functioning as a unit.

The importance of plants as energy producers

Energy perspectives are relevant to the ecosystem concept mentioned above. The process of photosynthesis enables a plant to retain, as chemical energy, approximately 1 per cent of the sun's radiant energy falling on the particular leaf's surface. As the plant is consumed by primary consumers, approximately 90 per cent of the leaf energy is lost from the biomass (see p54), either by respiration in the primary consumer, by heat radiation from the primary consumer's body or as dead organic matter excreted by the primary consumer. This organic matter, when incorporated in the soil, remains usefully within the ecosystem.

The relative levels of the total biomass as against the total organic matter in an ecosystem are an important feature. This balance can be markedly affected by physical factors such as soil type, by climatic factors such as temperature, rainfall, and humidity, and also influenced by the management system operating in that ecosystem. For example, a temperate woodland on 'heavy' soil with 750 mm annual rainfall will maintain a relatively large soil organic matter content, permitting good nutrient retention, good water retention and resisting soil erosion even under extreme weather conditions. For these reasons, the ecosystem is seen to be relatively stable. On the other hand, a tropical forest on a sandy soil with 3000 mm rainfall will have a much smaller soil organic matter reserve, with most of its carbon compounds being used in the living plants and animal tissues. As a consequence, nutrient and moisture retention and resistance to soil erosion are usually low; serious habitat loss can result when wind damage or human interference occurs. For temperate horticulturists, the main lesson to keep in mind is that high levels of soil organic matter are usually highly desirable, especially in sandy soils that readily lose organic matter.

Figure 3.3 **Mature woodland**

Succession

Communities of plants and animals change with time. Within the same habitat, the species composition will change, as will the number of individuals within each species. This process of change is known as 'succession'. Two types of succession are recognized.

- **Primary** succession is seen in a situation of uncolonized rock or exposed subsoil. Sand dunes, disused quarries and landslide locations are examples. Primary succession runs in parallel with the development of soils (see p300) or peat (see p328). It can be seen that plant and animal species from outside the new habitat will be the ones involved in colonization.

The term '**sere**' is often used instead of 'succession' when referring to a particular habitat. **Lithosere** refers to a succession beginning with uncolonized rock, **psammosere** to one beginning with sand (often in the form of sand dunes).

- **Secondary** succession is seen where a bare habitat is formed after vegetation has been burnt, or chopped down, or covered over with flood silt deposit. In this situation, there will often be plant seeds and animals which survive under the barren surface, which are able to begin colonization again by bringing topsoil, or at least some of its associated beneficial bacteria and other micro-organisms, to the surface. This kind of succession is the more common type in the British Isles. A **hydrosere** refers to succession occurring in a fresh water lake.

Influences on succession can come in two ways. '**Allogenic succession**' occurs when the stimulus for species change is an external one. For example, a habitat may have occasional flooding (or visits from grazing animals) which influence species change. In contrast, '**autogenic succession**' occurs when the stimulus for change is an internal one. For example, a gradual change in pH (or increased levels of organic matter) may lead to the species change.

Stages in succession

Referring now to secondary succession, there is commonly observed a characteristic sequence of plant types as a succession proceeds. The first species to establish are aptly called the '**pioneer** community'. In felled woodland, these may well be mosses, lichens, ferns and fungi. In contrast, a drained pond will probably have *Sphagnum* moss, reeds and rushes, which are adapted to the wetter habitat.

The second succession stage will see plants such as grasses, foxgloves and willow herb taking over in the ex-woodland area. Grasses and sedges are the most common examples seen in the drained pond. Such early colonizing species are sometimes referred to as **opportunistic**. They often show similar characteristics to horticultural weeds (see p184), having an extended seed germination period, rapid plant establishment, short time to maturity, and considerable seed production. They quickly cover over the previously bare ground.

The third succession stage involves larger plants, which, over a period of about five years, gradually reduce the opportunists' dominance. Honeysuckle, elder and bramble are often species that appear in ex-woodland, whilst willows and alder occupy a similar position in the drained pond. The term '**competitive**' is applied to such species.

The fourth stage introduces tree species that have the potential to achieve considerable heights. It may well happen that both the ex-woodland and the drained pond situation have the same tree species such as birch, oak and beech. These are described as **climax** species, and will dominate the habitat for a long time, so long as it remains undisturbed by natural or human forces. Within the climax community there often remain some specimens of the preceding succession stages, but they are now held in check by the ever-larger trees.

> A **climax association** is the community of plant species (and animals) present in a habitat at the end of the succession process.

This short discussion of succession has emphasized the plant members of the community. As succession progresses along the four stages described, there is usually an increase in **biodiversity** (an increase in numbers of plant species). It should also be borne in mind that for every plant species there will be several animal species dependent on it for food, and thus succession brings biodiversity in the plant, animal, fungal and bacterial realms. Not only is there an increase in species numbers in climax associations, but the food webs described below are also more complex, including important rotting organisms such as fungi which break down ageing and fallen trees.

Food chains

Charles Darwin is said to have told a story about a village with a large number of old ladies. This village produced higher yields of hay than the nearby villages. Darwin reasoned that the old ladies kept more cats than other people and that these cats caught more field mice which were important predators of wild bees. Since these bees were essential for the pollination of red clover (and clover improved the yield of hay), Darwin concluded that food chains were the answer to the superior hay yield. He was also highlighting the fact that inter-relationships between plants and animals can be quite complicated.

At any one time in a habitat, there will be a combination of animals associated with the plant community. A first example is a commercial crop, the strawberry, where the situation is relatively simple. The strawberry is the main source of energy for the other organisms, and is referred to, along with any weeds present, as the **primary producer** in that habitat. Any pest (e.g. aphid) or disease (e.g. mildew) feeding on the strawberries is termed a **primary consumer**, whilst a ladybird eating the aphid is called a **secondary consumer**. A habitat may include also tertiary and even quaternary consumers.

Food chains and webs

Any combination of species such as the above is referred to as a **food chain** and each stage within a food chain is called a **trophic level**. In the strawberry, this could be represented as:

$$\text{strawberry} \rightarrow \text{aphids} \rightarrow \text{ladybird}$$

In the soil, the following food chain might occur:

$$\textit{Primula}\ \text{root} \rightarrow \text{vine weevil} \rightarrow \text{predatory beetle}$$

In the pond habitat, a food chain could be:

$$\text{green algae} \rightarrow \textit{Daphnia}\ \text{crustacean} \rightarrow \text{minnow fish.}$$

Within any production horticulture crop, there will be comparable food chains to the ones described above. It is normally observed that in a monoculture such as strawberry, there will be a relatively short period of time (up to 5 years) for a complex food chain to develop (involving several species within each trophic level). However, in a long-term stable habitat, such as oak wood land or a mature garden growing perennials, there will be many plant species (primary producers), allowing many food chains to occur simultaneously. Furthermore, primary consumer species, e.g. caterpillars and pigeons, may be eating from several different plant types, whilst secondary consumers such as predatory beetles and tits will be devouring a range of primary consumers on several plant species. In this way, a more complex, interconnected community is developed, called a **food web** (see Figure 3.4).

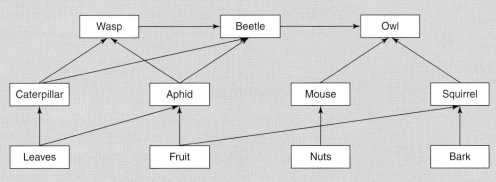

Figure 3.4 **Food web** in a woodland habitat

Decomposers

At this point, the whole group of organisms involved in the recycling of dead organic matter (called decomposers or detritovores) should be mentioned in relation to the food-web concept. The organic matter (see also Chapter 18) derived from dead plants and animals of all kinds is digested by a succession of species: large animals by crows, large trees by bracket fungi, small insects by ants, roots and fallen leaves by earthworms, mammal and bird faeces by dung beetles, etc. Subsequently, progressively smaller organic particles are consumed by millipedes, springtails, mites, nematodes, fungi and bacteria, to eventually create the organic molecules of humus that are so vital a source of nutrients, and a means of soil stability in most plant growth situations. It can thus be seen that although decomposers do not normally link directly to the food web they are often eaten by secondary consumers. They also are extremely important in supplying inorganic nutrients to the primary producer plant community.

Biomass

At any one time in a habitat, the amounts of living plant and animal tissue (**biomass**) can be measured or estimated. In production horticulture, it is clearly desirable to have as close to 100 per cent of this biomass in the form of the primary producer (crop), with as little primary consumer (pest or disease) as possible present. On the other hand, in a natural woodland habitat, the primary producer would represent approximately 85 per cent of the biomass, the primary consumer 3 per cent, the secondary consumer 0.1 per cent and the decomposers 12 per cent. This weight relationship between different trophic levels in a habitat (particularly the first three) is often summarized in graphical form as the 'pyramid of species'.

Countryside management utilizes these succession and food-web principles when attempting to strike a balance between the production of species diversity and the maintenance of an acceptably orderly managed area.

Succession to the climax stage is often quite rapid, occurring within 20 years from the occurrence of the bare habitat. Once established, a climax community of plants and animals in a natural habitat will usually remain quite stable for many years.

Garden considerations

When contemplating the distribution of our favourite species in the garden (ranging from tiny annuals to large trees), a thought may be given to their position in the succession process back in the natural habitat of their country of origin (see also p73). Some will be species commonly seen to colonize bare habitats. Most garden species will fall into the middle stages of succession. A few, whether they are trees, climbers, or low-light-requirement annuals or perennials, will be species of the climax succession. The garden contains plant species which compete in their native habitat. The artificial inter-planting of such species from different parts of the world (the situation found in almost all gardens), may give rise to unexpected results as this competition continues year after year. Such experiences are part of the joys, and the heartaches of gardening.

Companion planting

An increasingly common practice in some areas of horticulture (usually in small-scale situations) is the deliberate establishment of two or more plant species in close proximity, with the intention of deriving some cultural benefit from their association. Such a situation may seem at first sight to encourage competition rather than mutual benefit. Supporters of companion planting reply that plant and animal species in the natural world show more evidence of mutual cooperation than of competition.

Some experimental results have given support to the practice, but most evidence remains anecdotal. It should be stated, however, that whilst most commercial horticulturalists producers in Western Europe grow blocks of a single species, in many other parts of the world two or three different species are inter-planted as a regular practice.

Biological mechanisms are quoted in support of companion planting:

- **Nitrogen fixation**. Legumes such as beans convert atmospheric nitrogen to useful plant nitrogenous substances (see p366) by means of *Rhizobium* bacteria in their root nodules. Beans inter-planted with maize are claimed to improve maize growth by increasing its nitrogen uptake.
- **Pest suppression**. Some plant species are claimed to deter pests and diseases. Onions, sage, and rosemary release chemicals that mask the carrot crop's odour, thus deterring the most serious pest (carrot fly) from infesting the carrot crop. African marigolds (*Tagetes*) deter glasshouse whitefly and soil-borne nematodes by means of the chemical thiophene. Wormwood (*Artemisia*) releases methyl jasmonate as vapour that reduces caterpillar feeding, and stimulates plants to resist diseases such as rusts. Chives and garlic reduce aphid attacks.
- **Beneficial habitats**. Some plant species present a useful refuge for beneficial insects (see p271) such as ladybirds, lacewings and hoverflies. In this way, companion planting may preserve a sufficient level of these predators and parasites to effectively counter pest infestations. The following examples may be given: Carrots attract lacewings; yarrow (*Achillea*), ladybirds; goldenrod (*Solidago*), small parasitic wasps; poached-egg plant (*Limnanthes*), hoverflies. In addition, some plant species can be considered as traps for important pests. Aphids are attracted to nasturtiums, flea beetles to radishes, thus keeping the pests away from a plant such as cabbage.
- **Spacial aspects**. A pest or disease specific to a plant species will spread more slowly if the distance between individual plants is increased. Companion planting achieves this goal. For example, potatoes inter-planted with cabbages will be less likely to suffer from potato blight disease. The cabbages similarly would be less likely to be attacked by cabbage aphid.

Environmental factors and plant growth

Environmental stresses

Having dealt with the processes occurring in the natural habitat and in horticulture, it remains to mention some of the factors working against a diverse habitat. The main stresses to ecosystems in Britain and other parts of Western Europe are acidity, excess nutrients, high water tables and heavy metals.

Research suggests that in the last 25 years in Holland, for example, there has been a 25 per cent decrease in woodland species attributable to the increased acidity in the air. The same survey indicated species losses of 50 per cent, 6 per cent and 5 per cent in lakes due to excess nutrients, high water table and air acidity respectively.

Figure 3.5 Chimney

Plants are resilient organisms, but stresses imposed on habitats such as those near large towns, those in the wind-path of polluted air, those watered by rivers receiving industrial effluent, and agricultural fertilizers and pesticides are likely to lose indigenous species. The rapid increase in annual temperatures attributed to 'the greenhouse effect' is likely to change wild plant communities in as important a way as the environmental stresses mentioned above.

The effects of specific abiotic factors (pollutants) on plants

Acidity. Continuing increase in soil acidity reduces vital mycorrhizal activity, causes leaching of nutrients such as magnesium and calcium, and leaves phosphate in an insoluble form. In soils formed over limestone and chalk, the effects of acid rain are much less damaging.

Excess nutrient levels in water and soils (especially from fertilizers and farm silage) encourage increase in algae and a corresponding loss of dissolved oxygen. This process (called **eutrophication**) has a serious effect on plant and animal survival. It is seen most strikingly when fish in rivers and lakes are killed in this way.

Heavy metals may be released into the air or into rivers as by-products of chemical industries and the burning of fossil fuels. Cadmium, lead and mercury are three commonly discharged elements. While plants are more tolerant of these substances than animals, there is a slow increase within the plant cells, and more importantly the levels of chemicals increase dramatically as the plants are eaten and the chemicals move up the food chains (see also DDT p58).

Pesticides. Recent legislation has led to a greater awareness of pesticide effects on the environment. However, herbicide and insecticide leaching through sandy soils into watercourses continues to be a threat if application of the chemicals occurs near watercourses. A herbicide such as MCPA can kill algae, aquatic plants and fish.

A **high water table** can have a marked effect on a habitat if the effect is prolonged. The anaerobic conditions produced often lead to root death in all but the aquatic species present in the plant community.

Monitoring abiotic factors

There is constant monitoring by Government agencies for the factors mentioned above. This is especially so in National Parks, National Nature Reserves (**NNRs**), and Sites of Special Scientific Interest (**SSSIs**). Environmental scientists and laboratories have a range of techniques for assessing levels of these factors. Chemical tests for common nutrient substances and for pH can be performed in the field. More sophisticated analysis is required for heavy metals and pesticides. A common five day test for water quality called 'Biological Oxygen Demand' (**BOD**) enables a confident assessment of a water sample's pollution level. An unpolluted sample would register at about 3 mg of oxygen per unit volume per day whereas a sample polluted by fertilizers could be about 50 mg of oxygen per unit volume per day.

Plant modifications to extreme conditions (see also p80)

A survey of plants worldwide shows what impressive structural and physiological modifications they possess to survive in demanding habitats. A few examples of British species are described here.

A **xerophyte** is a plant adapted to living in a dry arid habitat.

Marram grass (*Ammophila arenaria*) living on sand dunes controls water-loss by means of leaf lamina which in cross-section is shown to be rolled up. It also possesses extremely long roots (see page 78).

A **hydrophyte** is a plant adapted to growing in water.

The yellow water lily (*Nupar lutea*) shows the following modifications: leaves with a thin cuticle (but with numerous stomata), large flat leaves and a stem with air sacs.

A **halophyte** is a plant adapted to living in a saline environment.

The coastal habitats provide many examples of species which must conserve water in salty, windy coastal conditions. Glasswort (*Salicornia stricta*) is a succulent with greatly reduced leaves which have a thick cuticle, and its stomata remain closed most of the day. The species is able to extract water from the seawater and is tolerant to internal salt concentrations that would kill most plant species (see plasmolysis, p123). An interesting halophyte is the sugar beet plant which was bred in France from a native coastal species. It is the only crop grown in Europe that receives salt (sodium chloride) as part of its fertilizer requirements.

Conservation

From the content of preceding paragraphs it can be seen that the provision of as extensive a system of varied habitats, each with its complex foodweb, in as many locations as possible, is increasingly being considered desirable in a nation's environment provision. In this way, a wide variety of species numbers (**biodiversity**) is maintained, habitats are more attractive and species of potential use to mankind are preserved. In addition, a society that bequeaths its natural habitats and ecosystems to future generations in an acceptably varied, useful and pleasant condition is contributing to the **sustainable development** of that nation.

The ecological aspects of natural habitats and horticulture have been highlighted in recent years by the **conservation movement**. One aim is to promote the growing of crops and maintain wildlife areas in such a way that the natural diversity of wild species of both plants and animals is maintained alongside crop production, with a minimum input of fertilizers and pesticides. Major public concern has focused on the effects of intensive production (monoculture) and the indiscriminate use by horticulturists and farmers of pesticides and quick-release fertilizers.

An example of wildlife conservation is the conversion of an area of regularly mown and 'weedkilled' grass into a wild flower meadow, providing an attractive display during several months of the year. The conversion of productive land into a wild flower meadow requires lowered soil fertility (in order to favour wild species establishment and competition), a choice of grass seed species with low opportunistic properties and a mixture of selected wild flower seed. The maintenance of the wild flower meadow may involve harvesting the area in July, having allowed time for natural flower seed dispersal. After a few years, butterflies and other insects become established as part of the wild flower habitat.

The horticulturist has three notable aspects of conservation to consider. Firstly, there must be no willful abuse of the environment in horticultural practice. Nitrogen fertilizer used to excess has been shown, especially in porous soil areas, to be washed into streams, since the soil has little ability to hold on to this nutrient (see p367). The presence of nitrogen in watercourses encourages abnormal multiplication of micro-organisms (mainly algae). On decaying these remove oxygen sources needed by other stream life, particularly fish (a process called eutrophication).

Secondly, another aspect of good practice increasingly expected of horticulturists is the intelligent use of pesticides. This involves a selection of those materials least toxic to man and beneficial to animals, and particularly excludes those materials that increase in concentration along a food chain. Lessons are still being learned from the widespread use of DDT in the 1950s. Three of DDT's properties should be noted. Firstly, it is long-lived (residual) in the soil. Secondly, it is absorbed in the bodies of most organisms with which it comes into contact, being retained in the fatty storage tissues. Thirdly, it increases in concentration approximately ten times as it passes to the next member of the food chain. As a consequence of its chemical properties, DDT was seen to achieve high concentrations in the bodies of secondary (and tertiary) consumers such as hawks, influencing the reproductive rate and hence causing a rapid decline in their numbers in the 1960s. This experience rang alarm bells for society in general, and DDT was eventually banned in most of Europe. The irresponsible action of allowing pesticide spray to drift onto adjacent crops, woodland or rivers has decreased considerably in recent years. This has in part been due to the Food and Environment Protection Act (FEPA) 1985, which has helped raise the horticulturist's awareness of conservation (see page 289).

A third aspect of conservation to consider is the deliberate selection of trees, features and areas which promote a wider range of appropriate species in a controlled manner. A golf course manager may set aside special areas with wild flowers adjacent to the fairway, preserve wet areas and plant native trees. Planting bush species such as hawthorn, field maple and spindle together in a hedgerow provides variety and supports a mixed population of insects for cultural control of pests. Tit and bat boxes in private gardens, an increasingly common sight, provide attractive homes for species that help in pest control. Continuous hedgerows will provide safe passage for mammals. Strips of grassland maintained around the edges of fields form a habitat for small mammal species as food for predatory birds such as owls. Gardeners can select plants for the deliberate encouragement of desirable species (nettles and *Buddleia* for butterflies; Rugosa roses and *Cotoneaster* for winter feeding of seed-eating birds; poached-egg plants for hoverflies).

It is emphasized that the development and maintenance of conservation areas requires continuous management and consistent effort to maintain the desired balance of species and required appearance of the area. As with gardens and orchards, any lapse in attention will result in invasion by unwanted weeds and trees. In a wider sense, the conservation movement is addressing itself to the loss of certain habitats and the consequent disappearance of endangered species such as orchids from their native areas. Horticulturists are involved indirectly because some of the peat used in growing media is taken from lowland bogs much valued for their rich variety of vegetation. Considerable efforts have been made to find alternatives to peat in horticulture (see p387) and protect the wetland habits of the British Isles.

Figure 3.6 **Pyracanthas** have good winter food for birds

Conservationists also draw attention to the thoughtless neglect and eradication of wild-ancestor strains of present-day crops; the gene-bank on which future plant breeders can draw for further improvement of plant species. There is also concern about the extinction of plants, especially those on the margins of deserts that are particularly vulnerable if global warming leads to reduced water supplies. In situ conservation mainly applies to wild species related to crop plants and involves the creation of natural reserves to protect habitats such as wild apple orchards and there is particular interest in preserving species with different ecological adaptions. Ex situ conservation includes whole plant collections in botanic gardens, arboreta, pineta and gene-banks where seeds, vegetative material and tissue cultures are maintained. The botanic gardens are coordinated by the Botanic Gardens Conservation International (BGCI), which is based at Kew Gardens, London, and are primarily concerned with the conservation of wild species.

Large national collections include the National Fruit Collection at Brogdale, Kent (administered by Wye College) and the Horticultural Research International at Wellesbourne, Birmingham, holds vegetables. The Henry Doubleday Heritage Seed Scheme conserves old varieties of vegetables which were once commercially available but which have been dropped from the National List (and so become illegal to sell). They encourage the exchange of seed. The National Council for the Conservation of Plants and Gardens (NCCPG) was set up by the Royal Horticultural Society at Wisley in 1978 and is an excellent example of professionals and amateurs working together to conserve stocks of extinction threatened garden plants, to ensure the availability of a wider range of plants and to stimulate scientific, taxonomic, horticultural, historical and artistic studies of garden plants. There are over 600 collections of ornamental plants encompassing 400 genera and some 5000 plants. A third of these are maintained in private gardens, but many are held in publicly funded institutions such as colleges, e.g. *Sarcococca* at Capel Manor College in North London, *Escallonia* at the Duchy College in Cornwall, *Penstemon* and *Philadelphus* at Pershore College and *Papaver orientale* at the Scottish Agricultural College, Auchincruive. Rare plants are identified and classified as 'pink sheet' plants.

Organic growing

The organic movement broadly believes that crops and ornamental plants should be produced with as little disturbance as possible to the balance of microscopic and larger organisms present in the soil and also in the above-soil zone. This stance can be seen as closely allied to the conservation position, but with the difference that the emphasis here is on the balance of micro-organisms. Organic growers maintain soil fertility by the incorporation of animal manures, or green manure crops such as grass–clover leys. The claim is made that crops receive a steady, balanced release of nutrients through their roots in a soil where earthworm activity recycles organic matter deep down; the resulting deep root penetration allows an effective uptake of water and nutrient reserves.

The use of most pesticides and quick-release fertilizers is said to be the main cause of species imbalance, and formal approval for licensed organic production may require soil to have been free from these two groups of chemicals for at least two years. Control of pests and diseases is achieved by a combination of resistant cultivars and 'safe' pesticides derived from plant extracts, by careful rotation of plant species, and by the use of naturally occurring predators and parasites. Weeds are controlled by mechanical and heat-producing weed controlling equipment, and by the use of mulches. The balanced nutrition of the crop is said to induce greater resistance to pests and diseases, and the taste of organically grown food is claimed to be superior to that of conventionally grown produce.

The organic production of food and non-edible crops at present represents about 5 per cent of the European market. The European Community Regulations (1991) on the 'organic production of agricultural products' specify the substances that may be used as 'plant-protection products, detergents, fertilizers, or soil conditioners' (see pps293 and 375). Conventional horticulture is, thus, still by far the major method of production and this is reflected in this book. However, it should be realized that much of the subsistence cropping and animal production in the Third World could be considered 'organic'.

Check your learning

1. Define the term 'ecosystem'.

2. Define the term 'semi-natural vegetation'.

3. Explain the importance of plants as energy producers within ecosystems.

4. Describe the stages in a named succession, giving plant species examples.

5. Describe the effects of three named pollutants on plant growth and development.

6. Describe three adaptations found in plants living in an extreme environment.

Further reading

Allaby, M. (1994). *Concise Oxford Dictionary of Botany*. Oxford University Press.

Ayres, A. (1990). *Gardening Without Chemicals*. Which/Hodder & Stoughton.

Baines, C. and Smart, J. (1991). *A Guide to Habitat Creation*. Packard Publishing.

Brown, L.V. (2008). *Applied Principles of Horticultural Science*. 3rd edn., Butterworth-Heinemann.

Caplan, B. (1992). *Complete Manual of Organic Gardening*. Headline Publishing.

Carson, R. (1962). *Silent Spring*. Hamish Hamilton.

Carr, S. and Bell, M. (1991). *Practical Conservation*. Open University/ Nature Conservation Council.

Dowdeswell, W.H. (1984). *Ecology Principles and Practice*. Heinemann Educational Books.

Innis, D.Q. (1997). *Intercropping and the Scientific Basis of Traditional Agriculture*. Intermediate Technology Publications.

Lampkin, N. (1990). *Organic Farming*. Farming Press.

Lisansky, S.G., Robinson, A.P., and Coombs, J. (1991). *Green Growers Guide*. CPL Scientific Ltd.

Mannion, A.M. and Bowlby, S.R. (eds) (1992). *Environmental Issues in the 1990s*. John Wiley & Sons.

Preston, C.D., Pearman, D.A., and Dines, T.D. (2002). *New Atlas of the British and Irish Flora*. Oxford University Press.

Tait, J. *et al.* (1998). *Practical Conservation*. Open University.

Chapter 4　Classification and naming

Summary

This chapter includes the following topics:

- **The major divisions of the plant kingdom**
- **Gymnosperms and angiosperms**
- **Monocotyledons and dicotyledons**
- **Nomenclature, the naming of plants**
- **Plant families**
- **Plant genera, species, subspecies**
- **Plant varieties and cultivars**

with additional information on the following:

- **Principles of classification**
- **Hybrids**
- **Identifying plants**
- **Geographical origins of plants**
- **Fungi**
- **Animals**
- **Bacteria**
- **Algae and lichens**
- **Viruses**

Figure 4.1 **A range of divisions of the plant kingdom**, including moss, fern, fungi, seed producing plants (pine needles) in natural habitat

The classification of plants

Any classification system involves the grouping of organisms or objects using characteristics common to members within the group. A classification can be as simple as dividing things by colour or size. Fundamental to most systems and making the effort worthwhile is that the classification meets a purpose; has a use. This is generally to make life simpler such as to find books in a library; they can be classified in different, but helpful ways, e.g. by subject or date or particular use.

Terms that are used in classification are:

- **taxonomy**, which deals with the principles on which a classification is based;
- **systematics** identifies the groups to be used in the classification;
- **nomenclature** deals with naming.

Various systems for organisms have been devised throughout history, but a seventeenth century Swedish botanist, **Linnaeus**, laid the basis for much subsequent work in the classification of plants, animals (and also minerals). The original divisions of the plant kingdom were the main groupings of organisms according to their place in evolutionary history. Simple single-celled organisms from aquatic environments evolved to more complex descendants, multicellular plants with diverse structures, which were able to survive in a terrestrial habitat, and develop sophisticated reproduction mechanisms.

The world of living organisms is currently divided into five kingdoms including:

- Plantae (plants)
- Animalia (animals)
- Fungi
- Bacteria (Prokaryotae)
- Protoctista (all other organisms that are not in the other kingdoms including algae and protozoa).

The organisms constituting the plant kingdom are distinguishable from animals in having sedentary growth, cellulose cell walls and polyploidy (see Chapter 10). They are able to change energy from one form (e.g. light) into organic molecules (autotrophic nutrition; see photosynthesis p110). Animals, amongst other things, have no cell walls and rely on eating ready-made organic molecules (heterotrophic nutrition).

Plant **divisions** (the names ending in -phyta) are further sub-divided into

- **class** (ending -psida);
- **order** (ending -ales);
- **family** (ending -aceae);
- **genus**;
- **species**.

A **species** is a group of individuals with the greatest mutual resemblance, which are able to breed amongst themselves.

Species is the basic unit of classification, and is defined as a group of individuals with the greatest mutual resemblance, which are able to breed amongst themselves. A number of species with basic similarities constitute a genus (plural genera), a number of genera constitute a family, and a number of families make up an order (see example given in Table 4.1). **Subspecies** are a naturally occurring variation within a species where the types are quite different from each other.

Table 4.1 Classification. The lettuce cultivar 'Little Gem' used to illustrate the hierarchy of the classification up to kingdom

Kingdom	Plantae	Plants
SubKingdom	Tracheobionta	Vascular plants
Superdivision	Spermatophyta	Seed plants
Division	Magnoliophyta	Flowering plants
Class	Magnoliopsida	Dicotyledons
Subclass	Asteridae	
Order	Asterales	
Family	Asteraceae	Aster family
Genus	*Lactuca*	lettuce
Species	*L. sativa*	garden lettuce
Cultivar	*L. sativa* 'Little Gem'	

In order to produce a universally acceptable system, the International Code of Botanical Nomenclature was formulated, which includes both non-cultivated plants and details specific to cultivated plants.

Kingdom Plantae

Plants

The first major breakdown of the plant kingdom is into divisions (see Table 4.2 below).

Table 4.2 Divisions of the plant kingdom

	Divisions:	Common name:
	Bryophyta	Mosses
	Hepatophyta	Liverworts
Vascular plants:		
	Equisetophyta	Horsetails
	Pteridophyta	Ferns
Seed plants:		
	Coniferophyta	Conifers
	Ginkgophyta	Ginkgo
	Magnoliophyta	Flowering plants

Divisions of the plant kingdom

Mosses and liverworts. Over 25 000 plant species which do not have a vascular system (see p92) are included in the divisions Bryophyta and Hepatophyta. They have distinctive vegetative and sexual reproductive structures, the latter producing spores that require damp conditions for survival. Many from both divisions are pioneer plants that play an important part in the early stages of soil formation. The low spreading carpets of vegetation also present a weed problem on the surface of compost in container-grown plants, on capillary benches and around glazing bars on greenhouse roofs.

Ferns and horsetails in the divisions Pteridophyta and Equisetophyta, have identifiable leaf, stem and root organs, but produce spores rather than seeds from the sexual reproduction process. Many species of ferns, e.g. maidenhair fern (*Adiantum cuniatum*), and some tropical horsetails, are grown for decorative purposes, but the common horsetail (*Equisetum arvense*), and bracken (*Pteris aquilina*) that spread by underground rhizomes are difficult weeds to control.

Seed-producing plants (Super-division – Spermatophyta) contain the most highly evolved and structurally complex plants. There are species adapted to most habitats and extremes of environment. Sexual reproduction produces a seed, which is a small, embryo plant contained within a protective layer.

Moss, Bryophyta

Liverwort, Hepatophyta

Ferns, Pteridophyta

Horsetail, Equisetophyta

Figure 4.2 **Four multicellular plant divisions**, moss, liverwort, fern, horsetail with horticultural significance

Angiosperms and gymnosperms

The subdivision into class brings about the gymnospermae, mostly consisting of trees and shrubs, and the angiospermae representing the greatest diversity of plants with adaptations for the majority of habitats. Structurally, the gymnosperms have much simpler xylem vessels than the more complex system in the angiosperms, and flowers are unisexual producing naked seeds. The angiosperms usually have hermaphrodite flowers, which produce complex seeds (see p103), inside a protective fruit.

Conifers (Coniferophyta) are a large division of many hundred species that include the pines (Order – Pinales) and yews (Order – Taxales). Characteristically they produce 'naked' seeds, usually in cones, the female organ. They show some primitive features, and often display structural adaptations to reduce water loss (see Figure 9.2). There are very many important conifers. Some are major sources of wood or wood pulp, but within horticulture many are valued because of their interesting plant habits, foliage shape and colours. The Cupressaceae, for example, includes fast growing species, which can be used as windbreaks, and small, slow growing types very useful for rock gardens. The yews are a highly poisonous group of plants that includes the common yew (*Taxus baccata*) used in ornamental hedges and mazes. The division Ginkgophyta is represented by a single surviving species, the maidenhair tree (*Ginkgo biloba*), which has an unusual slit-leaf shape, and distinctive bright yellow colour in autumn.

Flowering plants (Division – Magnoliophyta) have a flower structure for sexual reproduction producing seeds protected by fruits. This characteristic structure is used as the basis of their classification. There are estimated to be some 25 000 species, occupying a very wide range of habitats. Many in the division are important to horticulture, both as crop plants and weeds. This division is split into two main classes; the Liliopsida formerly the Monocotyledonae and generally known as the **monocotyledons,** and the Magnoliopsida, the **dicotyledons.** The main differences are given in Table 4.3.

Figure 4.3 **Yew (*Taxus baccata*) with berries** – a conifer

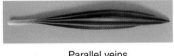

Parallel veins
(a) e.g. most monocotyledons

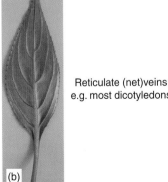

(b)

Reticulate (net)veins
e.g. most dicotyledons

Figure 4.4 **Leaf venation** in
(a) monocotyledon and (b) dicotyledon

Table 4.3 Differences between Monocotyledons and Dicotyledons

Monocotyledons	Dicotyledons
One seed leaf	Two seed leaves
Parallel veined leaves usually alternate and entire (see Figure 4.4)	Net veined leaves
Vascular bundles in stem scattered	Vascular bundles in stem in rings
Flower parts usually in threes, also three seed chambers in fruit	Flower parts usually in fours or fives, also four or five seed chambers in fruit
Except palms, are non-woody	Both woody and herbaceous species
Herbaceous plants	Woody stems showing annual rings

Monocotyledons include some important horticultural families, e.g. Arecaceae, the palms; Musaceae, the bananas; Cyperaceae, the sedges; Juncaceae, the rushes; Poaceae (formerly Graminae), the grasses; Iridaceae, the irises; Liliaceae, the lilies and the Orchidaceae, the orchids.

Dicotyledons has many more families significant to horticulture, including Magnoliaceae, the magnolias; Caprifoliaceae, the honeysuckles; Cactaceae, the cactuses; Malvaceae, the mallows; Ranunculaceae, the buttercups; Theaceae, the teas; Lauraceae, the laurels; Betulaceae, the birches; Fagaceae, the beeches; Solanaceae, the potatoes and tomato; Nymphaeaceae, the water lilies and Crassulaceae,

Figure 4.5 **Rose flower and hip** – a member of family Rosaceae with five petals, multiple sex organs and a succulent fruit (*see* page 104)

the stonecrops. Four of the biggest and most economically important families in this class have had a change of name. Fabaceae (formerly the Leguminosae), the pea and bean family, have five-petalled asymmetric or zygomorphic (having only one plane of symmetry) flowers, which develop into long pods (legumes) containing starchy seeds. The characteristic upturned umbrella-shaped flower head or **umbel** is found in the Apiaceae (formerly the Umbelliferae), the carrot family, and bears small white five-petalled flowers, which are wind-pollinated. The members of Asteraceae (formerly Compositae) have a characteristic flower head with many small florets making up the composite, regular (or actinomorphic) structure, e.g. chrysanthemum, groundsel. Members of the Brassicaceae (formerly Cruciferae) are characterized by their four-petal flower and contain the *Brassica* genus with a number of important crop plants such as cabbage, cauliflowers, swedes, Brussels sprouts, as well as the wallflower (*Cheiranthus cheiri*). Most of the brassicas have **a biennial** growth habit producing vegetative growth in the first season, and flowers in the second, usually in response to a cold stimulus (see vernalization). A number of weed species are found in this family, including shepherd's purse (*Capsella bursapastoris*), which is an **annual**. Many important genera, e.g. apples (*Malus*), pear (*Pyrus*) and rose (*Rosa*), are found within the Rosaceae family, which generally produces succulent fruit from a flower with five petals and often many male and female organs. Many species within this family display a **perennial** growth habit (see Table 4.4).

Nomenclature

The naming of cultivated plants

The binomial system

The name given to a plant species is very important. It is the key to identification in the field or garden, and also an international form of identity, which can lead to much information from books and the Internet. The common names which we use for plants, such as potato and lettuce, are, of course, acceptable in English, but are not universally used. A scientific method of naming can also provide more information about a species, such as its relationship with other species.

Linnaeus, working on classification and with the more detailed question of naming, formulated a system that he claimed should identify an individual plant type uniquely, by means of the composed **genus** name followed by the **species** name. For example, the chrysanthemum used for cut flowers is *Chrysanthemum* genus and *morifolium* species; note that the genus name begins with a capital letter, while the species has a small letter. Other examples are *Ilex aquifolium* (holly), *Magnolia stellata* (star-magnolia), *Ribes sangui-neum* (redcurrant).

Subspecies can evolve and display more distinct characteristics than the varieties detailed below, e.g. *Rhododendron arboreum* subsp.

cinnamomeum. The genus and species names must be written in *italics,* or underlined where this is not possible, to indicate that they are internationally accepted terms. However, these two words may not encompass all possible variations, since a species can give rise to a number of naturally occurring **varieties** with distinctive characteristics. In addition, cultivation, selection and breeding have produced variation in species referred to as cultivated varieties or **cultivars**.

The two terms, variety and cultivar are exactly equivalent, but the botanical variety name is referred to in Latin, beginning with a small letter, e.g. *Rhododendron arboreum* var *roseum,* while the cultivar is given a name often relating to the plant breeder who produced it, e.g. *Rhododendron arboreum* 'Tony Schilling'. There is no other significant difference in the use of the two terms, and therefore either is acceptable. However, the term cultivar will be used throughout this text. A cultivar name should be written in inverted commas and begin with a capital letter, after the binomial name or, when applicable, the common name. Examples include: *Prunus padus* 'Grandiflora', tomato 'Ailsa Craig', apple 'Bramley's seedling'.

If a cultivar name has more than one acceptable alternative, they are said to be **synonyms** (sometimes written syn.) e.g. *Phlox paniculata* 'Frau Alfred von Mauthner' syn. *P. paniculata* 'Spitfire'.

> A **variety** is a variation within a species which has arisen naturally.
>
> A **cultivar** is similarly a variation within a species which has arisen or has been bred in cultivation.

Hybrids

When **cross-pollination** occurs between two plants, hybridization results, and the offspring usually bear characteristics distinct from either parent. Hybridization can occur between different cultivars within a species, sometimes resulting in a new and distinctive cultivar (see Chapter 10), or between two species, resulting in an **interspecific hybrid**, e.g. *Prunus* $\times$ *yedoensis* and *Erica* $\times$ *darleyensis*. A much rarer hybridization between two different genera results in an **intergeneric hybrid**, e.g. $\times$ *Cupressocyparis leylandii* and $\times$ *Fatshedera lizei*. The names of the resulting hybrid types include elements from the names of the parents, connected or preceded by a multiplication sign ($\times$). A chimaera, consisting of tissue from two distinct parents, is indicated by a 'plus' sign, e.g. $+$ *Laburnocystisus adamii,* the result of a graft.

Further classifications of plants

Plants can be grouped into other useful categories. A classification based on their life cycle (ephemerals, annuals, biennials and perennials) has long been used by growers, who also distinguish between the different woody plants such as trees and shrubs. Growers distinguish between those plants that are able to withstand a frost (hardy) and those that cannot (tender); plants can be grouped according to their degree of hardiness. Table 4.4 brings together these useful terms, provides some definitions and gives some plant examples.

Table 4.4 Some commonly used terms that describe the life cycles, size and survival strategies of plants

Life cycles		
Ephemeral	A plant that has several life cycles in a growing season and can increase in numbers rapidly	e.g. groundsel (*Senecio vulgaris*)
Annual	A plant that completes its life cycle within a growing season	e.g. poached-egg flower (*Limnanthes douglasii*)
Biennial	A plant with a life cycle that spans two growing seasons	e.g. foxglove (*Digitalis purpurea*)
Perennial	A plant living through several growing seasons	
Herbaceous perennial	A perennial that loses its stems and foliage at the end of the growing season	e.g. Michaelmas daisy (Aster spp.) and hop (*Humulus lupulus*)
Woody plants		
Woody perennial	A perennial that maintains live woody stem growth at the end of the growing season	e.g. bush fruit, shrubs, trees, climbers (e.g. grape)
Shrub	A woody perennial plant having side branches emerging from near ground level. Up to 5 m tall	e.g. Lilac (*Syringa vulgaris*)
Tree	A large woody perennial unbranched for some distance above ground. Usually more than 5 m	e.g. Horse chestnut (*Aesculus hippocastanum*)
Deciduous	A plant that sheds all its leaves at once	e.g. Mock orange (*Philadelphus delavayi*)
Evergreen	A plant retaining leaves in all seasons	e.g. Aucuba (*Aucuba japonica*)
Hardiness		
Very hardy	A plant able to survive − 18°C	e.g. *Kerria japonica*
Moderately hardy	A plant able to survive − 15°C	e.g. *Camellia japonica*
Semi-hardy	A plant able to survive − 6°C	e.g. *Pittosporum crassifolium*
Tender	A plant not hardy below − 1°C	e.g. Pelargonium cvs

Identifying plants

A **flora** is a text written for the identification of flowering plant species. Some flora use only pictures to classify plants. More detailed texts use a more systematic approach where reference is made to a **key** of features that, by elimination, will lead to the name of a plant. Species are described in terms of their flowers, inflorescences, stems, leaves and fruit. This description will often include details of shape, size and colour of these plant parts.

Flowers. The number and arrangement of flower parts (see Figure 4.6) is the most important feature for classification and is a primary feature in plant identification. It can be described in shorthand using a floral formula or a floral diagram. For example, **the flora formula**, with the interpretation, for Wallflower (*Cheiranthus cheiri*), a member of the Cruciferae family is as follows:

$\oplus$	K4	C4	A2 + 4	G(2)
symmetrical flower	4 sepals in calyx	4 petals in corolla	6 anthers	2 ovaries joined together

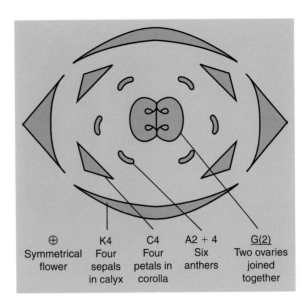

Figure 4.6 **Floral diagram of wallflower**

Figure 4.7 **Wallflower flower** (a) from above, (b) the side and (c) LS, illustrating the floral diagram above

Other examples of floral formulae include:

Sweet pea (Fabaceae):	·/·	K(5)	C5A(9) + 1	<u>G1</u>
Buttercup (Ranunculaceae):	⊕	K	C5A	G
Dead nettle (Labiatae):	·/·	K(5)	C(5)A4	<u>G(2)</u>
Daisy (Asteraceae):	⊕	K	C(5)A(5)	G(2)

The way that **flowers** are arranged on the plant is also distinctive in different families, e.g. raceme, common in the Fabaceae; corymb and capitulum found in the Asteraceae and umbel, very much associated with Apiaceae (see Chapter 7 for more detail).

Leaf form (see Figure 4.8) is a useful indicator when attempting to identify a plant and descriptions often include specific terms, a few are described below but many more are used in flora.

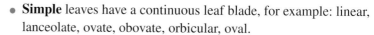

Figure 4.8 **Leaf forms**: (a) linear e.g. *Agapanths;* (b) lanceolate e.g. *Viburnum;* (c) oval e.g. *Garrya elliptica;* (d) peltate e.g. nasturtium; (e) hastate e.g. *Zantedeschia;* (f) lobed e.g. Geranium; (g) palmate e.g. lupin; (h) pinnate e.g. rose

- **Simple** leaves have a continuous leaf blade, for example: linear, lanceolate, ovate, obovate, orbicular, oval.
- **Margins** of leaves can be described: entire, sinuous, serrate, and crenate.
- **Leaf vein** arrangement also characterizes the plant: reticulate, parallel, pinnate and palmate.
- **Compound** leaves, such as compound palmate and compound pinnate, have separate leaflets each with an individual base on one leaf stalk (see p117 for leaf structure), but only the axillary bud is at the base of the main leaf stalk.

Most horticulturists yearn for stability in the **naming** of plants. Changes in names confuse many people who do not have access to up-to-date literature. On the other hand, the reasons for change are justifiable. New scientific findings may have shown that a genus or species belongs in a different section of a plant family, and that a new name is the correct way of acknowledging this fact. Alternatively, a plant introduced from abroad, maybe many years ago, may have mistakenly been given the incorrect name, along with all the cultivars derived from it.

Evidence from biochemistry, microscopy and DNA analysis is proving increasingly important in adding to the more conventional plant structural evidence for plant naming. There may be differing views whether a genus or species should be 'split' into smaller units, or several species be 'lumped' into an existing species or genus, or left unchanged. It seems likely that changes in plant names will continue to be a fact of horticultural life.

There has been a massive increase in communication across the world, especially as a result of the Internet. The level of information about plant names has improved. The International Code of Botanical Nomenclature (ICBN) has laid down an international system. Within Britain, the Royal Horticultural Society (RHS) has an advisory panel to help resolve problems in this area. An invaluable reference document 'Index

Kewensis' is maintained by Kew Gardens listing the first publication of the name for each plant species not having specific horticultural importance. Cultivated species are listed in the 'RHS Plant Finder', which also indicates where they can be sourced, is updated annually and can be viewed on the Internet. Further cooperation across Europe has led to the compilation of The International Plant Names Index with associated working parties formed from scientific institutions and the horticultural industry.

Geographical origins of plants

Gardens and horticultural units, from the tropics to more temperate climates, contain an astonishing variety of plant species from the different continents. Below is a brief selection of well-known plants, grown in Britain, illustrating this diversity of origin. It is salutary, when considering these far-flung places, to reflect on the sophisticated cultures, with skills in plant breeding and a passion for horticulture over the centuries that have taken wild plants and transformed them into the wonders that we now see in our gardens.

- **British Isles**; English Oak (*Quercus robur*), *Geranium robertianum*, foxglove, peppermint, *Pinus sylvestris*.
- **Far East** (China and Japan); cherry, cucumber, peach, walnut, *Clematis*, *Forsythia*, hollyhock, *Azalea*, rose.
- **India and South-East Asia**; mustard, radish.
- **Australasia**; *Acacia*, *Helichrysum*, *Hebe*.
- **Africa**; *Phaseolus*, pea, African violet, *Strelitzia*, *Freesia*, *Gladiolus*, *Impatiens*, *Pelargonium*, *Plumbago*.
- **Mediterranean**; asparagus, celery, lettuce, onion, parsnip, rhubarb, carnation, hyacinth, *Antirrhinum*, sweet pea, *Rosemarinus officinalis*.
- **Middle East and Central Asia**; apple, carrot, garlic, grape, leek, pear, spinach.
- **Northern Europe**; cabbage, *Campanula*, Crocus, forget-me-not, foxglove, pansy, *Primula*, rose, wallflower, parsley.
- **North America**; *Aquilegia*, *Ceonothus*, lupin, *Aster*, *Penstemon*, *Phlox*, sunflower.
- **Central and South America**; capsicum, maize, potato, tomato, *Fuchsia*, nasturtium, *Petunia*, *Verbena*.

Figure 4.9 **Oak Tree (*Quercus robur*)** – a native of the British Isles

Non-plant kingdoms

Fungi

Some **fungi** are single celled (such as yeasts) but others are multicellular, such as the moulds and the more familiar mushrooms and toadstools. Most are made up of a **mycelium**, which is a mass of thread-like filaments (**hyphae**). Their cell walls are made of chitin. Their energy and supply of organic molecules are obtained from other organisms (heterotrophic nutrition). They achieve this by secreting digestive enzymes on to their food source and absorbing the soluble products. They obtain their food directly from other living organisms, possibly causing disease (see Chapter 15), or from dead organic matter, so contributing to its breakdown in the soil (see Chapter 18).

Fungi are classified into three divisions:

- **Zygomycota** (mitosporic fungi) have simple asexual and sexual spore forms. Damping off, downy mildew, and potato blight belong to this group.
- **Ascomycota** have chitin cell walls, and show, throughout the group, a wide variety of asexual spore forms. The sexual spores are consistently formed within small sacs (asci), numbers of which may themselves be embedded within flask-shaped structures (perithecia), just visible to the naked eye. Black spot of rose, apple canker, powdery mildew, and Dutch elm disease belong to this group.
- **Basidiomycota** have chitin cell walls, and may produce, within one fungal species (e.g. cereal rust), as many as five different spore forms, involving more than one plant host. The fungi within this group bear sexual spores (basidiospores) from a microscopic club-shaped structure (basidium). Carnation rust, honey fungus, and silver leaf diseases belong to this group.

An artificially derived fourth grouping of fungi is included in the classification of fungi.

- The **Deuteromycota** include species of fungi that only very rarely produce a sexual spore stage. As with plants, the sexual structures of fungi form the most reliable basis for classification. But, here, the main basis for naming is the asexual spore, and mycelium structure. Grey mould (*Botrytis*), *Fusarium* patch of turf, and *Rhizoctonia* rot are placed within this group.

Figure 4.10 Fungi showing fruiting bodies

Animals

The **animal** kingdom includes a very large number of species that have a significant influence on horticulture mainly as pests (see Chapter 14) or as contributors to the recycling of organic matter (see Chapter 18).

Some of the most familiar animals are in the phylum **Chordata** that includes mammals, birds, fish, reptiles and amphibians. Mammal pest species include moles (see p200), rabbits (see p198), deer, rats and mice. Bird pest species are numerous including pigeons and bullfinches, but there are very many that are beneficial in that they feed harmful organisms such as tits that eat greenfly. Less familiar are important members of the phylum **Nematoda** (the round worms) that includes a very large number of plant disease causing organisms including Stem and Bulb Eelworm (see p229), Root Knot Eelworm (see p230), Chrysanthemum Eelworm (see p230) and Potato Root Eelworm. Phylum **Arthropoda** are the most numerous animals on earth and include insects, centipedes, millipedes and spiders; many of these are dealt with in the chapter on plant pests (Chapter 14), but it should be noted that there are many that are beneficial e.g. honey bees (see p136) and centipedes, which are carnivorous and many live on insect species that are harmful. Phylum **Annelida** (the segmented worms) includes earthworms, which are generally considered to be useful organisms especially when they are helping to decompose organic matter (see p321) or improving soil structure (see p311), but some species cause problems in fine turf when they produce worm casts. Phylum **Mollusca** is best known for the major pests: slugs and snails (see p203).

Bacteria

Bacteria are single-celled organisms sometimes arranged in chains or groups (colonial). They are autotrophic (can produce their own energy supply and organic molecules); some photosynthesize (see p110), but others are able to make organic molecules using the energy released from chemical reactions usually involving simple inorganic compounds. They have great importance to horticulture by their beneficial activities in the soil (see p321), and as causative organisms of plant diseases (see p251).

Algae and lichens

The **algae**, comprising some 18 000 species, are true plants, since they use chlorophyll to photosynthesize (see Chapter 8). The division Chlorophyta (green algae) contains single-celled organisms that require water for reproduction and can present problems when blocking irrigation lines and clogging water tanks. Marine algal species in Phaeophyta (brown algae) and Rhodophyta (red algae) are multicellular, and have leaf-like structures. They include the seaweeds, which accumulate mineral nutrients, and are therefore a useful source of compound fertilizer as a liquid feed. (The blue-green algae, which can cause problems in water because they produce unsightly blooms but are also toxic, have been renamed cyano-bacteria and placed in Kingdom Prokaryotae.)

Figure 4.11 **Lichen** – a combination of fungi and algae

Lichens

Classification is complex, since each lichen consists of both fungal and algal parts. Both organisms are mutually beneficial or symbiotic. The significance of lichens to horticulture is not great. Of the 15 000 species, one species is considered a food delicacy in Japan. However, lichens growing on tree bark or walls are very sensitive to atmospheric pollution, particularly to the sulphur dioxide content of the air. Different lichen species can withstand varying levels of sulphur dioxide, and a survey of lichen species can be used to indicate levels of atmospheric pollution in a particular area. Many contribute to the weathering of rock in the initial stages of soil development (see p300). Lichens are also used as a natural dye, and can form an important part of the diet of some deer.

Viruses

Viruses are not included in any of the kingdoms. They are visible only under an electron microscope, and do not have a cellular structure but consist of nucleic acid surrounded by an outer protein coat (a capsid). They do not have the cytoplasm, organelles and internal membranes found in the cells of living organisms (see p88). They cannot grow, move or reproduce without access to the cells of a host cell so they are not included in the classification of living things. Viruses survive by invading the cells of other organisms, modifying their behaviour and often causing disease e.g. arabis mosaic, chrysanthemum stunt, cucumber mosaic, leaf mosaic, plum pox, reversion, tomato mosaic and tulip break (see p253).

Check your learning

1. State the major divisions of the plant kingdom important to horticulture.

2. Describe the major differences between gymnosperms and angiosperms

3. Name the features of angiosperms which divide them into monocotyledons and dicotyledons.

4. Explain the reasons for plant nomenclature in horticulture.

5. Define the following terms: family, genus, species, subspecies, ephemeral, biennial, perennial, tender, half-hardy, hardy, herbaceous, woody, evergreen, semi-evergreen, and deciduous.

Further reading

Allaby, M. (1992). *Concise Oxford Dictionary of Botany*. Oxford University Press.

Baines, C. and Smart, J. (1991). *A Guide to Habitat Creation*. Packard Publishing.

Brown, L.V. (2008). *Applied Principles of Horticultural Science*. 3rd edn. Butterworth-Heinemann.

Dowdeswell, W.H. (1984). *Ecology, Principles and Practice*. Heinemann Educational Books.

Heywood, V.H. *et al.* (2007). *Flowering Plant Families of the World*. Firefly Books.

Hillier Gardener's Guide (2005). *Plant Names Explained*. David and Charles.

Ingram, D.S. *et al.* (eds) (2002). *Science and the Garden*. Blackwell Science Ltd.

Lord, T. (updated annually) *The RHS Plant Finder*. Dorling Kindersley.

Chapter 5 External characteristics of the plant

Summary

This chapter includes the following topics:

- **The external appearance or morphology of stem, root, leaf**
- **Their variation, adaptation and use in horticulture**

with additional information on:

- **The leaf form in gardening**
- **Plant size and growth rate**
- **Plant form in design**

Figure 5.1 The external appearance of the plant is often what is valued, e.g. flowers

Plant form

Most plant species at first sight appear very similar since all four organs, the **root, stem, leaf and flower**, are present in approximately the same form and have the same major functions. The generalized plant form for a dicotyledonous and a monocotyledonous plant can be seen in Figure 5.2.

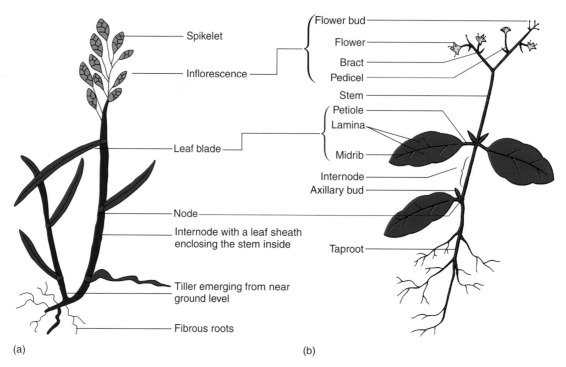

Figure 5.2 **Generalized plant form**: (a) monocotyledon; (b) dicotyledon

The development of the root and the stem from the seed is given in detail in Chapter 10. **Flowers** are the site of sexual reproduction in plants and their external appearance depends principally on the agents of pollination (see Chapters 7 and 10).

Roots

Root morphology (see Figure 5.3). The function of the root system is to take up water and mineral nutrients from the growing medium and to anchor the plant in that medium. Its major function involves making contact with the water in the growing medium. To achieve this it must have as large a surface area as possible. The root surface near to the tip where growth occurs (cell division in the meristem, see p93) is protected by the **root cap**. The root zone behind the root tip has tiny projections called **root hairs** reaching numbers of 200–400 per square millimetre, which greatly increase the surface area in this region (see Figure 5.4). Plants grown in hydroculture, e.g. NFT (p394), produce considerably fewer root hairs. The loss of root hairs during transplanting can check plant growth considerably, and the hairs can be points of entry of diseases such as club root (see Chapter 15). Figure 5.4 shows that the layer with the root hairs, the **epidermis**, is comparable with the epidermis of the stem (see stem structure); it is a single layer of cells which has a protective as well as an absorptive function.

Figure 5.3 **Germinated seed** showing primary and secondary roots growing from lower end of hypocotyl

A **taproot** is a single large root which will have many **lateral roots** growing out from it at intervals.

A **fibrous** root system consists of many roots growing out from the base of the stem.

Primary roots originate from the lower end of the hypocotyl.

Secondary roots are branches of the primary roots.

Adventitious roots grow in unusual places such as on the stem or other organ.

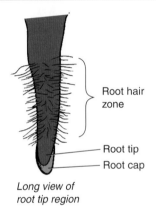

Long view of
root tip region

Root hair zone

Root tip

Root cap

Figure 5.4 **Root tip** showing the tip protected by root cap, and root hair zone

Two types of root system are produced; a **taproot** is a single large root which usually maintains a direction of growth in response to gravity (see geotropism) with many small lateral roots growing from it, e.g. in chrysanthemums, brassicas, dock. In contrast, a **fibrous root** system consists of many roots growing out from the base of the stem, as in grasses and groundsel (see Figure 5.3).

Stems

The **stem's** function is physically to support the leaves and flowers, and to transport water, minerals and food between roots, leaves and flowers (see p91 for stem structure). The leaf joins the stem at the **node** and has in its angle (axil) with the stem an **axillary bud**, which may grow out to produce a lateral shoot. The distance between one node and the next is termed the **internode**. In order to perform these functions, the stem produces tissues (see Chapter 6) specially formed for efficiency. It must also maintain a high water content to maintain turgor (see p122).

Buds

A bud is a condensed stem which is very short and has small folded leaves attached, both enclosing and protecting it. On the outside the leaves are often thicker and dark to resist drying and damage from animals and disease. A meristem is present at the tip of the stem, from which a flower or vegetative growth will emerge. A **terminal bud** is present at the tip of a main stem and will grow out to increase its length. Where leaves join the stem, **axillary buds** may grow into lateral shoots, or may remain dormant.

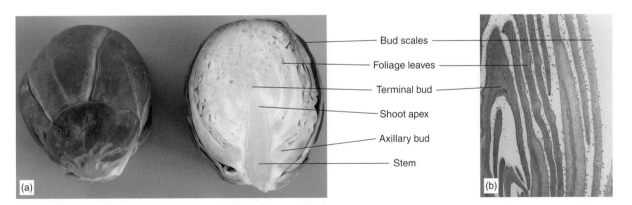

Bud scales
Foliage leaves
Terminal bud
Shoot apex
Axillary bud
Stem

(a) (b)

Figure 5.5 **Structure of a bud**: (a) Brussels sprout and (b) magnified image

Leaves

The **leaf**, consisting of the leaf blade (**lamina**) and stalk (**petiole**), carries out photosynthesis, its shape and arrangement on the stem depend on the water and light energy supply in the species' habitat. The arrangement of leaves and examples in different species, along the major differences between monocotyledons and dicotyledon described in Chapter 4. Leaf structure, as an organ of photosynth described in Chapter 8 (see p117).

Adaptations

The features of typical plants are given, but there are many variations on the basic form of the stem, root and leaves.

Adaptations to plant organs have enabled plants to compete and survive in their habitat. Plants adapted to dry areas (**xerophytes**), such as cacti, have leaves reduced to protective **spines** and stems capable of photosynthesis. **Thorns**, which are modified branches growing from axillary buds, also have a protective function, e.g. hawthorn (*Crataegus*). **Prickles** are specialized outgrowths of the stem epidermis, which not only protect but also assist the plant in scrambling over other vegetation, as in wild roses. Several species possess leaves modified specifically for climbing in the form of **tendrils**, as in many members of the Leguminosae family, and Clematis climb by means of a sensitive, **elongated leaf stalk**, which twists round their support. In runner beans, honeysuckle (*Lonicera*) and *Wisteria*, twining stems wind around other uprights for support. Others are able to climb with the help of adventitious roots such as ivies, and Virginia Creeper (*Parthenocissus*). **Epiphytes** are physically attached to aerial parts of other plants for support; they absorb and sometimes store water in aerial roots, as in some orchids.

To survive an environment with very low nutrient levels, such as the sphagnum peat moor (see p328), some plants have evolved methods of trapping insects and utilizing the soluble products of their decomposed prey. These **insectivorous** plants include the native sundew (*Drosera*) and butterwort (*Pinguicula spp.*), which trap their prey with sticky glands on their leaves. Pitcher plants (*Sarracenia spp.*) have leaves that

Figure 5.6 **Variations in structure** of plant parts as adaptations to modes of growth: (a) thorns of *Berberis* and *Pyracantha*; (b) prickles on the stem of rose, *Rosa sericea pteracantha* grown as an ornamental for its large red thorns; (c) tendrils of passion flower; (d) elongated leaf stalks of *Clematis*; (e) adventitious root formed on stems of ivy (*Hedera helix*); (f) twining stems of *Wisteria*

form containers which insects are able to enter into, but are prevented from escaping by slippery surfaces or barriers of stiff hairs. The Venus flytrap (*Dionaea muscpula*) has leaves that are hinged so that they can snap shut on their prey when it alights on one of the trigger hairs.

Plants found growing in coastal areas have adaptations that allow them to withstand high salt levels, e.g. salt glands as found in the cord grass (*Spartina spp.*) or *succulent* tissues in 'scurvy grass' (*Cochlearia*), both inhabitants of coastal areas.

Other modifications in plants are dealt with elsewhere. This includes the use of stems and roots as food and water storage organs (see vegetative propagation).

Leaf adaptations

Whilst remaining essentially the organ of photosynthesis and transpiration, the leaf takes on other functions in some species. The most notable of these is the climbing function. Tendrils are slender extensions of the leaf, and are of three types. In *Clematis spp*, the leaf petiole curls round the stems of other plants or garden structures in order to support the climber (see Figure 5.6(d)). In sweet pea (*Lathyrus odoratus*), the plant holds on with tendrils modified from the end-leaflets of the compound leaf. In the monocotyledonous climber, *Smilax china*, the support is provided by modified stipules (found at the base of the petiole). In cleavers (*Galium aparine*), both the leaf and stipules, borne in a whorl, have prickles that allow the weed to sprawl over other plant species.

Buds and bulbs are composed mainly of leaf tissue. In the former, the leaves (called scales) are reduced in size, hard, and brown rather than green. They tightly overlap each other, giving protection to the delicate meristematic tissues inside the bud. In a bulb, the succulent, light-coloured scale leaves contain all the nutrients and moisture necessary for the bulb's emergence. The scales are packed densely together around the terminal bud, minimizing the risk that might be caused by extremes of climate, or by pests such as eelworms or mice. In the houseplant *Bryophyllum daigremontianum*, the succulent leaf bears adventitious buds that are able to develop into young plantlets.

Leaf form

The novice gardener may easily overlook the importance that the shape, texture, venation, colour and size of leaves can contribute to the general appearance of a garden, as they focus more on the floral side of things. Flowers are the most striking feature, but they are often short-lived. It should be emphasized that the dominant theme in most gardens is the foliage and not the flowers (see Figures 5.7 and 5.8). The possibilities for contrast are almost endless when these five leaf aspects are considered.

In Chapter 4 (see p22) the range of leaf forms is described. Consider leaf shape first. The large linear leaves of *Phormium tenax* (New Zealand

Figure 5.7 Leaf form: shape, e.g. (a) *Phormium tenax*, (b) *Gunnera manicata*, (c) hostas and ferns; texture, e.g. (d) woolly leaves of silver mint, (e) variegation in ivy (*Hedera helix*) leaf

Figure 5.8 Leaf colours shown by examples *Helleborus, Berberis* and *Ajuga*

Flax) are a well-known striking example. In contrast are the large palmate leaves of *Gunnera manicata*. On a smaller scale, the shade-loving *Hostas*, with their lanceolate leaves, mix well with the pinnate-leaved *Dryopteris filix-mas* (Male fern).

Secondly, leaf texture is also important. Most species have quite smooth textured leaves. Notably different are *Verbascum olympicum, Stachys byzantina* (Lamb's tongue) and the alpine *Leontopodium alpinum* (Edelweiss) which all are woolly in texture. Glossy-leaved species such as *Ilex aquifolium* (Holly), and *Pieris japonica* provide a striking appearance.

Thirdly, the plant kingdom exhibits a wide variety of leaf colour tones (see Figure 5.8). The conifer, *Juniperus chinensis* (Chinese juniper), shrubs of the *Ceanothus* genus, and *Helleborus viridus* (Christmas rose) are examples of dark-leaved plants. Notable examples of plants with light-coloured leaves are the tree *Robinia pseudoacacia* (false acacia), the climber *Humulus lupulinus* 'Aureus' (common hop) and the creeping herbaceous perennial, *Lysimachia nummularia* 'Aurea' (Creeping Jenny). Plants with unusually coloured foliage may also be briefly mentioned: the small tree *Prunus* 'Shirofugen' (bronze-red), the sub-shrub *Senecio maritima* (silver-grey) and the shade perennial *Ajuga reptans* 'Atropurpurea' (bronze-purple).

Variegation (the presence of both yellow and green areas on the leaf) gives a novel appearance to the plant (see Figures 5.6 and 10.11). Example species are *Aucuba japonica* (Laurel), *Euonymus fortunei* and

Glechoma hederacea (Ground Ivy). Fourthly, in autumn, the leaves of several tree, shrub and climber species change from green to a striking orange-red colour. *Acer japonicum* (Japanese maple), *Euonymus alatus* (Winged spindle), and *Parthenocissus tri-cuspidata* (Boston ivy) are examples.

Plant size and growth rate

It is important for anyone planning a garden that they recognize the eventual size (both in terms of height and of width) of trees, shrubs and perennials. This vital information is quite often ignored or forgotten at the time of purchase. The impressive *Ginkgo* (Maidenhair tree) really can grow to 30 m in height (at least twice the height of a normal house) and is, therefore, not the plant to put in a small bed. Similarly, *X Cupressocyparis leylandii* (Leyland Cypress), seemingly so useful in rapidly creating a fine hedge, can also grow to 30 m, and reach 5 m in width, to the consternation of even the most friendly of neighbours.

The eventual size of a plant is recorded in plant encyclopaedias, which should be carefully scrutinized for this vital statistic. It may also be wise to contact a specialist nursery which deals with this important aspect on a day-to-day basis, and will give advice to potential buyers. It should be remembered that the eventual size of a tree or shrub may vary considerably in different parts of the country, and may be affected within a garden by factors such as aspect, soil, shade, and wind. Attention should also be given to the rate at which a plant grows; *Taxus* (Yew) or *Magnolia stellata* (Star Magnolia) are two notable examples of slow growing species.

Note that **trees** are large woody plants that have a main stem with branches appearing some distance above ground level. **Shrubs** are smaller, usually less than 3 m in height, but with branches developing at or near ground level to give a bushy appearance to the plant.

Plant form in design

Plant form as individual plants or in groups is the main interest for many in horticulture who use plants in the garden or landscape. Contrasts in plant shapes and sizes can be combined to please the eye of the observer.

The dominant plant within a garden feature is usually a tree or shrub chosen for its **special** striking appearance, or specimen plant. In a large feature, it may be a *Betula pendula* (silver birch) tree growing up to 20 m in height with a graceful form, striking white bark, and golden autumn colour. In a smaller feature, *Euphorbia characias* provides a very special effect with its 1 m high evergreen foliage, and springtime yellow blooms. Such plants can form a focal point in a garden or landscape.

Figure 5.9 **Flower of *Euphorbia characias ssp wulfenii*** can grow to 1 m

Spaced around these specimen plants, there may be included species providing a visually supportive **background or skeletal** form to the decorative feature. *Garrya elliptica*, a 4 m shrub with elliptical, wavy edged evergreen leaves and mid-winter catkins fits naturally into this category against a larger special plant. At 2.5 m, the evergreen shrub *Choisya ternata* (Mexican Orange Blossom) bearing fragrant white flowers in spring is a popular background species in decorative borders. *Jasminum nudifolium* (winter jasmine) is an example of a climber fulfilling this role. Such framework plants not only provide a suitable background, but also can provide continuity or unity through the garden or landscape and ensure interest all the year round.

Fitting further into the mosaic of plantings are the numerous examples of **decorative** species, exhibiting particular aspects of general structure or of flowering, and often having a deciduous growth. An example is the 0.3 m tall *Cytisus x kewensis* (a broom with a prostrate habit) with its downy arching stems and profuse creamy-white spring flowers. A contrasting example is the 2 m clump-forming grass species, *Cortaderia selloana*, producing narrow leaves and feathery late summer flowering panicles. Climbing species from the *Rosa* and *Clematis* genera also fit into the decorative category. Garden designers are also able to call on a very wide range of leaf forms to create textural or architectural interest in the border.

A host of deciduous **pretty** herbaceous and evergreen perennials are available for filling the decorative feature, fitting around the above-mentioned three categories. *Delphiniums* (up to 2 m), *Lupins* up to 1.5 m), *Asters* (up to 1.5 m), *Sedums* (up to 0.5 m), and *Alchemillas* (up to 0.5 m) are five examples illustrating a range of heights.

Finally, **infill** species either as bulbs (e.g. *Tulipa*, *Narcissus* or *Lilium*), perennials (e.g. *Saxifraga*, *Campanula*), or annuals (e.g. *Nicotiana* or *Begonia*), may be placed within the feature, sometimes for a relatively short period whilst other perennials are growing towards full-size. They are also used in colourful bedding displays.

Colour in flowers

The use of different flower colours in the garden has been the subject of much discussion in Britain over the last three hundred years. Many books have been written on the subject, and authorities on the subject will disagree about what combination of plants creates an impressive border. Some combinations are mentioned here, and Figure 5.10 illustrates one example of the harmony created by blue flowers placed next to yellow ones. Other combinations such as blue and white, e.g. *Ceanothus* 'Blue Mound' and *Clematis montana*, yellow and red, e.g. *Euphorbia polychroma* and *Geum rivale*, yellow and white, e.g. *Verbascum nigrum* and *Tanacetum parthenium*, purple and pale yellow, e.g. *Salvia x superba* and *Achillea* 'Lucky Break', red and lavender, e.g. *Rosa gallica* and *Clematis integrifolia*.

Figure 5.10 **Flower border** showing the use of flower colour: light blue flowers of *Brunnera macrophylla* contrast with yellow of *Asphodeline lutea* and dark blue flowered *Anchusa azurea*

Check your learning

1. Define the terms primary root, secondary root, tap root, lateral root, fibrous root, adventitious root.

2. Describe the structure of the root tip.

3. Explain the function of the root cap and root hairs.

4. Describe the position of the different types of buds on the plant.

5. Explain the structure of the stem in relation to its functions.

6. Describe examples of leaf variation in terms of size, form and colour.

Further reading

Bell, A.D. (1991). *Plant Form: an Illustrated Guide to Flowering Plant Morphology*. OUP.

Brickell, C. (ed.) (2003). *A–Z Encyclopaedia of Garden Plants*. 2 vols, Dorling Kindersley.

Clegg, C.J. and Cox, G. (1978). *Anatomy and Activities of Plants*. John Murray.

Hattatt, L. (1999). *Gardening with Colour*. Paragon.

Ingram, D.S. *et al.* (eds) (2002). *Science and the Garden*. Blackwell Science Ltd.

Mauseth, J.P. (1998). *Botany – An Introduction to Plant Biology*. Saunders.

Chapter 6   Plant cells and tissues

Summary

This chapter includes the following:

- **Plant cells**
- **Cell division**
- **Plant tissues**
- **Stem and root anatomy**
- **Secondary growth**

with additional information on:

- **Meristems**
- **Cell differentiation**

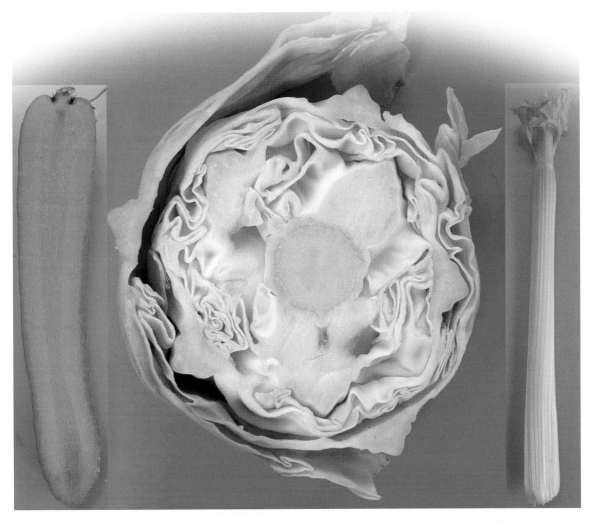

Figure 6.1 **Cabbage, carrot and celery** with examples of plant tissues, showing strengthening, structural and transport tissues

The anatomy of the plant

A **tissue** is a collection of specialized cells carrying out a particular function.

A close examination of the internal structure (anatomy) of the plant with a microscope will reveal how it is made up of different tissues. Each **tissue** is a collection of specialized cells carrying out one function, such as xylem conducting water. An **organ** is made up of a group of tissues carrying out a specific function, such as a leaf producing sugars for the plant. In the following chapter, the anatomy of the stem is illustrated and there is a comparable discussion of leaf structures in Chapter 8.

The cell

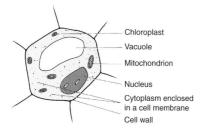

— Chloroplast
— Vacuole
— Mitochondrion
— Nucleus
— Cytoplasm enclosed in a cell membrane
— Cell wall

Figure 6.2 An unspecialized plant cell showing the organelles responsible for the life processes

Without the use of a microscope, the horticulturist will not be able to see cells, since they are very small (about a twentieth of a millimetre in size). They are very complex and scientific studies continue to discover more of the organization displayed in this fundamental unit.

A simple, unspecialized cell of parenchyma (see Figure 6.2) consists of a cellulose cell wall and contents (**protoplasm**) enclosed in a **cell membrane**, which is selective for the passage of materials in and out of the cell. The cellulose in the cell wall is laid down in a mesh pattern, which allows for stretching as the cell expands. Within the mesh framework are many apertures that, in active cells such as parenchyma, allow for strands of cytoplasm (called **plas-modesmata**) connecting between adjacent cells. These strands carry nutrients and hormones between cells, and are able to control the speed at which this movement takes place. When a plant wilts, its cells become smaller, but the plasmodesmata normally retain their links with adjacent cells. In the situation of 'permanent wilting' (see p342) or plasmolysis (see p123), there is a breakage of these strands and the plant is not able to recover.

Suspended in the jelly-like cytoplasm are small structures (organelles) each enclosed within a membrane and having specialized functions within the cell. In all tissues, the cell walls of adjoining cells are held together by calcium pectate (a glue-like substance which is an important setting ingredient in 'jam-making'). Some types of cell (e.g. xylem vessels) do not remain biochemically active, but die in order to achieve their usefulness. Here, the first wall of cellulose becomes thickened by additional cellulose layers and lignin, which is a strong, impervious substance.

The cell is made up of two parts, the nucleus and the cytoplasm. The nucleus coordinates the chemistry of the cell. The long chromosome strands that fill the **nucleus** (see also p138) are made up of the complex chemical deoxyri-bonucleic acid, usually known as DNA. In addition to its ability to produce more of itself for the process of cell division, DNA is also constantly manufacturing smaller but similar RNA (ribonucleic acid) units, which are able to pass through the nucleus membrane and attach themselves to other organelles. In this way, the nucleus is able to transmit instructions for the assembling, or destruction, of important chemicals within the cell.

There are six main types of organelle in the cytoplasm. The first, the **vacuole** is a sac containing dilute sugar, nutrients and waste products. It may occupy the major volume of a cell, and its main functions are storage and maintaining cell shape. The **ribosomes** make proteins from amino acids. Enzymes, which speed up chemical processes, are made of protein. The **Golgi apparatus** is involved in modifying and storing chemicals being made in the cell before they are transported where they are required. **Mitochondria** release energy in a controlled way, by the process of respiration, to be used by the other organelles. The energy is transferred via a chemical called ATP (adenosine triphosphate). The meristem areas of the stem, root and flower have cells with the highest number of mitochondria in order to help the rapid cell division and growth in these areas. **Plastids** such as the chloroplasts are involved in the production of sugar by the process of photosynthesis, and in the short-term storage of condensed sugar (in the form of starch). Lastly, the **endoplasmic reticulum** is a complex mesh of membranes that enables transport of chemicals within the cell, and links with the plasmodesmata at the cell surface. Ribosomes are commonly located on the endoplasmic reticulum. The whole of the living matter of a cell, nucleus and cytoplasm, are collectively called **protoplasm**.

Cell division

When a plant grows the cell number increases in the growing points or apical and lateral meristems of the stems and roots (see p93). The process of **mitosis** involves the division of one cell to produce two new ones.

> **Mitosis** is the process of cell division which results in a replication of cells.

The genetic information in the nucleus is reproduced exactly in the new cells to maintain the plant's characteristics. Each chromosome in the parent cell produces a duplicate of itself, thus producing sufficient material for the two new daughter cells (see Figure 6.3). A delicate, spindle-shaped structure ensures the separation of chromosomes, one complete set into each of the new cells. A dividing cell wall forms across the old cell to complete the division.

Tissues of the stem

Dicotyledonous stem

The internal structure of a dicotyledonous stem, as viewed in cross-section, is shown in Figure 6.4. Three terms, epidermis, cortex and pith are used to broadly describe the distribution of tissues across the stem. The **epidermis** is present as an outer protective layer of the stem, leaves and roots. It consists of a single layer of cells; a small proportion of them are modified to allow gases to pass through an otherwise impermeable layer (see stomata). The second general term, **cortex**, describes the zone of tissues found inside the epidermis and reaching inwards to the inner edge of the vascular bundles. A third term, **pith**,

1. INTERPHASE

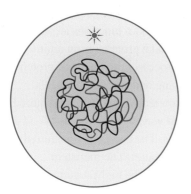

Before division, nucleus
is visible with a surrounding
membrane

2. PROPHASE

Nuclear envelope disappears, each
chromosome forms a duplicate
and arranges into a ball

3. METAPHASE

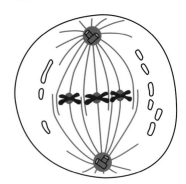

A spindle forms, chromosomes
attach to spindle and move to
centre of the cell

4. ANAPHASE

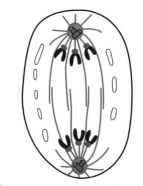

Chromosome pairs are pulled apart
and form two groups at either
end of the cell

5. TELOPHASE

Chromosomes form tight groups at
ends of cell; nuclear membrane
reforms; new cell wall forms to
divide cell into two identical
daughter cells

Figure 6.3 **Diagram to show the process of mitosis** (cell division)

refers to the central zone of the stem, which is mainly made up of
parenchyma cells.

Collenchyma and **sclerenchyma cells** are usually found to the inside
of the epidermis and are responsible for **support** in the young plant.
Both tissues have cells with specially thickened walls. When a cell
is first formed it has a wall composed mainly of **cellulose** fibres. In
collenchyma cells the amount of cellulose is increased to provide
extra strength, but otherwise the cells remain relatively unspecialized.
In sclerenchyma cells, the thickness of the wall is increased by the
addition of a substance called **lignin**, which is tough and causes the
living contents of the cell to disappear. These cells, which are long and
tapering and interlock for additional strength, consist only of cell walls.

The cortex of the stem contains a number of tissues. Many are made up of unspecialized cells such as **parenchyma**. In these tissues, the cells are thin walled and maintained in an approximately spherical shape by osmotic pressure (see p122). The mass of parenchyma cells (surrounding the other tissues) combine to maintain plant shape. Lack

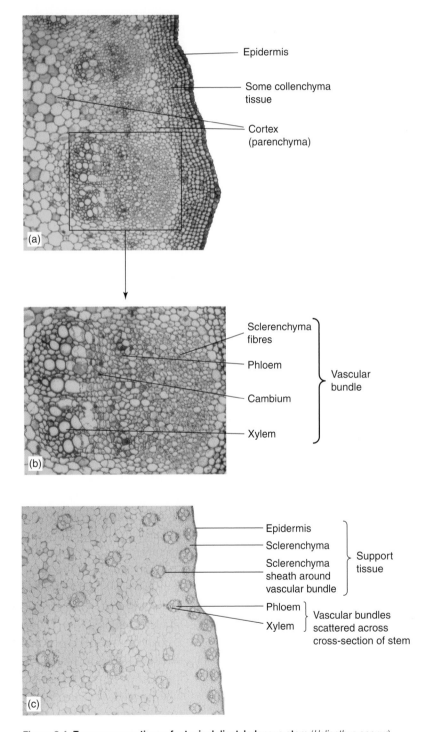

Figure 6.4 **Transverse sections of a typical dicotyledonous stem** (*Helianthus annuus*)
(a) and (b), a typical monocotyledonous stem (c) (*Zea*) and diagrams

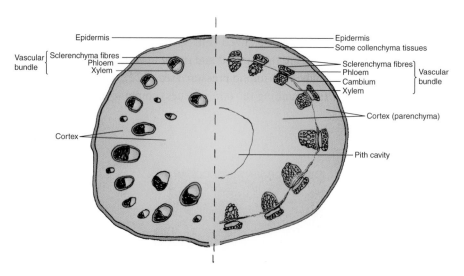

Figure 6.4 (d) A comparison between the stem tissue of monocotyledons (*Zea mays*) on the left and dicotyledonous (*Helianthus*) on the right

of water results in the partial collapse of the parenchyma cells and this becomes apparent as wilting. Parenchyma cells also carry out other functions, when required. Many of these cells contain chlorophyll (giving the stems their green colour) and so are able to photosynthesize. They release energy, by respiration, for use in the surrounding tissues. In some plants, such as the potato, they are also capable of acting as food stores (the potato tuber which stores starch). They are also able to undergo cell division, a useful property when a plant has been damaged. This property has practical significance when plant parts such as cuttings are being propagated, since new cells can be created by the parenchyma to heal wounds and initiate root development.

Contained in the cortex are **vascular bundles**, so named because they contain two vascular tissues that are responsible for transport. The first, **xylem**, contains long, wide, open-ended cells with very thick lignified walls, able to withstand the high pressures of water with dissolved minerals which they carry. The second vascular tissue, **phloem**, consists again of long, tube-like cells, and is responsible for the transport of food manufactured in the leaves carried to the roots, stems or flowers (see translocation). The phloem tubes, in contrast to xylem, have fairly soft cellulose cell walls. The end-walls are only partially broken down to leave sieve-like structures (sieve tubes) at intervals along the phloem tubes. Alongside every phloem tube cell, there is a small companion cell, which regulates the flow of liquids down the sieve tube. The phloem is seen on either side of the xylem in the marrow stem, but is found to the outside of the xylem in most other species. Phloem is penetrated by the stylets of feeding aphids (see Chapter 14). Also contained within the vascular bundles is the **cambium** tissue, which contains actively dividing cells producing more xylem and phloem tissue as the stem grows.

Figure 6.5 **A shredded leaf of *Phormium tenax*** showing fibres of xylem tissue

Stem growth

Growth of stems is initiated in the **apical**, or terminal, **bud** at the end of the stem (the apex). Deep inside the apical bud is a tiny mass of small, delicate jelly-like cells, each with a conspicuous nucleus but no cell vacuole. This mass is the **apical meristem** (see Figure 6.6). Here, cells divide frequently to produce four kinds of meristematic tissues. The first, at the very tip, continues as meristem cells. The second (protoderm) near the outside develops into the epidermis. The third (procambium) becomes the vascular bundles. The fourth (ground meristem) turns into the parenchyma, collenchyma, and sclerenchyma tissues of the cortex and pith. In addition to its role in tissue formation, the apical meristem also gives rise to small leaves (bud scales) that collectively protect the meristem. These scales and the meristem together form the **bud** (see p79). It should be noted that any damage to the sensitive meristem region by aphids, fungi, bacteria or herbicides would result in distorted growth. A fairly common example of such a distortion is **fasciation**, a condition that resembles a number of stems fused together. Buds located lower down the stem in the angle of the leaf (the axil) are called **axillary buds**; they contain a **lateral meristem** and often give rise to side branches.

In some plant families, e.g. the Graminae, the meristem remains at the base of the leaves, which are therefore protected against some herbicides, e.g. 2,4-D (see Chapter 13). This also means that grasses re-grow from their base after animals have grazed them. The new blades of grass grow from meristems between the old leaf and the stem. This means grasses can be **mown** which enables us to create lawns. The process of cutting back the grass also leads to it sending up several shoots instead of just one. This process of **tillering** helps thicken up the turf sward to make it such a useful surface for sport, as well as decoration. Mowing kills the dicotyledonous plants that have their stems cut off at the base and lose their meristems. However, many species are successful lawn weeds by growing in **prostrate** form; the foreshortened stem (very short internodes) creates a **rosette** of leaves that helps to conserve water, shades out surrounding plants and the growing point stays below the cutting height of the mower.

Elongation of the plant stem takes place in two stages. Firstly, cell division, described above, contributes a little. The second phase is cell **expansion**, which occurs at the base of the meristem. Here, the tiny unspecialized mer-istem cells begin to take in water and nutrients to form a cell vacuole. As a result, each cell elongates, and the stem rapidly grows. In the expansion zone, other developments begin to occur.

Cell differentiation

Most importantly, the cells begin to create their cell walls, and the connections between cells (plasmodesmata). The exact shape and chemical composition of the wall is different for each type of tissue cell, since it has a particular function to perform as described earlier;

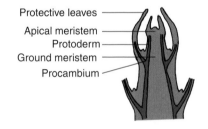

Protective leaves
Apical meristem
Protoderm
Ground meristem
Procambium

Figure 6.6 The tip of a dicotyledonous stem showing the four meristematic areas

sclerenchyma and collenchyma cells have walls thickened with lignin and cellulose while xylem and phloem vessels have developed walls and structures for transport. Leaf tissues similarly develop from parenchyma cells and form specialized tissues to carry out the process of photosynthesis. Leaf structure is described in Chapter 8.

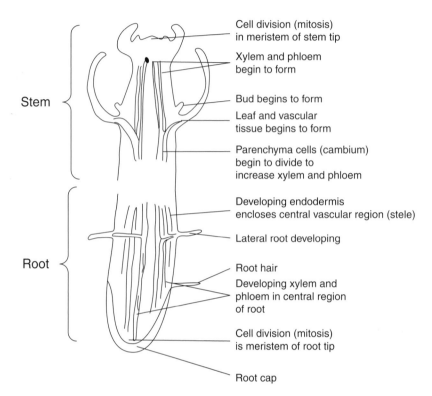

Figure 6.7 **Diagram of stem and root** showing areas of differentiation.

Secondary growth

Secondary growth results in the thickening of stems and roots and, in many cases, the production of wood.

As the stem length increases, so width also increases to support the bigger plant and supply the greater amount of water and minerals required.

The process in dicotyledons is called **secondary growth** (see Figure 6.8). Additional phloem and xylem are produced on either side of the cambium tissue, which now forms a complete ring. As these tissues increase towards the centre of the stem, so the circumference of the stem must also increase. Therefore a secondary ring of cambium (cork cambium) is formed, just to the inside of the epidermis, the cells of which divide to produce a layer of corky cells on the outside of the stem. This layer will increase with the growth of the tissue inside the stem, and will prevent loss of water if cracks should occur. As more secondary growth takes place, so more phloem and xylem tissue is produced but the phloem tubes, being soft, are squashed as the more numerous and very hard xylem vessels occupy more and more of the cross-section of the stem. Eventually, the majority of the stem consists of secondary xylem that forms the **wood**.

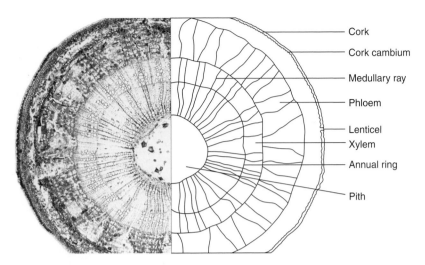

Figure 6.8 **Cross-section of lime** (*Tilia europea*) stem showing tissues produced in secondary growth

The central region of xylem sometimes becomes darkly stained with gums and resins (**heartwood**) and performs the long-term function of support for a heavy trunk or branch. The outer xylem, the **sapwood**, is still functional in transporting water and nutrients, and is often lighter in colour. The xylem tissue produced in the spring has larger diameter vessels than autumn-produced xylem, due to the greater volume of water that must be transported; a distinct ring is therefore produced where the two types of tissue meet. As these rings will be formed each season, their number can indicate the age of the branch or trunk; they are called **annual rings**. The phloem tissue is pushed against the cork layers by the increasing volume of xylem so that a woody stem appears to have two distinct layers, the wood in the centre and the **bark** on the outside.

If **bark** is removed, the phloem also will be lost, leaving the vascular cambium exposed. The stem's food transport system from leaves to the roots is thus removed and, if a trunk is completely **ringed** (or 'girdled'), the plant will die. Rabbits or deer in an orchard may cause this sort of damage. 'Partial ringing', i.e. removing the bark from almost the whole of the circumference, can achieve a deliberate reduction in growth rate of vigorous tree fruit cultivars and woody ornamental species. Initially, the **bark** is smooth and shiny, but with age it thickens and the outer layers accumulate chemicals (including suberin) that make it an effective protection against water loss and pest attack. This part of the bark (called **cork**) starts to peel or flake off. This is replaced from below and the cork gradually takes on its characteristic colours and textures. Many trees such as silver birch, London Plane, *Prunus serrula*, *Acer davidii* and many pines and rhododendrons have attractive bark and are particularly valued for winter interest (see Figure 6.9).

Since the division of cells in the cambium produces secondary growth, it is important that when **grafting** a **scion** (the material to be grafted) to a **stock**, the vascular cambium tissues of both components be positioned as close to each other as possible (see p92). The success of a graft depends very much on the rapid **callus** growth derived from the cambium, from

Figure 6.9 **Examples of the decorative effects of tree bark** (a) *Myrtus luma* (b) *Euonymus alatus* (c) *Eucalytus parvifolia* (d) *Quercus agrifolia* (e) *Caucasian Wing-nut (Pterocarya fraxinifolia)* (f) *Eucalyptus urnigera* (g) *Pinus nigra ssp. Salzmannii* (h) *Betula utilis jacquemontii* (i) White Willow (*Salix alba*) (j) *Prunus serula tibetica* (k) Date Plum (*Diospyros lotus*) (l) Black Walnut (*Juglans nigra*)

which new cambial cells form and subsequently from which the new xylem and phloem vessels form to complete the union. The two parts then grow as one to carry out the functions of the plant stem.

A further feature of a woody stem is the mass of lines radiating outwards from the centre, most obvious in the xylem tissues. These are **medullary**

Figure 6.10 **Bark of silver birch** showing lenticels

rays, consisting of parenchyma tissue linking up with small areas on the bark where the corky cells are less tightly packed together (**lenticels**). These allow air to move into the stem and across the stem from cell to cell in the medullary rays. The oxygen in the air is needed for the process of **respiration**, but the openings can be a means of entry of some diseases, e.g. Fireblight. Other external features of woody stems include the **leaf scars** which mark the point of attachment of leaves fallen at the end of a growing season, and can be a point of entry of fungal spores such as apple canker.

Monocotyledonous stem

This has the same functions as those of a dicotyledon; therefore the cell types and tissues are similar. However, the arrangement of the tissues does differ because increase in diameter by **secondary growth** does not take place. The stem relies on extensive sclerenchyma tissue for support that, in the maize stem shown in Figure 6.4, is found as a sheath around each of the scattered vascular bundles. Monocotyledonous stem structures are seen at their most complex in the palm family. From the outside, the trunk would appear to be made of wood, but an internal investigation shows that the stem is a mass of sclerified vascular bundles. The absence of secondary growth in the vascular bundles makes the presence of cambium tissue unnecessary.

Secondary thickening is found not only in trees and shrubs, but also in many herbaceous perennials and annuals that have woody stems. However, trees and shrubs do exhibit this feature to the greatest extent.

Tissues of the root

The layer with the root hairs, the **epidermis**, is comparable with the epidermis of the stem; it is a single layer of cells which has a protective as well as an absorptive function. Inside the epidermis is the parenchymatous **cortex** layer. The main function of this tissue is respiration to produce energy for growth of the root and for the absorption of mineral nutrients. The cortex can also be used for the storage of food where the root is an overwintering organ (see p79).

The cortex is often quite extensive and water must move across it in order to reach the transporting tissue that is in the centre of the root. This central region, called the **stele**, is separated from the cortex by a single layer of cells, the **endodermis**, which has the function of controlling the passage of water into the stele. A waxy strip forming part of the cell wall of many of the endodermal cells (the Casparian strip) prevents water from moving into the cell by all except the cells outside it, called **passage cells**.

Water passes through the endodermis to the **xylem** tissue, which transports the water and dissolved minerals up to the stem and leaves. The arrangement of the xylem tissue varies between species, but often appears in transverse section as a star with varying numbers of 'arms'. **Phloem** tissue is responsible for transporting carbohydrates from the leaves as a food supply for the production of energy in the cortex.

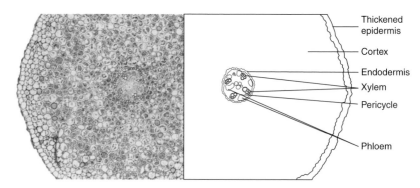

Figure 6.11 **Cross-section of *Ranunculus* root** showing thickened outer region, large area of cortex and central vascular region or stele enclosed in an endodermis.

A distinct area in the root inside the endodermis, the **pericycle**, supports cell division and produces lateral roots, which push through to the main root surface from deep within the structure. Roots age and become thickened with waxy substances, and the uptake rate of water becomes restricted.

Check your learning

1. Define the term 'tissue'.

2. List the main types of tissue found in the plant.

3. Explain the function of the tissues named.

4. State the tissues involved in the strengthening in monocotyledons.

5. State the function of the cell components.

6. Describe the role of meristems in cell division.

7. Describe the process of secondary growth in dicotyledons.

8. Describe how the internal structure of a root differs from that of the stem.

Further reading

Bowes, B.G. (1996). *A Colour Atlas of Plant Structure*. Manson Publishing Ltd.

Clegg, C.J. (2003). *Green Plants: the Inside Story*. III Adv Biology series. Hodder Murray.

Clegg, C.J. and Cox, G. (1978). *Anatomy and Activities of Plants*. John Murray.

Cutler, D.F. (1978). *Applied Plant Anatomy*. Longman.

Ingram, D.S. *et al.* (eds) (2002). *Science and the Garden*. Blackwell Science Ltd.

Mauseth, J.P. (1998). *Botany – An Introduction to Plant Biology*. Saunders.

Chapter 7   Plant reproduction

Summary

This chapter includes the following topics:

- **Types of inflorescence**
- **Flower structure**
- **Function of flower parts**
- **Characteristics of flowers**
- **Tepals**
- **Seeds and fruits**

with additional information on:

- **Parthenocarpy**

Figure 7.1 *Euphorbia cyparissias* 'Fens Ruby'

Figure 7.2 Range of flowers as organs of **sexual reproduction** having similar basic structure, but varying appearance having adapted for successful pollination or by plant breeding (*see* chapter 10) (a) *Iris chrysographes* 'Kew Black'; (b) *Eryngium giganteum*, ('Miss Willmott's ghost'); (c) *Trollius chinensis* 'Golden Queen'; (d) *Rosa* 'L.D.Braithwaite'; (e) *Hemerocallis* 'Rajah'; (f) *Aquilegia fragrans*; (g) *Oenothera* 'Apricot Delight; (h) *Helenium* 'Wyndley'; (i) *Helleborus xhybridus*; (j) *Nepeta nervosa*; (k) *Primula vialii*

Flowers

Types of inflorescence

The organs of sexual reproduction in the flowering plant division are flowers, and variation in their arrangement can be identified and named:

- **spike** is an individual, unstalked series of flowers on a single flower stalk, e.g. Verbascum;

Figure 7.3 Inflorescence types, (a) spike; *Verbascum* (b) raceme; Foxglove and (c) *Veronica*; (d) corymb, *Achillea*; (e) umbel, Hogweed; (f) capitulum, *Inula*

- **raceme** consists of individual stalked flowers, the stalks all the same length again spaced out on a single undivided main flower stalk, e.g. foxglove (see Figure 7.3), hyacinth, lupin, wallflower;
- **compound racemes** have a number of simple racemes arranged in sequence on the flower stalk, e.g. grasses;
- **corymb** is similar to a raceme except that the flower stalks, although spaced out along the main stalk, are of different lengths so that the flowers are all at the same level, e.g. *Achillea* (see Figure 7.3). A very common sight in hedgerows;
- **umbel** has stalked flowers reaching the same height with the stalks seeming to start at the same point on the main stem, e.g. hogweed (see Figure 7.3);
- **capitulum** or composite flower forms a disc carrying flower parts radiating out from the centre, as if compressed from above, e.g. *Inula* (see Figure 7.3), daisy, chrysanthemum.

The number and arrangement of flower parts are the most important features for classification and are a primary feature in plant identification (see Chapter 4, p67).

Flower structure

The flower structure is shown in Figure 7.4.

The flower is initially protected inside a flower bud by the **calyx** or ring of **sepals**, which are often green and can therefore photosynthesize. The development of the flower parts requires large energy expenditure by the

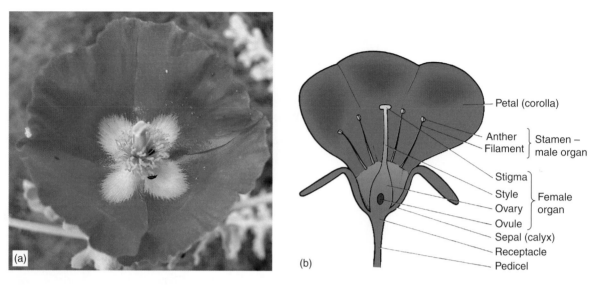

Figure 7.4 **Flower structure**, e.g. (a) flower of *Glaucium corniculatum* and (b) diagram of typical flower to show structures involved in the process of sexual reproduction

Figure 7.5 **Tulip 'Attila'**, e.g. of tepals – outer layers of flower are similar

A plant possessing flowers with both male and female organs is **hermaphrodite**.

Species with separate male and female flowers on the same plant are **monoecious**.

Species which produce male and female flowers on different plants are **dioecious**.

plant, and therefore vegetative activities decrease. The **corolla** or ring of **petals** may be small and insignificant in **wind-pollinated flowers**, e.g. grasses (see p134), or large and colourful in **insect-pollinated species** (see p135). The colour and size of petals can be improved in cultivated plants by breeding, and may also involve the multiplication of the petals or **petalody**, when fewer male organs are produced.

The flower may include other parts:

- Tepals, where the outer layers of the flower have a similar appearance, making the sepals and petals indistinguishable. They are common in monocotyledons such as tulips (see Figure 7.5) and lilies.
- Androecium, the male organ, consists of a **stamen which** bears an **anther** that produces and discharges the **pollen grains**.
- Gynaecium, the female organ, is positioned in the centre of the flower and consists of an **ovary** containing one or more **ovules** (egg cells). The **style** leads from the ovary to a **stigma** at its top where pollen is captured.
- The flower parts are positioned on the **receptacle**, which is at the tip of the **pedicel** (flower stalk).
- **Nectaries** may develop on the receptacle, at the base of the petals; these have a secretory function, producing substances such as nectar which attract pollinating organisms.
- Associated with the flower head or **inflorescence** are leaf-like structures called **bracts**, which can sometimes assume the function of insect attraction, e.g. in *Poinsettia*.

The flowers of many species have both male and female organs (hermaphrodite), but some have separate male and female flowers (monoecious), e.g. *Cucurbita*, walnut, birch (*Betula*), whereas others produce male and female flowers on different plants (dioecious), e.g. holly, willows, *Skimmia japonica* and ***Ginkgo biloba***.

The seed

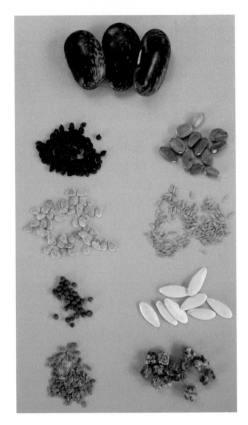

Figure 7.6 **Seeds**: a range of species, from top – runner bean, left to right – leek, artichoke, tomato, lettuce, Brussels sprout, cucumber, carrot, beetroot

The seed, resulting from sexual reproduction, creates a new generation of plants that bear characteristics of both parents. The plant must survive often through conditions that would be damaging to a growing vegetative organism. The seed is a means of protecting against extreme conditions of temperature and moisture, and is thus the **overwintering stage**.

Seed structure

The basic seed structure is shown in Figure 7.7. The main features of the seed are:

- **embryo**, in order to survive the seed must contain a small immature plant protected by a seed coat;
- **testa**, the seed coat, is formed from the outer layers of the ovule after **fertilization**;
- **micropyle**, a weakness in the testa, marks the point of entry of the pollen tube prior to fertilization;
- **hilum**, this is the point of attachment to the fruit.

The embryo consists of a **radicle**, which will develop into the primary root of the seedling, and a **plumule**, which develops into the shoot system, the two being joined by a region called the **hypocotyl**. A single seed leaf (**cotyledon**) will be found in monocotyledons, while two are present as part of the embryo of dicotyledons. The cotyledons may occupy a large part of the seed, e.g. in beans, to act as the food store for the embryo.

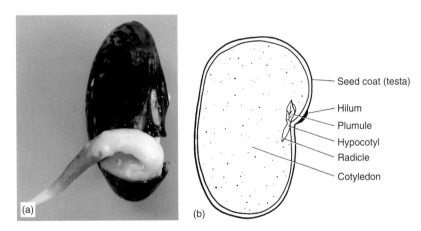

Figure 7.7 **The structure of the seed**, (a) runner bean seed just beginning to germinate and showing developing radicle showing a geotropic response (*see* page 000), (b) long section of bean seed showing structure

The **seed** is formed from the ovule of the flower and is the result of the reproductive process.

In some species, e.g. grasses and Ricinus (castor oil plant), the food of the seed is found in a different tissue from the cotyledons. This tissue is called **endosperm** and is derived from the fusion of extra cell nuclei, at the same time as fertilization. Plant food is usually stored as the carbohydrate, starch, formed from sugars as the seed matures, e.g. in peas and beans. Other seeds, such as sunflowers, contain high proportions of fats and oils, and proteins are often present in varying proportions. The seed is also a rich store of nutrients that it requires when a seedling, such as phosphate (see p367).

The seed structure may be specialized for wind dispersal, e.g. members of the Asteraceae family, including groundsel, dandelion and thistle, which have parachutes, as does *Clematis* (Ranunculaceae). Many woody species such as lime (*Tilia*), ash (*Fraxinus*), and sycamore (*Acer*) produce winged fruit. Other seed-pods are explosive, e.g. balsam and hairy bittercress. Organisms such as birds and mammals distribute hooked fruits such as goosegrass and burdock, succulent types (e.g. tomato, blackberry, elderberry), or those that are filled with protein (e.g. dock). Dispersal mechanisms are summarized in Table 7.1.

Seeds are contained within fruits which provide a means of protection and, often, dispersal.

The fruiting plant

The development of the true fruit involves either the expansion of the ovary into a juicy **succulent** structure, or the tissues becoming hard and dry. In false fruits other parts, such as the inflorescence, e.g. pineapple and mulberry, and the receptacle, as in apple, become part of the structure

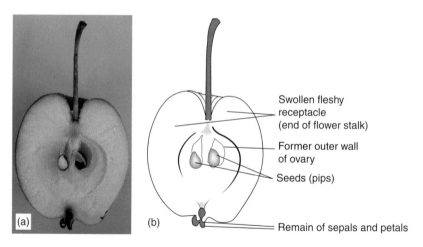

Figure 7.8 **Apple – a** false fruit (pome) (a) LS of crab apple and (b) showing structure

The **fruit** is the protective and distributary structure for the seed and forms from the ovary after fertilization.

The succulent fruits are often eaten by animals, which help seed dispersal, and may also bring about chemical changes to break dormancy mechanisms (see p151). Some fruits (described as being **dehiscent**), release their seeds into the air. They do this either by an explosive method as seen in brooms and poppies; or by tiny feathery parachutes, seen in willow herb and groundsel. Dry fruits may rot away gradually to release their seeds by an **indehiscent** action. Different adaptations of fruit, many of which are of economic import ance, and the methods by which seeds are dispersed are summarized in Table 7.1 and illustrated in Figure 7.9.

A **false fruit** is formed from parts other than, or as well as, the ovary wall.

Table 7.1 Fruits and the dispersal of seeds

True Fruits (formed from the ovary wall after fertilization):		
Succulent (indehiscent)	Drupes	Cherry, plum Blackberry (collection of drupes)
	Berries	Gooseberry, Marrow, banana
Dry indehiscent	Schizocarps	Sycamore
	Samara	Trefoil
	Lomentum	Hogweed
	Cremocarb	Hollyhocks
	Carcerulus	Acorn, rose, strawberry
	Achenes (nuts)	
Dry dehiscent	Capsules	Poppy, Violet, Campanula
	Siliquas	Wallflowers, stocks
	Siliculas	Shepherd's purse, honesty
	Legumes	Pea, bean, lupin
	Follicles	Delphinium, monkshood
False Fruits (formed from parts other than, or as well as, the ovary wall):		
From inflorescence	Pineapple, mulberry	
From receptable	Apple, pear	
Method of seed dispersal	**Type of fruit**	**Examples**
Animals	Succulent	Elderberry, blackberry – eaten by birds
		Mistletoe, yew – stick to beaks
	Hooked	Burdock, goose-grass – catch on fur
Wind	Winged	Ash, sycamore, lime, elm
	Parachutes	Dandelion, clematis, thistles
	Censer (dry capsules)	Poppy, campion, antirrhinum
Explosion	Pods	Peas, lupins, gorse, vetches, geranium

Fruit set

The process of pollination, in most species, stimulates fruit set. The hormones, in particular gibberellins, carried in the pollen, trigger the production of auxin in the ovary, which causes the cells to develop. In species such as cucumber, the naturally high content of auxin enables fruit production without prior fertilization, i.e. **parthenocarpy**, a useful phenomenon when the object of the crop is the production of seedless fruit. Such activity can be simulated in other species, especially when poor conditions of light and temperature have caused poor fruit set in species such as tomato and peppers. Here, the flowers are sprayed with an auxin-like chemical, but the quality of fruit is usually inferior. Pears can be sprayed with a solution of gibberellic acid to replace the need for pollination. Fruit ripening occurs as a result of hormonal changes and

Fruits

Succulent

(a) Drupe, e.g. Sloe   (b) Berry, e.g. *Viburnum*

Dry indehiscent

Samara, e.g. Sycamore   Lomentum, e.g. Trefoil   Cremocarb, e.g. Hogweed

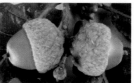

Carcerulus, e.g. Hollyhock   Achene, e.g. Acorn

Dry dehiscent

Capsule, e.g. Poppy   Siliqua   Silicula, e.g. Honesty

Legume, e.g. Lupin   Follicle, e.g. Monkshood

Seed dispersal

Eaten by animals e.g. Blackberry   Hooked, e.g. Burdock   Winged, e.g. Ash

Parachute, e.g. Dandelion   Censer, e.g. Antirrhinum   Pod, e.g. *Geranium*

Figure 7.9 **Fruit types** and seed dispersal

involves in tomatoes a change in the sugar content, i.e. at the crucial stage called **climacteric**. After this point, fruit will continue to ripen and also respire after removal from the plant. Ethylene is released by ripening fruit, which contributes to deterioration in store. Early ripening can be brought about by a spray of a chemical, e.g. ethephon, which stimulates the release of ethylene by the plant, e.g. in the tomato.

Reproduction in simple multicellular green plants

The seed-producing plants represent the most important division of the plant kingdom in horticulture. Other, simpler multicellular green plants reproduce sexually, but also asexually. **Alternation of generations** exists when two stages of quite distinct types of growth occur. In ferns (Pteridophyta), a vegetative stage produces a spore forming body on the underside of leaves (see Figure 7.10). Spores are released and, with suitable damp conditions, germinate to produce a sexual leafy stage in which male and female organs develop and release cells which fertilize and develop in the body of the plant. These spores then germinate while nourished by the sexual leafy stage and develop in turn into a new vegetative plant. Ferns can be produced in cultivation by spores if provided with damp sterile conditions to allow the tiny spores to germinate without competition (see Figure 7.11). Vegetative propagation by division of plants or rhizomes is common.

Figure 7.10 **Fern spores** on underside of leaves, *Dryopteris erythrosora* and *Phyllitis scolopendrium Cristata*

Many plants are able to reproduce both sexually and asexually by vegetative propagation. This is described in detail in Chapter 12.

Figure 7.11 **Germinating fern spores** and plantlets

Check your learning

1. Describe the main types of inflorescence of flowering plants.
2. Describe the structure and function of the parts of a typical dicotyledonous flower.
3. Describe the structure of a seed.
4. Define the terms: monoecious, dioecious, hermaphrodite, seed, fruit, false fruit.
5. Describe the main types of fruit, namely dry, hard, fleshy, indehiscent and give an example of each.

Further reading

Bleasdale, J.K.A. (1983). *Plant Physiology in Relation to Horticulture*. Macmillan.

Leopold, A.C. (1977). *Plant Growth and Development*. 2nd edn. McGraw-Hill.

Raven, P.H. *et al.* (2005). *Biology of Plants*. W.H. Freeman.

Wareing, P.F. and Phillips, I.D.J. (1981). *The Control of Growth and Differentiation in Plants*. Pergamon.

Chapter 8   Plant growth

Summary

This chapter includes the following topics:

- **Photosynthesis**
- **Leaf structure**
- **Respiration**

with additional information on:

- **Lighting of crops**
- **Storage**
- **Carbon chemistry**

Figure 8.1 **Pumpkins** representing the large amount of growth a plant can produce

In any horticultural situation, growers are concerned with controlling and even manipulating plant growth. They must provide for the plants the optimum conditions to produce the most efficient growth rate and the end product required. Therefore the processes that result in growth are explored in order that the most suitable or economic growth can be achieved. Photosynthesis is probably the single most important process in plant growth. Respiration is the process by which the food matter produced by photosynthesis is converted into energy usable for growth of the plant. Photosynthesis and respiration make these processes possible, and the balance between these results in growth. It must be emphasized that growth involves the plant in hundreds of chemical processes, occurring in the different organs and tissues throughout the plant.

Growth is a difficult term to define because it really encompasses the totality of all the processes that take place during the life of an organism. However, it is useful to distinguish between the processes which result in an increase in size and weight, and those processes which cause the changes in the plant during its life cycle, which can usefully be called development, described in Chapters 7 and 11.

Photosynthesis

The following environmental requirements for photosynthesis are explained in detail below:

- carbon dioxide
- light
- adequate temperature
- water (see also Chapters 9 and 10).

All living organisms require organic matter as food to build up their structure and to provide chemical energy to fuel their activities. Whilst photosynthesis is the crucial process, it should be remembered that a multitude of other processes are occurring all over the plant. Proteins are being produced, many of which are enzymes necessary to speed up chemical reactions in the leaf, the stem, the root, and later in the flower and fruit. The complex carbohydrate, cellulose, is being built up as cell walls of almost every cell. Nucleo-proteins are being provided in meristematic areas to enable cell division. These are three examples of many, to show that growth involves much more than just photosynthesis and respiration.

Photosynthesis is the process in the chloroplasts of the leaf and stem by which green plants manufacture food in the form of high energy carbohydrates such as sugars and starch, using light as energy.

All the complex organic compounds, based on carbon, must be produced from the simple raw materials, water and carbon dioxide. Many organisms are unable to manufacture their own food, and must therefore feed on already manufactured organic matter such as plants or animals. Since large animals predate on smaller animals, which themselves feed on plants, all organisms depend directly or indirectly on photosynthesis occurring in the plant as the basis of a **food web** or chain (see p53).

A summary of the process of photosynthesis is given in Table 8.1 as a word formula and as a chemical equation. This apparently simple equation represents, in reality, two different stages in the production of glucose. The first, the 'light reaction' occurs during daylight, and splits water into hydrogen and oxygen. The second, the 'dark reaction' occurring at night, takes the hydrogen and joins it to carbon dioxide to make glucose.

Table 8.1 Two ways to represent the chemistry involved in photosynthesis

a. Written in a conventional way, the process can be expressed in the following way:
carbon dioxide plus water plus light gives rise to glucose plus oxygen (when in the presence of chlorophyll)

b. Written in the form of a chemical equation, which represents molecular happenings at the sub-microscopic level, the above sentence becomes:
6 C_2O molecules + 6 H_2O molecules plus light give rise to 1 $C_6H_{12}O_6$ molecule + 6 O_2 molecules (when in the presence of chlorophyll)

Most plant species follow a **'C–3'** process of photosynthesis where the intermediate chemical compound contains three carbon atoms (C–3) before producing the six-carbon glucose molecule (C–6). Many C–3 plants are not able to increase their rate of photosynthesis under very high light levels.

In contrast, a **'C–4'** process is seen in many tropical families, including the maize family, where plants which use an intermediate compound containing four carbon atoms (C4) are able to continue to respond to very high levels of light, thus increasing their productivity.

A third process, called **'CAM'** (Crassulacean Acid Metabolism) was first discovered in the stonecrop family. Here, the intermediate chemical is a different four-carbon compound, malic acid. This third process has more recently been found in several other succulent plant families (including cacti), all of which need to survive conditions of drought. Such plants need to keep their stomata closed during the heat of the day, but this prevents the entry of carbon dioxide. During the night, carbon dioxide is absorbed and stored as malic acid, ready for conversion to glucose the next day. CAM plants do not normally grow very fast because they are not able to store large quantities of this malic acid, and thus their potential for glucose production is limited.

Carbon chemistry

All living organisms, from viruses to whales, have the element carbon at the centre of their chemistry. The study of this element's chemical activity is called **organic** chemistry. Originally, it was thought that all organic compounds came from living organisms, but modern chemistry has brought synthetic urea fertilizer and DDT, which are organic.

Unusually in the range of chemical elements, carbon (like silicon) has a combining power, or valency, of four (see p387 for **basic chemistry**). This means that each carbon atom can react with four other atoms, whether they are atoms of other elements such as hydrogen, or with more carbon atoms. Since the four chemical bonds from the carbon atom point in diametrically opposite directions, it can be seen that molecules containing carbon are three dimensional, a feature which is very important in the chemistry of living things.

Carbon normally forms **covalent** (non-ionic) bonds in its molecules. Here, the bond shares the electrons, and so there is no chemical charge associated with the molecule (see also ionic bonds). Two of the simplest molecules to contain carbon are carbon dioxide and methane (see Figure 8.2).

Carbon dioxide, or CO_2, is a constituent of the air we breathe and the sole source of carbon for almost all plants. In this compound, the carbon attaches to two oxygen atoms, each with a valency of two.

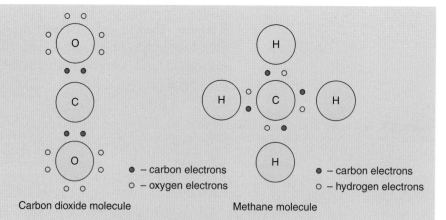

● – carbon electrons
○ – oxygen electrons

● – carbon electrons
○ – hydrogen electrons

Carbon dioxide molecule

Methane molecule

Figure 8.2 **Carbon dioxide and methane**

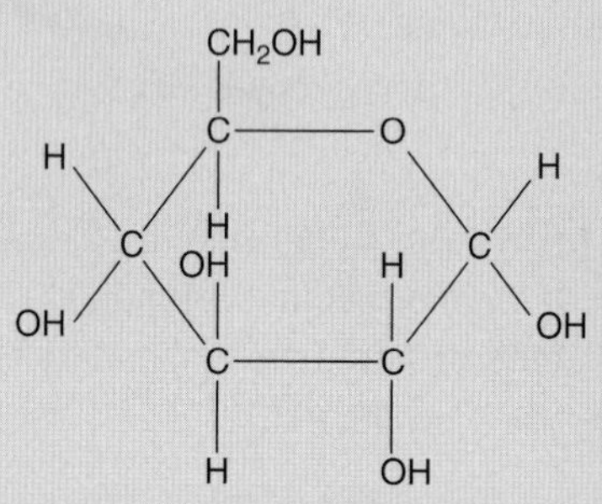

CH₂OH

Figure 8.3 **Sugar molecule**

Methane, or CH_4, is often referred to as 'marsh gas' and is the major constituent of the North Sea gas supply. In this compound, the carbon atom is bonded with four hydrogen atoms, each of which has a valency of one. Most fuels, such as petrol, are chemically related to methane, but with more carbon atoms in the molecule. This family of chemicals (containing progressively more carbons in the molecule are methane, butane, propane and octane) is collectively called the **hydrocarbons**. The molecules in petrol mainly contain eight carbon atoms, hence the term 'octane'.

Slightly more complicated is the glucose molecule ($C_6H_{12}O_6$). Here there is a circular molecule and Figure 8.3 indicates its three-dimensional structure. Glucose is the molecule produced by photosynthesis. It is the starting point for the synthesis of all the many molecules used by the plant, i.e. starch, cellulose (see Figure 8.4), proteins, pigments, auxins, DNA, etc. These molecules may contain chains with hundreds of carbon atoms joined in slightly different ways, but the basic chemistry is the same. All the valency (see p388) requirements of carbon, oxygen and hydrogen are still met.

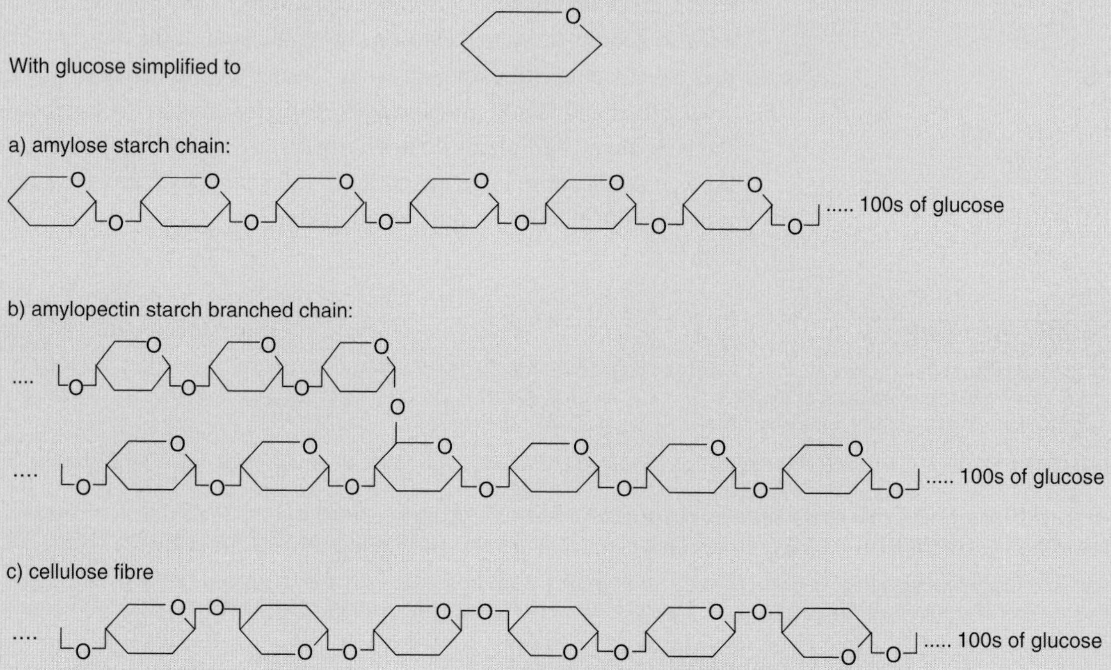

With glucose simplified to

a) amylose starch chain:

..... 100s of glucose

b) amylopectin starch branched chain:

....

.... 100s of glucose

c) cellulose fibre

.... 100s of glucose

Figure 8.4 Starch and cellulose molecules

Requirements for photosynthesis

Carbon dioxide

In order that a plant may build up organic compounds such as sugars, it must have a supply of carbon which is readily available. **Carbon dioxide** is present in the air in concentrations of 330 ppm (parts per million) or 0.03 per cent, and can diffuse into the leaf through the stomata. Carbon dioxide gas moves ten thousand times faster in air than it would in solution through the roots. The amount of carbon dioxide in the air immediately surrounding the plant can fall when planting is very dense, or when plants have been photosynthesizing rapidly, especially in an unventilated greenhouse.

This reduction will slow down the rate of photosynthesis, but a grower may supply additional carbon dioxide inside a greenhouse or polythene tunnel to **enrich** the atmosphere up to about three times the normal concentration, or an optimum of 1000 ppm (0.1 per cent) in lettuce. Such practices will produce a corresponding increase in growth, provided other factors are available to the plant. If any one of these is in short supply, then the process will be slowed down. This principle, called the law of limiting factors, states that the factor in least supply will limit the rate of the process, and applies to other non- photosynthetic processes in the plant. It would be wasteful, therefore, to increase the carbon dioxide concentration artificially, e.g. by burning propane gas, or releasing pure carbon dioxide gas, if other factors were not proportionally increased.

Figure 8.5 **CO₂ burner** enriches the glasshouse environment, so helping provide an optimum growing environment

Light

Light is a factor required for photosynthesis to occur. In any series of chemical reactions where one substance combines with another to form a larger compound, energy is needed to fuel the reactions. Energy for photosynthesis is provided by light from the sun or from artificial lamps. As with carbon dioxide, the amount of light energy present is important in determining the rate of photosynthesis – simply, the more light or greater **illuminance** (intensity) absorbed by the plant, the more photosynthesis can take place. Light energy is measured in joules/square metre, but for practical purposes the light for plant growth is measured according to the light falling on a given area, that is lumens per square metre (lux). More recently, the unit 'microwatts per sq. metre' has been introduced. One lux, in natural sunlight, is equal to microwatts per sq. metre. Whilst the measurement of illuminance is a very useful tool for the grower, it is difficult to state the plant's precise requirements, as variation occurs with species, age, temperature, carbon dioxide levels, nutrient supply and health of the plant.

However, it is possible to suggest approximate limits within which photosynthesis will take place; a minimum intensity of about 500–1000 lux enables the plant's photosynthesis rate to keep pace with **respiration**, and thus maintain itself. The maximum amount of light many plants can usefully absorb is approximately 30 000 lux, while good growth in many plants will occur at 10 000–15 000 lux. Plant species adapted to shade conditions, however, e.g. *Ficus benjamina* , require only 1000 lux. Other shade-tolerant plants include Taxus spp., Mahonia and Hedera. In summer, light intensity can reach 50 000–90 000 lux and is therefore not limiting, but in winter months, between November and February, the low natural light intensity of about 3000–8000 lux is the limiting factor for plants actively growing in a heated greenhouse or polythene tunnel. Care must be taken to maintain clean glass or polythene, and to avoid condensation that restricts light transmission. Intensity can be increased by using artificial lighting, which can also extend the length of day, which is short during the winter, by **supplementary lighting**. This method is used for plants such as lettuce, bedding plants and brassica seedlings.

Total replacement lighting

Growing rooms which receive no natural sunlight at all use controlled temperatures, humidities, and carbon dioxide levels, as well as light. Young plants which can be grown in a relatively small area, and which are capable of responding well to good growing conditions in terms of growth rate, are often raised in a growing room.

The type of lamp

Lamps are chosen for increasing intensity, and therefore more photosynthesis. All such lamps must have a relatively high efficiency of conversion of electricity to light, and only **gas discharge** lamps are able to do this. Light is produced when an electric arc is formed across the gas filament enclosed under pressure inside an inner tube. Light, like other forms of energy, e.g. heat, X-rays and radio waves, travels in the form of waves, and the distance between one wave peak and the next is termed the wavelength. Light wavelengths are measured in nanometres (nm); 1 nm = one thousandth of a micrometre. Visible light wavelengths vary from 800 nm (red light – in the long wavelength area) to 350 nm (blue light – in the short wavelength area), and a combination of different wavelengths (colours) appears as white light. Each type of lamp produces a characteristic wavelength range and, just as different coloured substances absorb and reflect varying colours of light, so a plant absorbs and reflects specific wavelengths of light.

Since the photosynthetic green pigment chlorophyll absorbs mainly red and blue light and reflects more of the yellow and green part of the spectrum, it is important that the lamps used produce a balanced wavelength spectrum to include as high a proportion of those colours as possible, in order that the plant makes most efficient use of the light provided. The gas included in a lamp determines its light characteristics. The two most commonly used gases for horticultural lighting are mercury vapour, producing a green blue light with no red, and sodium, producing yellow light. This limited spectrum may be modified by the inclusion of fluorescent materials in the inner tube, which allow the tube to re-emit wavelengths more useful to the plant emitted by the gas and re-emit the energy as a shorter wavelength. Thus, modified mercury lamps produce the desirable red light missing from the basic emission.

Figure 8.6 **High-pressure sodium lamp** is used for supplementary lighting

Low-pressure mercury-filled tubes produce diffuse light and, when suitably grouped in banks, provide uniform light close to plants. These are especially useful in a growing room, provided that they produce a broad spectrum of light as is seen in the 'full spectrum fluorescent tubes'. Gas enclosed at **high pressure** in a second inner tube produces a small, high intensity source of light. These small lamps do not greatly obstruct natural light entering a greenhouse and, while producing valuable uniform supplementary illumination at a distance, cause no leaf scorch. Probably the most useful lamp for supplementary lighting in a greenhouse is a high-pressure sodium lamp, which produces a high intensity of light, and is relatively efficient (27 per cent).

Carbon dioxide **enrichment** should be matched to artificial lighting in order to produce the greatest growth rate and most efficient use of both factors.

Temperature

The complex chemical reactions which occur during the formation of carbohydrates from water and carbon dioxide require the presence of chemicals called **enzymes** to accelerate the rate of reactions. Without these enzymes, little chemical activity would occur. Enzyme activity in living things increases with temperature from 0°C to 36°C, and ceases at 40°C. This pattern is mirrored by the effect of air temperature on the rate of photosynthesis. But here, the optimum temperature varies with plant species from 25°C to 36°C as optimum. It should be borne in mind that at very low light levels, the increase in photo-synthetic rate with increased temperature is only limited. This means that any input of heating into the growing situation during cold weather will be largely wasted if the light levels are low.

Integrated environmental control in a greenhouse is a form of computerized system developed to maintain near-optimum levels of the main environmental factors (light, temperature and carbon dioxide) necessary for plant growth. It achieves this by frequent monitoring of the greenhouse using carefully positioned sensors. Such a system is able to avoid the low temperature/light interaction described above. The beneficial effects to plant growth of lower night temperatures compared with day are well known in many species, e.g. tomato. The explanation is inconclusive, but the accumulation of sugars during the night appears to be greater, suggesting a relationship between photosynthesis and respiration rates. Such responses are shown to be related to temperature regimes experienced in the areas of origin of the species.

Temperature adaptations

Adaptation to extremes in temperature can be found in a number of species; for example resistance to high temperatures above 40°C in **thermophiles**; resistance to **chilling injury** is brought about by lowering the freezing point of cell constituents. Both depend on the stage of development of the plant, e.g. a seed is relatively resistant,

Figure 8.7 **The glasshouse environment** may be controlled by means of (a) a computer situated in the glasshouse, while (b) conditions are monitored throughout the glasshouse

but the hypocotyl of a young seedling is particularly vulnerable. Resistance to chilling injury is imparted by the cell membrane, which can also allow the accumulation of substances to prevent freezing of the cell contents. **Hardening off** of plants by gradual exposure to cold temperatures can develop a change in the cell membrane, as in bedding plants and peas. Examples of plant hardiness are found in Table 4.3.

Water

Water is required in the photosynthesis reaction but this represents only a very small proportion of the total water taken up by the plant (see transpiration). Water supply through the xylem is essential to maintain leaf turgidity and retain fully open stomata for carbon dioxide movement into the leaf. In a situation where a leaf contains only 90 per cent of its optimum water content, stomatal closure will prevent carbon dioxide entry to such an extent that there may be as much as 50 per cent reduction in photosynthesis. A visibly wilting plant will not be photosynthesizing at all.

Minerals

Minerals are required by the leaf to produce the **chlorophyll** pigment that absorbs most of the light energy for photosynthesis. Production of chlorophyll must be continuous, since it loses its efficiency quickly. A plant deficient in iron, or magnesium especially, turns yellow (**chlorotic**) and loses much of its photosynthetic ability. **Variegation** similarly results in a slower growth rate.

The leaf

The leaf is the main organ for photosynthesis in the plant, and its cells are organized in a way that provides maximum efficiency. The upper epidermis is transparent enough to allow the transmission of light into the lower leaf tissues. The sausage-shaped palisade mesophyll cells are packed together, pointing downwards, under the upper epidermis. The sub-cellular **chloroplasts** within them carry out the photosynthesis process. The absorption of light by chlorophyll occurs at one site and the energy is transferred to a second site within the chloroplast where it is used to build up carbohydrates, usually in the form of insoluble starch. The spongy mesophyll, below the palisade mesophyll, has a loose structure with many air spaces. These spaces allow for the two-way diffusion of gases. The carbon dioxide from the air is able to reach the palisade mesophyll; and oxygen, the waste product from photosynthesis, leaves the leaf. The numerous stomata on the lower leaf surface are the openings to the outside by which this gas movement occurs. The numerous small vascular bundles (veins) within the leaf structure contain the xylem vessels that provide the water for the photosynthesis reaction. The phloem cells are similarly present in the vascular bundles, for the removal of sugar to other plant parts. Figure 8.8 shows the

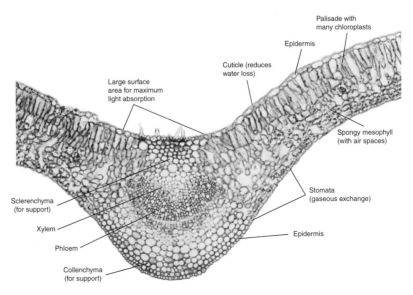

Figure 8.8 **Cross-section of *Ligustrum*** leaf showing its structure as an efficient photosynthesizing organ

structure of the leaf and its relevance to the process of photosynthesis. A newly expanded leaf is most efficient in the absorption of light, and this ability reduces with age. The movement of the products of photosynthesis is described in Chapter 9.

Pollution

Gases in the air, which are usually products of industrial processes or burning fuels, can cause damage to plants, often resulting in scorching symptoms of the leaves. Fluoride can accumulate in composts and be present in tap water, so causing marginal and tip scorch in leaves of susceptible species such as *Dracaena* and *Gladiolus*. Sulphur dioxide and carbon dioxide may be produced by faulty heat exchangers in glasshouse burners, especially those using paraffin. Scorch damage over the whole leaf is preceded by a reddish discolouration.

Respiration

Respiration is the process by which sugars and related substances are broken down to yield energy, the end-products being carbon dioxide and water.

In order that growth can occur, the food must be broken down in a controlled manner to release energy for the production of useful structural substances such as cellulose, the main constituent of plant cell walls, and proteins for enzymes. This energy is used also to fuel cell division and the many chemical reactions that occur in the cell.

The energy requirement within the plant varies, and reproductive organs can respire at twice the rate of the leaves. Also, in apical meristems, the processes of cell division and cell differentiation require high inputs of energy. In order that the breakdown is complete, oxygen is required in the process of aerobic respiration. A summary of the process is given in Table 8.2.

Table 8.2 A summary of the process of aerobic respiration

a. Written in a conventional way, the process can be expressed in the following way:
glucose plus oxygen gives rise to carbon dioxide plus water plus energy in the mitochondria of the cell.

b. Written in the form of a chemical equation, which represents molecular happenings at the sub-microscopic level, the above sentence becomes:
1 $C_6H_{12}O_6$ molecule plus 6 O_2 molecules give rise to 6 CO_2 molecules plus 6 H_2O molecules plus energy in the mitochondria of the cell.

It would appear at first sight that respiration is the reverse of photosynthesis (see p111). This supposition is correct in the sense that photosynthesis creates glucose as an energy-saving strategy, and respiration breaks down glucose as an energy releasing mechanism. It is also correct in the sense that the simple equations representing the two processes are mirror images of each other.

It should, however, be emphasized that the two processes have two notable differences. The first is that respiration in plants (as in animals) occurs in all living cells of all tissues at all times in leaves, stems, flowers, roots and fruits. Photosynthesis occurs predominantly in the

palisade mesophyll tissue of leaves. Secondly, respiration takes place
in the torpedo-shaped organelles of the cell called mitochondria.
Photosynthesis occurs in the oval-shaped chloroplasts. Details of
biochemistry, beyond the scope of this book, would reveal how
different these processes are, in spite of their superficial similarities.
In the absence of oxygen, inefficient anaerobic respiration takes place
and incomplete breakdown of the carbohydrates produces alcohol
as a waste product, with energy still trapped in the molecule. If a
plant or plant organ such as a root is supplied with low oxygen
concentrations in a waterlogged or compacted soil, the consequent
alcohol production may prove toxic enough to cause root death.
Over-watering, especially of pot plants, leads to this damage and
encourages damping-off fungi.

Storage of plants

The actively growing plant is supplied with the necessary factors
for photosynthesis and respiration to take place. Roots, leaves or
flower stems removed from the plant for sale or planting will cease
to photosynthesize, though respiration continues. Carbohydrates and
other storage products, such as proteins and fats, continue to be broken
down to release energy, but the plant reserves are depleted and dry
weight reduced. A reduction in the respiration rate should therefore be
considered for stored plant material, whether the period of storage is a
few days, e.g. tomatoes and cut flowers, or several months, e.g. apples.

Attention to the following factors may achieve this aim:

- **Temperature**. The enzymes involved in respiration become
  progressively less active with a reduction in temperatures from 36°C
  (optimum) to 0°C. Therefore, a cold store employing temperatures
  between 0°C and 10°C is commonly used for the storage of materials
  such as cut flowers, e.g. roses; fruit, e.g. apples; vegetables, e.g.
  onions; and cuttings, e.g. chrysanthemums, which root more readily
  later. Long-term storage of seeds in **gene banks** (see Chapter 10) uses
  liquid nitrogen at 20°C.
- **Oxygen and carbon dioxide**. Respiration requires oxygen in
  sufficient concentration; if oxygen concentration is reduced, the
  rate of respiration will decrease. Conversely, carbon dioxide is a
  product of the process and as with many processes, a build-up of a
  product will cause the rate of the process to decrease. A controlled
  environment store for long-term storage, e.g. of top fruit, is
  maintained at 0° C–5°C according to cultivar, and is fed with inert
  nitrogen gas to exclude oxygen. Carbon dioxide is increased by up to
  10 per cent for some apple cultivars.
- **Water loss**. Loss of water may quickly desiccate and kill stored
  material, such as cuttings. Seeds also must not be allowed to lose
  so much water that they become non-viable, but too humid an
  environment may encourage premature germination with equal loss of
  viability.

Check your learning

1. State an equation in words which describes the process of photosynthesis.
2. Explain how optimum levels of temperature, carbon dioxide, water and light sustain maximum photosynthesis.
3. Describe the anatomy of a typical leaf as an organ of photosynthesis.

4. State an equation in words which describes the process of respiration.
5. Explain how optimum levels of oxygen, water and temperature affect the rate of respiration.

Further reading

Attridge, T.H. (1991). *Light and Plant Responses*. CUP.

Bickford, E.D. *et al.* (1973). *Lighting for Plant Growth*. Kent State University Press.

Bleasdale, J.K.A. (1983). *Plant Physiology in Relation to Horticulture*. Macmillan.

Brown, L.V. (2008). *Applied Principles of Horticultural Science*. 3rd edn. Butterworth-Heinemann.

Capon, B. (1990). *Botany for Gardeners*. Timber Press.

Grow Electric Handbook No. 2 (1974). *Lighting in Greenhouses*, Part 1. Electricity Council.

Hopkins, W.G. and Huner, N.P.A. (2004). *Introduction to Plant Physiology*. John Wiley & Sons.

Ingram, D.S. *et al.* (eds) (2002). *Science and the Garden*. Blackwell Science Ltd.

MAFF (1978). *Carbon Dioxide Enrichment for Lettuce*, HPG51, HMSO.

Sutcliffe, J. (1977). *Plants and Temperature*. Edward Arnold.

Thompson, A.K. (1996). *Postharvest Technology of Fruit and Vegetables*. Blackwell Science Ltd.

Chapter 9 Transport in the plant

Summary

This chapter includes the following topics:

- **Water movement in the plant**
- **Diffusion and osmosis**
- **Xylem and phloem**
- **Transpiration and water loss**
- **Effect of evaporation, temperature and humidity**
- **Stomata**
- **Minerals in the plant**

with additional information on:

- **Sugar movement in plants**

Figure 9.1 **Apple** (Cox on M1) excavated at 16 years to reveal distribution of roots. Note the vigorous main root system near the surface with some penetrating deeply (courtesy of Dr E.G. Coker)

Water

Water is the major constituent of any living organism and the maintenance of a plant with optimum water content is a very important part of plant growth and development (see Soil Water, Chapter 19). Probably more plants die from lack of water than from any other cause. **Minerals** are also raw materials essential to growth (see p126), and are supplied through the root system.

Functions of water

The plant consists of about 95 per cent water, which is the main constituent of **protoplasm** or living matter. When the plant cell is full of water, or **turgid**, the pressure of water enclosed within a membrane or vacuole acts as a means of support for the cell and therefore the whole plant, so that when a plant loses more water than it is taking up, the cells collapse and the plant may wilt. Aquatic plants are supported largely by external water and have very little specialized support tissue. In order to survive, any organism must carry out complex chemical reactions, which are explained, and their horticultural application described, in Chapter 8. Raw materials for these chemical reactions must be transported and brought into contact with each other by a suitable medium; water is an excellent solvent. One of the most important processes in the plant is photosynthesis, and a small amount of water is used up as a raw material in this process.

- **Diffusion** is a process whereby molecules of a gas or liquid move from an area of high concentration to an area where there is a relatively lower concentration of the diffusing substance, e.g. sugar in a cup of tea will diffuse through the tea without being stirred – eventually! If the process is working against a concentration gradient, energy is needed.
- **Osmosis** can therefore be defined as the movement of water from an area of low salt concentration to an area of relatively higher salt concentration, through a partially permeable membrane. The greater the osmotic pressure then the faster water moves into the root cells, a process which is also affected by increased temperature.

Movement of water

Water moves into the plant through the roots, the stem, and into the leaves, and is lost to the atmosphere. Water vapour moves through the **stomata** (see p117) by diffusion from inside the leaf into the air immediately surrounding the leaf where there is a lower relative humidity (see p41).

The pathway of water movement through the plant falls into three distinct stages:

- water uptake;
- movement up the stem;
- transpiration loss from the leaves.

Water uptake

The movement of water into the roots is by a special type of diffusion called **osmosis**. Soil water enters root cells through the **cell wall** and **membrane**. Whereas the cell wall is permeable to both soil water and the dissolved inorganic minerals, the cell membrane is permeable to water, but allows only the smallest molecules to pass through, somewhat like a sieve. Therefore the cell membrane is considered to be a **partially permeable membrane**.

A greater concentration of minerals is usually maintained inside the cell compared with that in the soil water. This means that, by osmosis, water will move from the soil into the cell where there is relatively lower concentration of water, as there are more inorganic salts and sugars. The greater the difference in concentration of inorganic salts the faster water moves into the root cells.

If there is a build-up of salts in the soil, either over a period of time or, for example, where too much fertilizer is added, water may move out of the roots by osmosis, and the cells are then described as **plasmolyzed**. Cells that lose water this way can recover their water content if the conditions are rectified quickly, but it can lead to permanent damage to the cell interconnections (see p88). Such situations can be avoided by correct dosage of fertilizer and by monitoring of conductivity levels in greenhouse soils and NFT systems (see Chapter 22).

Movement of water in the roots

It is the function of the root system to take up water and mineral nutrients from the growing medium and it is constructed accordingly, as described in Chapter 6. Inside the epidermis is the parenchymatous cortex layer. The main function of this tissue is respiration to produce energy for growth of the root and for the absorption of mineral nutrients. The cortex can also be used for the storage of food where the root is an overwintering organ (see p92).

The cortex is often quite extensive and water moves across it in order to reach the transporting tissue that is in the centre of the root. Movement is relatively unrestricted as it moves through the intercell spaces and the lattice work of the cell wall (see p88). The central region, the **stele**, is separated from the cortex by a single layer of cells, the **endodermis**, which has the function of controlling the passage of water into the stele. A waxy strip forming part of the cell wall of many of the endodermal cells (the Casparian strip) prevents water from moving into the cell by all except the cells outside it, called **passage cells**. In this way, the volume of water passing into the stele is restricted. If such control did not occur, more water could move into the transport system than can be lost through the leaves. In some conditions, such as in high air humidity (see p41), more water moves into the leaves than is being lost to the air, and the more delicate cell walls in the leaf may burst. This condition is known as **oedema**, and commonly occurs in *Pelargonium* as dark green patches becoming brown, and also in weak-celled plants such as lettuce, when it is known as **tipburn**, because the margins of the leaves in particular will appear scorched. **Guttation** may occur when liquid water is forced onto the leaf surface.

Water passes through the endodermis to the **xylem** tissue (see p 92), which transports the water and dissolved minerals up to the stem and leaves. The arrangement of the xylem tissue varies between species, but often appears in transverse section as a star with varying numbers of 'arms' (see Figure 6.11).

> Xylem tissue transports the water and dissolved minerals up to the stem and leaves.

A distinct area in the root inside the endodermis, the **pericycle**, supports cell division and produces lateral roots, which push through to the main root surface from deep within the structure. Roots, as with stems, age and become thickened with waxy substances, and the uptake rate of water becomes restricted. Root anatomy is described in Chapter 6.

Movement of water up the stem

Three factors contribute to water movement up the stem:

- **root pressure** by which osmotic forces (see p122) push water up the stem to a height of about 30 cm. This can provide a large proportion of the plant's water needs in smaller annual species;
- **capillary action** (attraction of the water molecules for the sides of the xylem vessels), which may lift water a few centimetres, but which is not considered a significant factor in water movement;
- **transpiration pull** is the major process that moves soil water to all parts of the plant.

Transpiration

> Transpiration is the loss of water vapour from the leaves of the plant.

Any plant takes up a lot of water through its roots; for example, a tree can take up about 1000 litres (about 200 gallons) a day. Approximately 98 per cent of the water taken up moves through the plant and is lost by transpiration; only about 2 per cent is retained as part of the plant's structure, and a yet smaller amount is used up in photosynthesis.

The seemingly extravagant loss through leaves is due to the unavoidably large pores in the leaf surface (**stomata**) essential for carbon dioxide diffusion (see Figure 8.8). However, two other points should be considered here:

- water vapour diffuses outward through the leaf stomata more quickly than carbon dioxide (to be used for photosynthesis) entering. However, the plant is able to partially close the stomata to reduce water loss without causing a carbon dioxide deficiency in the leaf;
- the diffusion rate of water vapour through the stomata leads to a leaf cooling effect enabling the leaf to function whilst being exposed to high levels of radiation.

The plant is able to reduce its transpiration rate because the **cuticle** (a waxy waterproof layer) protects most of its surface and the stomata are able to close up as the cells in the leaf start to lose their turgor (see leaf structure p117). The stomatal pore is bordered by two sausage-shaped guard cells, which have thick cell walls near to the pore. When the guard cells are fully turgid, the pressure of water on the thinner walls causes the cells to buckle and the pore to open. If the plant begins to lose more water, the guard cells lose their turgidity and the stomata close to prevent any further water loss. Stomata also close if carbon dioxide concentration in the air rises above optimum levels.

A remarkable aspect of transpiration is that water can be pulled ('sucked') such a long way to the tops of tall trees. Engineers have long known that columns of water break when they are more than about 10 m long, and yet even tall trees such as the giant redwoods pull water up a hundred metres from ground level. This apparent ability to flout the laws of nature is probably due to the small size of the xylem vessels, which greatly reduce the possibility of the water columns collapsing.

A further impressive aspect of the plant structure is seen in the extreme ramifications of the xylem system in the veins of the leaf. This fine network ensures that water moves by transpiration pull right up to the spongy mesophyll spaces in the leaf (see p117), and avoids any water movement through living cells, which would slow the process down many thousand times.

If the air surrounding the leaf becomes very humid, then the **diffusion** of water vapour will be much reduced and the rate of transpiration will decrease. Application of water to greenhouse paths during the summer, **damping down** (see p15), increases relative humidity and reduces transpiration rate. While the air surrounding the leaf is moving, the humidity of air around the leaf is low, so that transpiration is maintained and greater water loss is experienced.

Windbreaks (see p38) reduce the risk of desiccation of crops. Ambient temperatures affect the rate at which liquid water in the leaf evaporates and thus determines the transpiration rate (see p124).

A close relationship exists between the daily fluctuation in the rate of transpiration and the variation in solar radiation. This is used to assess the amount of water being lost from cuttings in mist units (see misting p176); a light-sensitive cell automatically switches on the misting. In artificial conditions, e.g. in a florist shop, transpiration rate can be reduced by providing a cool, humid and shaded environment. **Plasmolyzed** leaf cells can occur if highly concentrated sprays cause water to leave the cells and result in scorching (see p123).

The evaporation of water from the cells of the leaf means that in order for the leaf to remain turgid, which is important for efficient photosynthesis, the water lost must be replaced by water in the xylem. Pressure is created in the xylem by the loss from an otherwise closed system and water moves up the petiole of the leaf and stem of the plant by suction (see **transpiration pull**). If the water in the xylem column is broken, for example when a stem of a flower is cut, air moves into the xylem and may restrict the further movement of water when the cut flower is placed in vase water. However, by cutting the stem under water

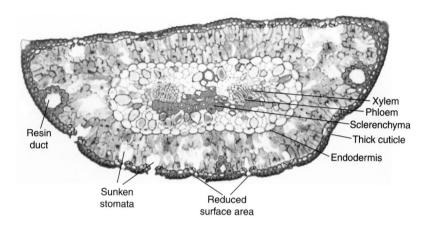

Figure 9.2 **Cross-section of pine leaf** (*Pinus*) showing some adaptations to reduce water loss

the column is maintained and water enters at a faster rate than if the plant was intact with a root system.

Anti transpirants are plastic substances which, when sprayed onto the leaves, will create a temporary barrier to water loss over the whole leaf surface, including the stomata. These substances are useful to protect a plant during a critical period in its cultivation; for example, conifers can be treated while they are moved to another site.

Structural adaptations to the leaf occur in some species to enable them to withstand low water supplies with a reduced surface area, a very thick cuticle and sunken guard cells protected below the leaf surface (see Figures 9.2 and 9.3). Compare this cross-section with that of a more typical leaf shown in Figure 8.8. In extreme cases, e.g. cacti, the leaf is reduced to a spine, and the stem takes over the function of photosynthesis and is also capable of water storage, as in the stonecrop (*Sedum*). Other adaptations are described on p81.

Minerals

Essential minerals are those inorganic substances necessary for the plant to grow and develop normally. They can be conveniently divided into two groups. The **major nutrients (macronutrients)** are required in relatively large quantities whereas the **micronutrients (trace elements)** are needed in relatively small quantities, usually measured in parts per million, and within a narrow concentration range to avoid deficiency or toxicity. The list of essential nutrients is given in Table 9.1.

Non-essential minerals, such as sodium and chlorine, appear to have a role in the plants but not as a universal requirement for growth and development. Sodium is made use of in many plants, notably those of estuarine origins, but whilst it does not appear to be essential there is an advantage in using agricultural salt on some crops such as beet or carrots. Aluminium plays an important part in the colour of *Hydrangea*

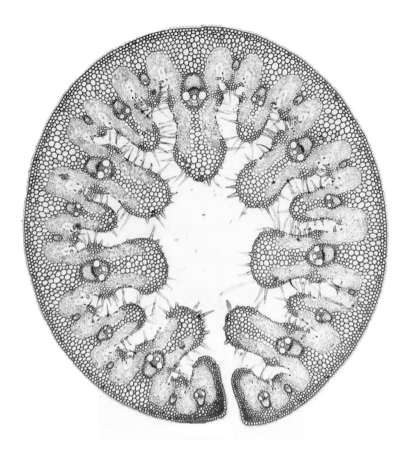

Figure 9.3 **Transverse section of Marram Grass leaf**, showing adaptations to prevent water loss; outer thick cuticle, curling by means of hinge cells to protect inner epidermis, stomata sunken into surface to maintain high humidity

Table 9.1 Nutrient requirements

Macronutrients (major nutrients)		Micronutrients (trace elements)	
N	Nitrogen	Fe	Iron
P	Phosphate	Bo	Boron
K	Potassium	Mn	Manganese
Mg	Magnesium	Cu	Copper
Ca	Calcium	Zn	Zinc
S	Sulphur	Mo	Molydenum

flowers (see p84) and silicon occurs in many grasses to give them a cutting edge or sharp ridges on their leaves.

Functions and deficiency symptoms of minerals in the plant

Many essential minerals have very specific functions in the plant cell processes. When in short supply (**deficient**) the plant shows certain characteristic symptoms, but these symptoms tend to indicate an extreme deficiency. To ensure optimal mineral supplies, growing media analysis or plant tissue analysis (see p377) can be used to forecast low nutrient levels, which can then be addressed.

Nitrogen is a constituent of **proteins**, nucleic acids and chlorophyll and, as such, is a major requirement for plant growth. Its compounds comprise about 50 per cent of the dry matter of protoplasm, the living substance of plant cells.

Deficiency causes slow, spindly growth in all plants and yellowing of the leaves (chlorosis) due to lack of chlorophyll. Stems may be red or purple due to the formation of other pigments. The high mobility of nitrogen in the plant to the younger, active leaves leading to the old leaves showing the symptoms first.

Phosphorus is important in the production of nucleic acid and the formation of adenosine triphosphate (see ATP p89). Large amounts are therefore concentrated in the meristem. Organic phosphates, so vital for the plant's respiration, are also required in active organs such as roots and fruit, while the seed must store adequate levels for germination. Phosphorus supplies at the seedling stage are critical; the growing root has a high requirement and the plant's ability to establish itself depends on the roots being able to tap into supplies in the soil before the reserves in the seed are used up (see p367).

Deficiency symptoms are not very distinctive. Poor establishment of seedlings results from a general reduction in growth of stem and root systems. Sometimes a general darkening of the leaves in dicotyledonous plants leads to brown leaf patches, while a reddish tinge is seen in monocotyledons. In cucumbers grown in deficient peat composts or NFT, characteristic stunting and development of small young leaves leads to brown spotting on older leaves.

Potassium. Although present in relatively large amounts in plant cells, this mineral does not have any clear function in the formation of important cell products. It exists as a cation and acts as an osmotic regulator, for example in guard cells (*see* p125), and is involved in resistance to chilling injury, drought and disease.

Deficiency results in brown, scorched patches on leaf tips and margins (see Figure 21.4), especially on older leaves, due to the high mobility of potassium towards growing points. Leaves may develop a bronzed appearance and roll inwards and downwards.

Magnesium is a constituent of chlorophyll. It is also involved in the activation of some enzymes and in the movement of phosphorus in the plant.

Deficiency symptoms appear initially in older leaves because magnesium is mobile in the plant. A characteristic interveinal chlorosis appears (see Figure 21.5), which subsequently become reddened and eventually necrotic (dead) areas develop.

Calcium is a major constituent of plant cell walls as calcium pectate, which binds the cells together. It also influences the activity of meristems especially in root tips. Calcium is not mobile in the plant so the deficiency symptoms tend to appear in the younger tissues first. It causes weakened cell walls, resulting in inward curling, pale young leaves, and sometimes death of the growing point. Specific disorders include 'topple' in tulips, when the flower head cannot be supported by the top of the stem, 'blossom end rot' in tomato fruit, and 'bitter pit' in apple fruit.

Sulphur is a vital component of many proteins that includes many important enzymes. It is also involved in the synthesis of chlorophyll. Consequently a deficiency produces a chlorosis that, due to the relative immobility of sulphur in the plant, shows in younger leaves first.

Iron and **manganese** are involved in the synthesis of chlorophyll; although they do not form part of the molecule they are components of some enzymes required in its synthesis. Deficiencies of both minerals result in leaf chlorosis. The immobility of iron causes the younger leaves to show interveinal chlorosis first. In extreme cases, the growing area turns white.

Boron affects various processes, such as the translocation of sugars and the synthesis of gibberellic acid in some seeds (*see* dormancy p131). Deficiency causes a breakdown and disorganization of tissues, leading to early death of the growing point. Characteristic disorders include 'brown heart' of turnips, and 'hollow stem' in brassicas. The leaves may become misshapen, and stems may break. Flowering is often suppressed, while malformed fruit are produced, e.g. 'corky core' in apples, and 'cracked fruit' of peaches.

Copper is a component of a number of enzymes. Deficiency in many species results in dark green leaves, which become twisted and may prematurely wither.

Zinc, also involved in enzymes, produces characteristic deficiency symptoms associated with the poor development of leaves, e.g. 'little leaf' in citrus and peach, and 'rosette leaf' in apples.

Molybdenum assists the uptake of nitrogen, and although required in very much smaller quantities, its deficiency can result in reduced plant nitrogen levels. In tomatoes and lettuce, deficiency of molybdenum can lead to chlorosis in older leaves, followed by death of cells between the veins (interveinal necrosis) and leaf margins. Tissue browning and infolding of the leaves may occur and in *Brassicae*, the 'whiptail' leaf symptom involves a dominant midrib and loss of leaf lamina.

Mineral uptake

Minerals are absorbed to form the soil solution (see Chapter 21). The plants take up only water-soluble material so all supplies of nutrients including fertilizers and manures must be in the form of **ions** (charged particles). The movement of the elements in the form of **ions** occurs in the direction of root cells containing a higher mineral concentration

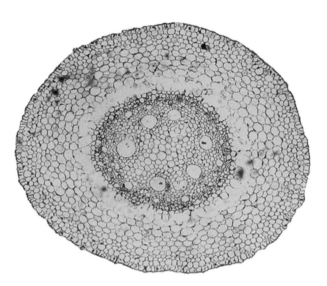

Figure 9.4 **Cross-section of *Zea mais* root** showing its structure in the absorption and transport of water and minerals

than the soil, i.e. **against a concentration gradient**. The passage in the water medium across the root cortex is by simple diffusion, but transport across the endodermis requires a supply of energy from the root cortex. The process is therefore related to temperature and oxygen supply (see respiration p118).

Nutrients are taken up predominantly by the extensive network of fine roots that grow in the top layers of the soil (see Figure 9.1). Damage to the roots near the soil surface by cultivations should be avoided because it can significantly reduce the plant's ability to extract nutrients. It is recommended that care should be taken to ensure that trees and shrubs are planted so their roots are not buried too deeply and many advocate that the horizontally growing roots should be set virtually at the surface to give the best conditions for establishment.

The surface thickening that occurs in the ageing root does not significantly reduce the absorption ability of most minerals, e.g. potassium and phosphate, but calcium is found to be principally taken up by the young roots.

Sugars

Movement of sugars in the plant

Phloem tissue is responsible for transporting carbohydrates from the leaves as a food supply for the production of energy in the cortex.

The product of photosynthesis (see p110) in most plants is starch (some plants produce sugars only), which is stored temporarily in the chloroplast or moved in the phloem to be more permanently stored in the seed, the stem cortex or root, where specialized storage organs such as **rhizomes** and **tubers** may occur (see p158).

The movement or **translocation** of materials around the plant in the phloem and xylem is a complex operation. **Phloem** is principally responsible for the transport of the products of photosynthesis as soluble sugars, usually sucrose, which move under pressure to areas of need, such as roots, flowers or storage organs. Each phloem **sieve-tube cell** (see p92) has a smaller companion cell that has a high metabolic rate. Energy is thus made available to the protoplasm at the end of each sieve-plate, which is able to 'pump' dilute sugar solutions around the plant. The flow can be interrupted by the presence of disease organisms such as club root (see p244).

Check your learning

1. Define the term transpiration and describe the environmental factors which affect it.

2. Describe the pathway and plant tissues involved in water and mineral movement through the plant.

3. Define the terms diffusion and osmosis.

4. Explain how evaporation and consequent water loss can be controlled in the plant by stomata and structural adaptations.

Further reading

Clegg, C.J. and Cox, G. (1978). *Anatomy and Activities of Plants*. John Murray.

Ingram, D.S. *et al.* (eds) (2002). *Science and the Garden*. Blackwell Science Ltd.

Mauseth, J.P. (1998). *Botany – An Introduction to Plant Biology*. Saunders.

Moorby, J. (1981). *Transport Systems in Plants*. Longman.

Scott Russell, R. (1982). *Plant Root Systems: Their Function and Interaction with the Soil*. McGraw-Hill.

Sutcliffe, J. (1971). *Plants and Water*. Edward Arnold.

Sutcliffe, J.F. and Baker, D.A. (1976). *Plant and Mineral Salts*. Edward Arnold.

Chapter 10 Pollination and fertilization

Summary

This chapter includes the following topics:

- Pollination
- Fertilization
- Compatability and incompatibility
- Parthenocarpy
- Polyploids, including triploids
- F1 hybrid breeding

with additional information on the following:

- The genetic code
- Cell division
- Inheritance of characteristics
- Other breeding programmes
- Monohybrid and dihybrid crosses
- Mutations
- New breeding technology
- Breeders' rights
- Gene Banks

Figure 10.1 **Bees and other pollinating insects** are attracted to large, colourful flowers

Introductory principles

Ever since growers first selected seed for their next crop, they have influenced the genetic make-up and potential of succeeding crops. A basic understanding of plant breeding principles is useful if horticulturists are to understand the potential and limits of what plant cultivars can achieve, and so that they can make realistic requests to the plant breeder for improved cultivars. Plant breeding now supplies a wide range of plant types to meet growers' specific needs. The plant breeder's skill relies on their knowledge of flower biology, cell biology and genetics. Desirable plant characters such as yield, flower colour and disease resistance are selected and incorporated by a variety of methods. This chapter attempts to give a background to the principles used by plant breeders.

Plant breeding follows two scientific findings. Firstly, characteristics of a species are commonly passed on from one generation to the next (**heredity**). Secondly, sexual reproduction is also able to generate different characteristics in the offspring (**variation**). A plant breeder relies on the principles of heredity to retain desirable characteristics in a breeding programme, while new characteristics are introduced in several ways to produce new cultivars.

The processes of pollination and fertilization are first discussed as they are the plant processes that lead to the genetic make-up of the plant that follows.

Pollination

The flower's function is to bring about sexual reproduction (the production of offspring following the fusion of male and female nuclei). The male and female nuclei are contained within the pollen grain and ovule respectively and pollination is the transfer process. Cross-pollination ensures that variation is introduced into new generations of offspring.

Self pollination occurs when pollen comes from the same flower (or a different flower on the same plant) as the ovule, common in Fabaceae (bean family).

Cross-pollination occurs when pollen comes from a flower of a different plant, with a different genetic make-up from the ovule, common in Brassicaceae (cabbage family).

Natural agents of cross-pollination are mainly wind and insects.

Pollination is the plant process whereby a male pollen grain is transferred from the anther to the stigma of the flower, thus enabling fertilization to take place.

Wind-pollinated flowers. The characteristics of wind-pollinated flowers are their small size, their green appearance (lacking coloured petals), their absence of nectaries and scent production, and their production of large amounts of pollen which is intercepted by large stigmas. They also often have proportionally large stigmas that protrude from the flower to maximize the chances of intercepting pollen grains in the air.

Figure 10.2 **Wind-pollinated species** have small, inconspicuous flowers, e.g. (a) *Stipa calamagrostis* (b) *Cyperus chira* (c) *Luzula nivea*

The commonest examples of wind-pollinated plants are the grasses, and trees with catkins such as *Salix* (willow), *Betula* (birch), *Corylus* (hazel), *Fagus* (beech), *Quercus* (oak). The Gymnosperma (conifers) also have wind-pollination from the small male cones.

Figure 10.3 **Insect-pollinated flowers**, e.g. (a) Day Lily (*Hemerocallis*) (b) *Digitalis stewartii* (c) *Verbascum* 'Cotswold Queen', are brightly coloured and sometimes have guidelines in the petals to attract insects

Insect-pollinated flowers. The characteristics of insect-pollinated flowers are brightly coloured petals (and scent production) to attract insects, and the presence of nectaries to entice insects with sugary food. Insects such as bees and flies collect the pollen on their bodies as they fly in and out and carry it to other flowers. In tropical countries, slugs, birds, bats and rodents are also pollinators. Some floral mechanisms, e.g. in snapdragon and clover, physically prevent entry of smaller non-pollinating insects and open only when heavy bees land on the flower. Other plant species, such as *Arum* lily, trap pollinating insects for a period of time to give the best chance of successful fertilization.

Certain *Primula* spp. have stigma and stamens of differing lengths to encourage cross-pollination; in thrum-eyed flowers, the anthers emerge further from the flower than the stigma, so that insects rub against them when reaching into the flower tube; in pin-eyed flowers, the stigma protrudes from the flower and will catch the pollen from the same place on the insect body, so ensuring cross-pollination (see Figure 10.4).

Figure 10.4 Structural mechanisms to encourage **cross-pollination** in insect-pollinated flowers are shown in (a) and (b) snapdragon flowers where the flower only opens to the weight of the bee on the lower petal and (c) and (d) *Primula* flowers where the stamens and style are arranged differently

Bees in pollination

The well-known social insect, the honey bee (*Apis mellifera)*, is helpful to horticulturists. The female worker collects pollen and nectar in special pockets (honey baskets) on its hind legs. This is a supply of food for the hive and, in collecting it, the bee transfers pollen from plant to plant. Several crops, such as apple and pear, do not set fruit when self-pollinated. The bee therefore provides a useful function to the fruit grower. In large areas of fruit production the number of resident hives may be insufficient to provide effective pollination, and in cool, damp or windy springs, the flying periods of the bees are reduced.

It may therefore be advantageous for the grower to introduce beehives into the orchards during blossom time, as an insurance against bad weather. One hive is normally adequate to serve 0.25 ha of fruit. Blocks of four hives placed in the centre of a 1 ha area require foraging bees to travel a maximum distance of 70 m. In addition to honey bees, wild species, e.g. the potter flower bee (*Anthophora retusa*) and red-tailed bumble-bee (*Bombus lapidarius)*, increase fruit set, but their numbers are not high enough to dispense with the honey bee hives.

All species of bee are killed by broad spectrum insecticides, e.g. deltamethrin, and it is important that spraying of such chemicals be restricted to early morning or evening during the blossom time period when hives have been introduced.

Figure 10.5 Bumble-bee boxes provided in glasshouse for pollination of tomatoes

Fertilization. The union of male and female gametes to produce a zygote (fertilized ovule).

In commercial greenhouses, the pollination of crops such as tomatoes and peppers is commonly achieved by in-house nest-boxes of bumble-bees, *Bombus terrestris* (see Figure 10.5). Plant breeders may use blowflies in glasshouses to carry out pollination. They also perform mechanical transfer of pollen by means of small brushes.

When a pollen grain arrives at the stigma of the same plant species, it absorbs sugar and moisture from the stigma's surface and then germinates to produce a **pollen tube**. The pollen tube contains the 'male' nucleus (and also an extra 'second nucleus'). These nuclei are carried in the pollen tube as they grow down inside the style and into the ovary wall.

Fertilization

After entering the ovule, the male nucleus fuses with the female nucleus, their chromosomes becoming intimately associated. The term '**gamete**' is used to describe the agents, both male and female, that are involved in fertilization. In animals, the gametes are the eggs and sperms. In plants, they are the ovules and pollen.

Incompatible, in relation to fertilization, is a genetic mechanism that prevents self fertilization, thus encouraging cross-pollination, e.g. in Brassicaceae. The mechanism operates by inhibiting any of the following four processes:

- pollen germination;
- pollen tube growth;
- ovule fertilization;
- embryo development.

Compatible, in relation to fertilization , is a genetic mechanism that allows self fertilization, thus encouraging self pollination, e.g. in Fabaceae.

Parthenocarpy, where fertilization does not occur before fruit formation, is a useful phenomenon when the object of the crop is the production of seedless fruit, as in cucumber. It is usually accompanied by a high level of auxin in the plant and may be induced in pears by a spray of gibberellic acid.

Parthenocarpy is the formation of fruit without prior fertilization.

The fertilized ovule (**zygote**) undergoes repeated cell division of its young unspecialized cells before beginning to develop tissues through differentiation (see p93), that form the embryo within the seed.

Additionally, however, it should be noted that there is often a second fertilization within the ovule, which has led to the term 'double fertilization'. The second fertilization involves the 'second nucleus' of the pollen tube (mentioned above) fusing with two extra ('polar') nuclei present in the ovule itself. The resulting tissue consequently contains three sets of chromosomes (triploid, see p146) and is called **endosperm**. Endosperm is a short-term food supply used by the embryo to help its growth. Endosperm is found in the seeds of many plant families, but is best developed in the grass family. In maize seed, for example, the endosperm often represents more than half the seed volume. Anyone making popcorn will be eating 'exploded endosperm'.

Not all parts of a seed are derived from embryo or endosperm origins. The outer coat (testa) is formed from the outer layers of the ovule and is thus maternal in origin. Also, the ovary, which contained the ovules in the flower before fertilization, develops and expands to form the fruit of the plant (see seed structure, p103).

In the previous sections of this chapter, descriptions have been given of the plant processes that lead to the development of a seed. Genetics also requires some knowledge of the microscopic details of reproductive cells and of the biological processes in which they are involved, since this knowledge explains how plant characteristics are passed on from generation to generation.

The genetic code

All living plant cells contain a nucleus which controls every activity in the cell (see p00). Within the nucleus is the chemical deoxyribonucleic acid (DNA), a very large molecule made up of thousands of atoms (see also carbon chemistry, p111). DNA contains hundreds of sub-units (nucleotides), each of which contains a chemically active zone called a 'base'. There are four different bases: guanine, cytosine, thymine and adenine. The sequence of these bases is the method by which genetic information is stored in the nucleus, and also the means by which information is transmitted from the nucleus to other cell organelles (this sequence is called the **genetic code**). A change in the base sequencing of a plant's code will lead to it developing new characteristics. These very long molecules of DNA are called **chromosomes**. Each species of plant has a specific number of chromosomes. The cells of tomato (*Lycopersicum esculentum)* contain 24 chromosomes, the cells of *Pinus* and *Abies* species 24 and onions 16 (human beings have 46). Each chromosome contains a succession of units, called **genes**, containing many base units. Each gene usually is the code for a single characteristic such as flower colour or disease resistance. Scientists have been able to correlate many gene locations with plant characteristics that they control. Microscopic observation of cells during cell division reveals two similar sets of chromosomes, e.g. in tomatoes a total of 12 similar or homologous pairs. The situation in a nucleus where there are two sets of chromosomes is termed the **diploid** condition. A gene for a particular characteristic, such as flower colour, has a precise location on one chromosome, and on the same location of the homologous chromosome. For each characteristic, therefore, there are at least two alleles (alternative forms of the gene), one on each chromosome in the homologous pair, which provide genetic information for that characteristic. The fact that every living plant cell which has a nucleus has a complete set of all genetic information (**totipotency**) means that cells have the information to become any specialized cell in the plant. Therefore, when organs are removed from their usual place, as in vegetative propagation, they are able to develop new parts, such as adventitious roots, using this information. Vegetative propagation is described in detail in Chapter 12.

Cell division

When a plant grows, the cell numbers increase in the growing points of the stems and roots, the division of one cell producing two new ones. Genetic information in the nucleus is reproduced exactly in the new cells to maintain the plant's characteristics. The process of **mitosis** achieves this (see p89). Each chromosome in the parent cell produces a duplicate of itself, thus producing sufficient material for the two new daughter cells. A delicate, spindle-shaped structure ensures the separation of chromosomes, one complete set into each of the new cells. A dividing cell wall forms across the old cell to complete the division.

Meiosis

In the anthers and ovaries (parts of the plant producing the sex cells; the pollen and ovules respectively) cell division needs to be radically different. Sexual reproduction involves the fusion of genetic material contributed by the sex cells of each parent (see fertilization). Half of the chromosomes in the cells of an offspring are therefore inherited from the male parent, and half from the female. To ensure that the chromosome number in the offspring is equal to that of the parents, the number of chromosomes in the male and female sex cells (pollen and ovule) must be halved. This halving is achieved by a special division process, meiosis, in the anthers and ovaries. It ensures the separation of each homologous chromosome from its partner so that each sex cell contains only one complete set of chromosomes. This cell condition is termed **haploid**.

Inheritance of characteristics

As mentioned above, genetic information is passed from parent to offspring when material from male and female parent comes together by fusion of the sex cells. Genes from each parent can, in combination, produce an intermediate form, a mixture of the parents' characteristics in the offspring; e.g. a gene for red flowers inherited from the male parent, combined with a gene for white flowers from the female parent, could produce pink-flowered offspring, if both conditions are equal (see below). If one gene, however, was completely dominant over the other, e.g. if the red gene inherited from the male parent was **dominant** over the white female gene, all offspring would produce red flowers (see Figure 10.6). The non-dominant (**recessive**) white gene will still be present as part of the genetic make-up of the offspring cells and can be passed on to the next generation. If it then were to combine at fertilization with another white gene, the offspring would be white-flowered.

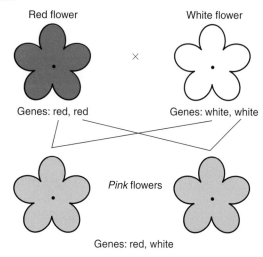

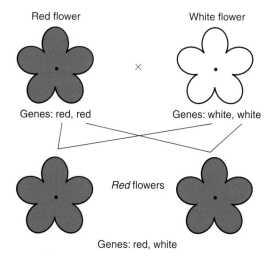

Figure 10.6 The pattern of inheritance of **genes**

The example above considers the inheritance of **one** pair of genes by a single offspring (**monohybrid cross**). However, many sex cells are produced from one flower in the form of pollen grains and ovules, and these give rise to many seeds in the next generation. The plant breeder tries to **predict** the ratio (or percentage) of offspring with each option (in this case, the ratio of red- flowered offspring to white-flowered offspring).To understand this prediction process, consider now the same example of flower colour in a little more detail.

If a red-flowered plant, containing two genes for red (described as **pure**), is carefully fertilized (**crossed**) with a 'pure' white-flowered plant, the red-flowered plant supplies in this case pollen as the male parent, and the white-flowered plant supplies ovule as the female parent. As both parents are pure, the male parent can produce only one type of sex cell, containing the 'red' version (**allele**) of the gene, and the female parent only the white (**allele**). Since all pollen grains will carry 'red' genes and all ovules 'white' , then in the absence of dominance, the only possible combination for the first generation (or **F1**) is pink offspring, each containing an allele for red and an allele for white (i.e. **impure**). Figure 10.7 illustrates this inheritance by using letters to describe genes, **R** to represent a red-inducing gene, and **r** to represent a white-inducing gene.

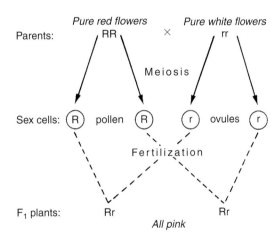

Figure 10.7 **Simple inheritance**: production of the F1 generation

The **genotype** (the genetic make-up of a cell) can be represented by using letters, e.g. Rr. The **phenotype** (the outward appearance e.g. red, pink or white flower) results from the genotype's action. If these plants from the F1 generation were now used as parents and crossed (or perhaps self-pollinated), then the results in the second or **F2** generation would be as shown in Figure 10.8, i.e. 25 per cent of the population would have the phenotype of red flowers, 50 per cent pink flowers and 25 per cent white flowers (a ratio of 1:2:1). The plant breeder, by analysing the ratios of each colour, would be able to calculate which colour genes were present in the parents, and whether the 'colour' was pure.

More usually, one of the alleles at the gene site exhibits dominance and the other is recessive. As a result, when offspring inherit both alleles (dominant and recessive), there is no intermediate phenotype; only the dominant gene expresses itself (the recessive is masked). It was the monk Gregor Mendel working in a monastery in what is now Brno in

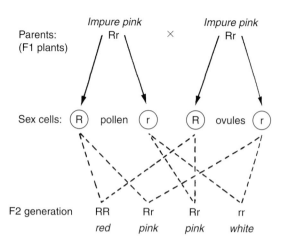

Figure 10.8 **Simple inheritance**: production of the F2 generation

the Czech Republic in the middle of the eighteenth century who laid the foundations of genetics with his breeding experiments.

He used peas and worked with single characteristics such as the height of the plant. He crossed tall with short plants and found that in the second or F2 generation the plants were either tall or short; there were no middle-sized phenotypes. They were found to be in the ratio of three tall to one short plant. Similarly, he crossed pure (homozygous) round pea parents (RR) with pure (homozygous) wrinkled pea parents (rr). When crossing the two pure parents (RR × rr), the genotype of the first generation, or F1, follows the pattern as shown in Figure 10.7, but all the plants produced round peas, i.e. the round pea allele was dominant in peas.

When the second or F2 generation was produced the genotype followed the pattern shown in Figure 10.8, but there were three round peas to every one wrinkled pea type. Both RR and Rr genotypes produced the same phenotype; only a double recessive (rr) produced plants with wrinkled peas. He went on to look at other simple 'single gene characteristics' including seed colour (yellow dominant and green recessive) and flower colour (purple dominant and white recessive). In all these cases he found that the ratio of phenotypes in the second or F2 generation was 3:1. By observation he established the **Principle of Segregation** which states that the phenotype is determined by the pair of alleles in the genotype and only one allele of the gene pair can be present in a single gamete (i.e. passed on by a parent).

Dihybrid cross. Mendel then went on to investigate the crossing of plants that differed in **two** contrasting characters. The results of crossing tall purple-flowered peas with short white-flowered ones produced all tall purple-flowered peas. As Mendel showed, they all had the same genotype (TtPp) and the same phenotype. The F2 generation produced from these parents is illustrated in Figure 10.9, where the combinations are shown in a Punnet Square (or 'checkerboard' – a useful way of showing the genotypes produced in a cross). Note that each gene has behaved independently; there are 12 tall to four short (ratio 3:1) and 12 purple to four white-flowered plants (3:1).

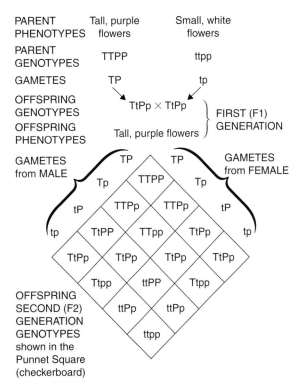

Figure 10.9 **Punnet square** showing a dihybrid cross

This example illustrates Mendel's **Principle of Independent Assortment** which states that the alleles of **unlinked** genes (i.e. genes not on the same chromosome) behave independently at meiosis. The cross involving two independent genes produces combinations of phenotypes in the ratio 9:3:3:1.

For the cross TtPp × TtPp, the genotypes produced are shown in Figure 10.9 and the phenotypes are as follows:

9 Tall, purple-flowered plants

1	TTPP
2	TTPp
2	TtPP
4	TtPp

3 Tall, white-flowered plants

2	TTpp
1	Ttpp

3 Short, purple-flowered plants

2	ttPP
1	ttPp

1 Short, white-flowered plants

1	ttpp

Not all genes act independently in the way shown above. Some are **linked** by being on the same chromosome and so they will usually appear together. The number of grouped genes is equal to the number of chromosome pairs for the species (seven for peas).

F1 hybrid breeding

Breeders may choose to produce seeds for the commercial market which are all F1 offspring or sometimes called F1 hybrids. F1 hybrid seeds are important to the grower since, given a uniform environment, all plants of the same cultivar will produce a uniform crop because they are all genetically identical (see Figure 10.10). Crops grown from F1 hybrid seed such as cabbage, Brussels sprouts and carrots can be harvested at one time and they have similar characteristics of yield. Similarly, F1 hybrid flower crops will have uniformity of colour and flower size.

Figure 10.10 **Uniformity** of crop in **F1 hybrid** *Poinsettias*

Another feature of F1 hybrids is **hybrid vigour**. Plants crossed from parents with quite different characteristics will display the feature to a marked extent, giving outstanding growth, especially in good growing conditions. The desirable characteristics of the two parents, such as disease resistance, good plant habit, high yield and good fruit or flower quality, may be incorporated along with established characteristics of successful commercial cultivars by means of the F1 hybrid breeding programme.

F1 hybrid seed production first requires suitable parent stock, which must be pure for all characteristics. In this way, genetically identical offspring are produced, as described in Figure 10.7. The production of pure parent plants involves repeated self pollination (selfing) and selection, over eight to twelve generations, resulting in suitable **inbred** parent lines. During this and other self pollination programmes, vigour is lost (the parent plants do not look impressive) but, of course, the vigour is restored by hybridization.

The parent lines must now be cross-pollinated to produce the F1 hybrid seed. It is essential to avoid self pollination at this stage, therefore one of the lines is designated the male parent to supply pollen. The anthers in the flowers of the other line, the female parent, are removed, or treated to prevent the production of viable pollen. The growing area must be isolated to exclude foreign pollen, and seed is collected only from the female parent. This seed is more expensive than most other commercial seed, due to the complex breeding programme requiring intensive labour. Seed collected from the planted commercial F1 hybrid crop represents the F2, and will produce plants with very diverse characteristics (Figure 10.8). Some F2 seed, however, is deliberately produced by breeders for flowering plants, such as geraniums and fuchsias, where a variety of colour and habit is required for bedding plant display.

Other breeding programmes

In addition to F1 hybrid breeding, where specific improvements are achieved, plant breeders may wish to bring about more general improvements to existing cultivars, or introduce characteristics such as disease resistance. Programmes are required for crops which self-pollinate (**inbreeders**), or those which cross-pollinate (**outbreeders**), and two of these strategies are described.

Pedigree breeding is the most widely used method in plant breeding by both amateurs and professionals. Two plants with different desirable characteristics are crossed to produce an F1 population. These F1 plants of very similar genotype are then crossed (called '**selfed**') and any offspring with useful characteristics are selected for further selfing to produce a line of desirable plants. After repeated selfing and selection, the characteristics of the new lines are compared with existing cultivars and assessed for improvements. Further field trials will determine a new type's suitability for submission and possible registration as a new cultivar. If a plant breeder wants to produce a strain adapted to particular conditions (e.g. a cultivar with hardiness), exposure of plants of a selected cultivar to the desired conditions will eliminate unsuitable plants, allowing the hardy plants to set seed. Repetition of this process gradually adapts the whole population. Selection for other characteristics such as earliness may be selected by harvesting seed early.

Disease resistance breeding (see also p290). A genetic characteristic enables the plant to combat fungal attack. The disease organism may itself develop a corresponding genetic capacity to overcome the plant's resistance by mutation. The introduction of disease resistance into existing cultivars often requires a **backcross breeding** programme. This involves a commercial cultivar lacking resistance crossed with a wild plant which exhibits resistance. For example, a lettuce cultivar may lack resistance to downy mildew (*Bremia lactucae*), or a tomato cultivar lack tomato mosaic virus resistance.

This commercial cultivar is crossed with a resistant wild plant to produce an F1, then an F2. From this F2, plants having both the characteristics of the commercial cultivar and also disease resistance are selected. The process continues with backcrossing of these selected plants with the original commercial parent to produce an F1, from which an F2 is produced. More commercial characteristics may be incorporated by further backcrossing and selection over a number of generations, until all the characteristics of the commercial cultivar are restored, but with the additional disease resistance.

Polyploids

Polyploidy occurs when duplication of chromosomes fails to result in mitotic cell division.

Polyploids are plants with cells containing more than the diploid number of chromosomes; e.g. a triploid has three times the haploid number, a tetraploid four times and the polyploid series continues in many species up to octaploid (eight times haploid). An increase in size of cells, with a resultant increase in roots, fruit and flower size of many species of chrysanthemums, fuchsias, strawberries, turnips and grasses, is the result of polyploidy. There is a limit to the number of chromosomes that a species can contain within its nucleus. Polyploidy occurs when duplication of chromosomes (see mitosis) fails to result in mitotic cell division. The multiplication of a polyploid cell within a meristem may form a complete polyploid shoot that, after flowering and fertilization, may produce polyploid seed. Polyploidy can occur spontaneously, and has led to many variant types in wild plant populations. It can be artificially induced by the use of a mitosis inhibitor, colchicine.

Triploids

The crossing of a tetraploid and a diploid gives rise to a triploid. Triploids, having an odd number of chromosomes which are unable to pair up during meiosis, are often infertile in nature. But there are a few important examples in horticulture, notably 'Bramley's Seedling' apple cultivar. Pollen from such cultivars is sterile, being derived from an irregular meiosis division in the anthers. The presence nearby of suitable pollinator cultivars such as 'James Grieve' and 'Grenadier' provide suitable viable pollen at the same flowering time as 'Bramley's Seedling'. Thus two pollinators are required, i.e. one to pollinate the Bramley's Seedling and a second to pollinate the pollinator. An alternative strategy for a private gardener is the inclusion of a pollinator onto the triploid tree by means of a suitable graft (see p176), the result is sometimes called a 'family tree'. Polyploidy only becomes significant in the plant when the mutated cell is part of a meristem.

Mutations

Spontaneous changes in the content or arrangement of chromosomes (**mutations**), whether in the cells of the vegetative plant or in the reproductive cells, occur in nature at the rate of approximately one cell in one million. These changes to the plant DNA are one of the most important causes of new alleles (see p141) leading to changes in the characteristics of the individual. Extreme chromosome alterations result in malformed and useless plants, but slight rearrangements may provide horticulturally desirable changes in flower colour or plant habit. Such desirable mutations have been seen in plants such as chrysanthemum, *Dahlia* and *Streptocarpus*. Mutation breeding also produces these variations, but using irradiation treatments with X-rays, gamma rays or mutagenic chemicals increases the mutation rate. In both situations (natural mutations and induced mutations), the mutation only becomes significant in the plant when the mutated cell originates in a meristem, where it proceeds to create a mass of novel genetic tissues (and organs).

Figure 10.11 Chimaera (distinct genetical tissues) in variegated horseradish

When a shoot with a different coloured flower or leaf arises, it is often referred to as a **sport**. A more extreme example of a mutation is a **chimaera**. This occurs when organs (and even whole plants) have two or more genetically distinct kinds of tissues existing together. This often results in variegation of the leaves, as seen in some *Acer* and *Pelargonium* species. Horticulturists use one form or other of vegetative propagation to preserve and increase the genetic novelty. These useful mutations may give rise to potential new cultivars in just one generation. (See Figure 10.11)

Recombinant DNA technology

For the plant breeder it has historically been difficult to predict whether the progeny from a breeding programme would show the desired characteristics. The term **recombinant DNA technology** refers to a modern method of breeding that enables novel sources of DNA to be integrated with greater certainty into a plant's existing genotype. Two new techniques have appeared in the last few years that have enabled this major shift in breeding practice.

The first technique is **marker-assisted breeding** . Breeders are now able to analyze chromosome material and establish what DNA sequence is present on the chromosome. Some plant characters such as disease resistance are hard to evaluate in newly bred plants, as infection may be difficult to achieve under test conditions. Since the breeders are now able to recognize the chromosome DNA sequence for plant resistance, they can apply this knowledge by analyzing newly bred plants for this desirable character. Whilst resistance to a disease may be complex, involving several genes acting together, the marker-assisted technique has proved a powerful form of assistance in this area.

The second technique is **genetic modification** (now known as **GM)**, or genetic engineering. By this method, genes derived from other plant species can be incorporated into the species in question. The commonest technique involves the bacterium *Agrobacterium tumifasciens*. This organism (see also p263) causes crown gall disease on plants such as apple. The bacterium contains a circular piece of DNA (plasmid) that on entering plant cells can integrate its DNA into that of the infected plant cell. Breeders are able to develop strains of *A. tumifasciens* in large numbers. The new strains can be induced to accept, in their plastids, a desirable gene taken from other variants of the same plant species, or taken from other species. Wounded plants infected by a bacterial strain begin to multiply the newly acquired gene by integrating it into the cells of the plant. Tissues developing around the point of infection can then be used for micro-propagation of the new genetically-modified cultivar.

Confirmation of successful genetic change can be achieved most easily when the newly introduced gene is already linked in the bacterial plasmid by a marker gene. Two common kinds of marker were used initially, resistance to an antibiotic and resistance to a herbicide. In this way, the breeder was able to test whether incorporation of a desirable new character was successful by exposing it to the antibiotic or herbicide concerned. Alternative methods to the use of antibiotic markers have been sought. There seems little doubt that major advances in the quantity and quality of horticultural crops could follow GM methods of breeding. However, there are fears that such methods could result in deterioration of food quality or pose a threat to the environment.

The Plant Varieties and Seeds Act, 1964

This Act protects the rights of producers of new cultivars. The registration of a new cultivar is acceptable only when its characteristics are shown to be significantly different from any existing type. Successful registration enables the plant breeder to control the licence for the cultivar's propagation, whether by seed or vegetative methods. Separate schemes operate for the individual genera of horticultural and agricultural crops, but all breeding activities may benefit from the 1964 Act. Producers of licenced plants pay a royalty fee to the breeder.

Gene banks

As new cultivars are produced and grown for use in modern horticulture, old cultivars and wild sources of variation (which could be a source of valuable characteristics and be useful in future breeding programmes) are being lost. Since initiatives in 1974, there continues to be much interest in gene conservation and several gene banks have been established. A gene bank provides a means of storing large quantities of seed of diverse origins at low temperatures, while some plant material (i.e. that which cannot be stored as seed) is maintained by tissue culture (see p177).

Check your learning

1. Define the terms, pollination, self pollination and cross-pollination.

2. Describe two characteristics of wind-pollinated flowers and two characteristics of insect pollinated flowers.

3. Define, in relation to fertilization, the terms compatible and incompatible.

4. Define the term parthenocarpy and explain its importance in horticulture.

5. Describe the characteristics of F1 hybrids and explain the meaning of hybrid vigour.

Further reading

Bos, I. and Calligari, P. (2007). *Selection Methods in Plant Breeding*. Kluwer.

George, R.A.T. (1999). *Vegetable Seed Production*. Longman.

Have, van der, D.J. (1979). *Plant Breeding Perspectives*, Wageningen. Centre for Agricultural Publishing and Documentation.

Ingram, D.S. *et al.* (eds) (2002). *Science and the Garden*. Blackwell Science Ltd.

North, C. (1979). *Plant Breeding and Genetics in Horticulture*. Macmillan.

Simmonds, N.W. (1979). *Principles of Crop Improvement*. Longman.

Watts, L. (1980). *Flower and Vegetable Breeding*. Grower Books.

Chapter 11   Plant development

Summary

This chapter includes the following:

- **Growth and development**
- **Seed germination**
- **Seed viability and dormancy**
- **Tropisms**
- **The vegetative plant**
- **Photoperiodism**

with additional information on:

- **Apical dominance**
- **Juvenility**
- **Pruning**
- **Extended flower life**
- **The ageing plant**

Figure 11.1 **Autumn colour** in the leaves of *Parthenocissus tricuspida* (Boston Ivy), developing in response to environmental changes

Growth and development

Growth is a difficult term to define because it really encompasses the totality of all the processes that take place during the life of an organism. However, it is useful to distinguish between growth as the processes which result in an increase in size and weight (described in Chapter 8), and those processes which cause the changes in the plant during its life cycle, which can usefully be called **plant development**. This is described here through the typical life cycle of plants, from seed to senescence.

Seeds

Seed germination

Details of the structure of seeds can be found in Chapter 7.

The main requirements for the successful germination of most seed are as follows:

> **Seed germination** is the emergence of the radicle through the testa, usually at the micropyle.

- **Water supply** to the seed is the first environmental requirement for germination. The water content of the seed may fall to 10 per cent during storage, but must be restored to about 70 per cent to enable full chemical activity. Water initially is absorbed into the structure of the testa in a way similar to a sponge taking up water into its air space, i.e. by **imbibition**. This softens the testa and moistens the cell walls of the seed so that the next stages can proceed. The cells of the seed take up water by **osmosis**, often assuming twice the size of the dry seed. The water provides a suitable medium for the activity of enzymes in the process of respiration. A continuous water supply is now required if germination is to proceed at a consistent rate, but the growing medium, whether it is outdoor soil or compost in a seed tray, must not be waterlogged, because **oxygen** essential for aerobic respiration would be withheld from the growing embryo. In the absence of oxygen, **anaerobic respiration** occurs and eventually causes death of the germinating seed, or suspended germination, i.e. **induced dormancy**.
- **Temperature** is a very important germination requirement, and is usually specific to a given species or even cultivar. It acts by fundamentally influencing the activity of the enzymes involved in the biochemical processes of respiration, which occur between 0°C and 40°C. However, species adapted to specialized environments respond to a narrow range of germination temperatures. For example, cucumbers require a minimum temperature of 15°C and tomatoes 10°C. On the other hand, lettuce germination may be inhibited by temperatures higher than 30°C and in some cultivars, at 25°C, a period of induced dormancy occurs. Some species, such as mustard, will germinate in temperatures just above freezing and up to 40°C, provided they are not allowed to dry out.
- **Light** is a factor that may influence germination in some species, but most species are indifferent. Seed of *Rhododendron, Veronica* and

Phlox is inhibited in its germination by exposure to light, while that of celery, lettuce, most grasses, conifers and many herbaceous flowering plants is slowed down when light is excluded. This should be taken into account when the covering material for a seedbed is considered (see tilth). The colour (wavelength) of light involved may be critical in the particular response created. Far red light (720 nm), occurring between red light and infra-red light and invisible to the human eye, is found to inhibit germination in some seeds, e.g. birch, while red light (660 nm) promotes it. A canopy of tall deciduous plants filters out red light for photosynthesis. Seeds of species growing under this canopy receive mainly far red light, and are prevented from germinating. When the leaves fall in autumn, these seeds will germinate in response to the now available red light and to the low winter temperatures.

Typical germination process

Seed dormancy

As soon as the embryo begins to grow out of the seed, i.e. **germinates**, the plant is vulnerable to damage from cold or drought. Therefore, the seed must have a mechanism to prevent germination when poor growing conditions prevail. **Dormancy** is a period during which very little activity occurs in the seed, other than a very slow rate of respiration. Seeds will not germinate until dormancy is broken.

A **thick testa** prevents water and oxygen, essential in germination, from entering the seed. Gradual breakdown of the testa, occurring through bacterial action or freezing and thawing, eventually permits germination following unsuitable conditions. The passing of fruit through the digestive system of an animal, such as a bird, may promote germination, e.g. in tomato, cotoneaster and holly. Many species, e.g. fat hen, produce seed with variable dormancy periods, to spread germination time over a number of growing seasons. Spring soil cultivations can break the seed coat and induce germination of weed seeds (see p185). This structural dormancy, in horticultural crops, may present germination problems in plants such as rose rootstock species and *Acacia*. Physical methods using sandpaper or chemical treatment with sulphuric acid (collectively known as **scarification**) can break down the seed coat and therefore the dormancy mechanism.

Chemical inhibitors may occur in the seed to prevent the germination process. **Abscisic acid** at high concentrations helps maintain dormancy while, as dormancy breaks, progressively lower levels occur, with a simultaneous increase in concentrations of growth promotors such as gibberellic acid and cytokinins. Inhibitory chemicals located just below the testa may be washed out by soaking in water.

Cold temperatures cause similar breaks of dormancy in other species (**stratification**), the exact temperature requirement varying with the period of exposure and the plant species. Many alpine plants require a 4°C stratification temperature while other species, e.g. *Ailanthus*, *Thuja*, ash and many other trees and shrubs, require both moisture and the chilling treatment. The chemical balance inside the seed may be changed in favour of germination by treatment with chemicals such as gibberellic acid and potassium nitrate.

An **undeveloped embryo** in a seed is incapable of germinating until time has elapsed after the seed is removed from the parent plant, i.e. the **after ripening** period has occurred, as in the tomato and many tropical species, such as palms. Some seeds such as *Acacia* are recorded to have a dormancy of more than a hundred years.

The practical implications of the above are considered in detail in Chapter 12.

Seed viability

There are a number of essential germination requirements for a successful seedling emergence to occur. A **viable seed** has the potential for germination, given the required external conditions. Its viability, therefore, indicates the activity of the seed's internal organs, i.e. whether the seed is 'alive' or not. Most seeds remain viable until the next growing season, a period of about eight months, but many can remain dormant for a number of years until conditions are favourable for germination. In general, viability of a batch of seed diminishes with time, its maximum viability period depending largely on the species. For example, celery seed quickly loses viability after the first season, but wheat has been reported to germinate after scores of years. The germination potential of any seed batch will depend on the storage conditions of the seed, which should be cool and dry, slowing down respiration and maintaining the internal status of the seed. These conditions are achieved in commercial seed stores by means of sensitive control equipment. Packaging of seed for sale takes account of these requirements and often includes a waterproof lining of the packet, which maintains constant water content in the seeds.

The Seeds Acts

In the UK the Seeds Acts control the quality of seed to be used by growers. A seed producer must satisfy the minimum requirements for species of vegetables and forest tree seed by subjecting a seed batch to a government testing procedure. A sample of the seed is subjected to standardized ideal germination conditions, to find the proportion that is viable (**germination percentage**). The germination and emergence under less ideal field conditions (**field emergence**), where tilth and disease factors are variable, may be much lower than germination percentage.

The sample is also tested for **quality** which provides information, available to the purchaser of the seed, covering trueness to type; that is, whether the characteristics of the plants are consistent with those of the named cultivar; the percentage of non-seed material, such as dust; the percentage of weed seeds, particularly those of a poisonous nature (see Weeds Act, Chapter 13). The precise regulations for sampling and testing, and requirements for specific species, have changed slightly since the 1920 Act, the 1964 Act (which also included the details of plant varieties), and the entry of Britain into the European Community. Some control under EC regulations is made of the provenance of forestry seed, as the geographical location of its source is important in relation to a number of factors, including response to drought, cold, dormancy, habit, and pest and disease susceptibility.

The seedling

Within the seed is a food store that provides the means to produce energy for germination. Once the food store has been exhausted, the seedling must rapidly become independent in its food supply and begin to photosynthesize. It must therefore respond to stimuli in its environment to establish the direction of growth. Such a response is termed **a tropism**, and is very important in the early survival of the seedling (Figure 11.2).

A **tropism** is a directional growth response to an environmental stimulus.

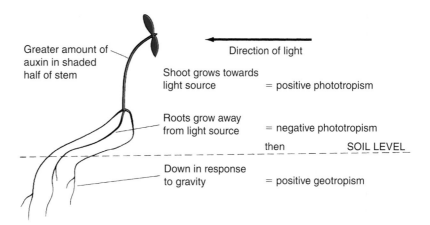

Figure 11.2 **Geotropism and phototropism** shown as mechanisms assisting the survival of the seedling

Geotropism is a directional growth response to gravity. The emergence of the radicle from the testa is followed by growth of the root system, which must take up water and minerals quickly so that the shoot system may develop. A seed germinating near the surface of a growing medium must not put out roots that grow on to the surface and dry out, but the roots must grow downwards to tap water supplies. Conversely, the plumule must grow away from the pull of gravity so the leaves develop in the light.

Etiolation is the type of growth which the shoot produces as it moves through the soil in response to gravity. The developing shoot is delicate and vulnerable to physical damage, and therefore the growing tip is often protected by being bent into a plumular hook. The stem grows quickly, is supported by the structure of the soil and therefore is very thin and spindly, stimulated by friction in the soil which causes release of ethylene. The leaves are undeveloped, as they do not begin to function until they move into the light. Mature plants that are grown in dark conditions also appear etiolated.

Seedling development

The emergence of the plumule above the growing medium is usually the first occasion that the seedling is subjected to light. This stimulus inhibits the extension growth of the stem so that it becomes thicker and stronger, but the seedling is still very susceptible to attack from pests and damping-off diseases. The leaves unfold and become green in response to light, which enables the seedling to photosynthesize and so support its development. The first leaves to develop, the cotyledons, derive from the seed and may emerge from the testa while still in the soil, as in peach and broad bean (**hypogeal** germination), or be carried with the testa into the air, where the cotyledons then expand (**epigeal** germination), e.g. in tomatoes and cherry.

> **Hypogeal** germination occurs when the cotyledons develop above ground outside the seed.
>
> **Epigeal** germination occurs when the testa merges above ground initially enclosing the cotyledons.

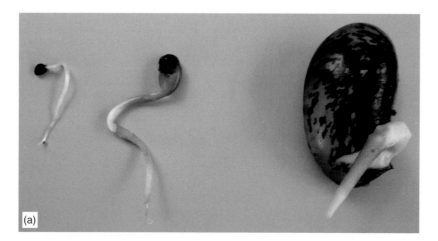

Figure 11.3 Seed germination: (a) **epigeal** germination on left in leek and tree lupin, **hypogeal** germination on right in runner bean (b) later stage showing hypogeal in bean on left and epigeal in tree lupin on right

Phototropism occurs so the shoot grows towards a light source that provides the energy for photosynthesis. A bend takes place in the stem just below the tip as cells in the stem away from the light grow larger than those near to the light source. A greater concentration of auxin in the shaded part of the stem causes the extended growth (see Figure 11.2). Roots display a negative phototropic response, growing away from light when exposed at the surface of the growing medium, e.g. on a steep bank. The growth away from light may supersede the root's geotropic response, and will cause the roots to grow back into the growing medium.

Hydrotropism is the growing of roots towards a source of water. The explanation of this tropism has not been found, but it can be shown to occur. The **cotyledons** that emerge from the testa contribute to the growth of the seedling in photosynthesis, but the **true leaves** of the plant, which often have a different appearance to the cotyledon, very quickly unfold.

Apical dominance

After the germination of the seed, the plumule establishes a direction of growth, due partly to the geotropic and phototropic forces acting on it. Often the terminal bud of the main stem sustains the major growth pattern, while the axillary buds are inhibited in growth to a degree that depends on the species.

In tomatoes and chrysanthemums, the lateral shoots have the potential to grow out, but are inhibited by a high concentration of auxin, which accumulates in these buds. The source of the chemical is the terminal bud, which maintains the inhibition. In commercial chrysanthemum production, the removal of the main shoot (stopping) is a common practice. It takes away the auxin supply to the axillary buds, which are then able to grow out to create a larger, more balanced inflorescence. Conversely, the practice of **disbudding** in chrysanthemums and carnations, takes out the axillary buds to allow the terminal bud to develop into a bigger bloom that benefits from the greater food availability.

Conditions for early plant growth

Many plant species are propagated in glasshouses. A few principles are described here to help ensure success. Seed trays should be thoroughly cleaned to prevent the occurrence of diseases such as 'damping off' (see p246). Fresh growing medium should be used for these tiny plants that have little resistance to disease. Compost low in soluble fertilizer is less likely to scorch young plants. Compost should be firmed down in containers to provide closer contact with the developing root system.

With very small seed there is a danger that too many seeds are sown together, with the result that the seedlings intertwine and are hard to separate. This problem can often be avoided by diluting batches of very small seed with some fine sand before sowing. Small seed samples need to be sown on the surface of the compost and then covered with only a fine sprinkling of compost. In this way, their limited food reserves are not overtaxed as they struggle through the compost to reach the light.

Water quality is important with young plants. Mains water is recommended, as it will be free from diseases. Water butts and reservoirs need particular scrutiny to avoid problems. Water that has been left to reach the ambient temperature of the glasshouse is less likely to harm seedlings. The compost in seed trays should be kept permanently moist (but not waterlogged), as seedling roots dry out easily. Glass or plastic covers placed over seed trays will help prevent moisture loss, and these can be removed when root establishment has occurred and seedlings are pushing against the covers.

As soon as seedlings have expanded their cotyledon leaves, they should be carefully transferred ('pricked off') from the seed tray and placed in another tray filled with compost having higher levels of fertility. The seedlings should be spaced at approximately 2.5 cm intervals, thus providing a root volume for increased growth. Later, plants will be transferred to pots ('potted-on') to allow for further growth.

Plants growing in glasshouses are tender. The cuticle covering leaves and stems is very thin. Growth is rapid and the stem's mechanical strength is likely to be dependant on tissues such as collenchyma and parenchyma rather than the sturdier xylem vessels (see p92). When a plant is transferred from a glasshouse to cooler, windier outside conditions (for example in spring), it may become stressed, lose leaves and stop growing. It is advisable to 'harden off' plants before this stressful exposure. Reducing heat and increasing ventilation in the glasshouse are two ways of achieving this aim. Traditionally plants were moved out into cold frames to gradually expose them to the conditions into which they are to be planted. Moving plants out during the day and back inside overnight for a number of weeks, is another strategy.

The vegetative plant

The role of the vegetative stage in the life cycle of the plant is to grow rapidly and establish the individual in competition with others. It must therefore photosynthesize effectively and be capable of responding to good growing conditions. Growing rooms with near-ideal conditions of light, temperature and carbon dioxide utilize this capacity that will reduce with the ageing of the plant (see Chapter 8).

Juvenility

The **juvenile** stage is a period after germination that is capable of rapid vegetative growth and is unlikely to flower.

The early growth stage of the plant, **juvenile** growth, is characterized by certain physical appearances and activities that are different from those found in the later stages or in **adult** growth. Often leaf shapes vary; e.g. the juvenile ivy leaf is three-lobed while the adult leaf is more oval, as shown in Figure 11.4. The habit of the plant is also different; the juvenile stem of ivy tends to grow horizontally and is vegetative in nature, while the adult growth is vertical and bears flowers. Other examples are common in conifer species where the complete appearance of the plant is altered by the change in leaf form, for example, *Chamaecyparis fletcheri* and many *Juniperus* species such as *J. chinensis*. In the genera *Chamaecyparis* and *Thuja*, the juvenile condition can be achieved permanently by repeated vegetative propagation producing plants called **retinospores**, which are used as decorative features.

Figure 11.4 **Juvenile** growth on left, showing adventitious roots and lobed leaf, **adult** growth on right showing flowers and entire leaf, in ivy (Hedera helix)

Leaf retention is also a characteristic of juvenility. It can be significant in species such as beech (see Figure 11.5), where the phenomenon is exaggerated, and the trees

Figure 11.5 **Leaf retention** in the lower **juvenile** branches of beech (*Fagus sylvatica*); compare with the bare adult branches

can be pruned back to the vegetative growth. This can create additional protection in **windbreaks** although the barrier created tends to be too solid to provide the ideal wind protection (see p38).

Many species that require an environmental change to stimulate flower initiation, such as the Brassicas that require a cold period, will not respond to the stimulus until the juvenile period is over; about eleven weeks in Brussels sprouts.

The adult stage essential for sexual reproduction is less useful for vegetative propagation than the responsive juvenile growth, a condition due probably to the hormonal balance in the tissues. Figure 11.4 shows the spontaneous production of adventitious roots on the ivy stem. Adult growth should be removed from **stock plants** (see p174) to leave the more successful juvenile growth for cutting.

The ability of the plant to reproduce vegetatively is widely used in horticulture and these methods, both natural and artificial, are detailed in Chapter 12.

Vegetative propagation

Although the life cycle of most plants leads to sexual reproduction, **all plants have the potential to reproduce asexually or by vegetative propagation, when pieces of the parent plant are removed and develop into a wholly independent plant**. All living cells contain a nucleus with a complete set of genetical information (*see* genetic code, Chapter 10), with the potential to become any specialized cell type. Only part of the total information is brought into operation at any one time and position in the plant.

If parts of the plant are removed, then cells lose their orientation in the whole plant and are able to produce organs in positions not found in the usual organization. These are described as **adventitious** and can, for example, be roots on a stem cutting, buds on a piece of root, or roots and buds on a piece of leaf used for vegetative propagation. Many plant species use the ability for vegetative propagation in their normal pattern of development, in order to increase the number of individuals of the species in the population. The production of these vegetative **propagules**, as with the production of seed, is often the means by which the plant survives adverse conditions (*see* overwintering), acting as a food store which will provide for the renewed growth when it begins. The stored energy in the swollen tap roots of dock and dandelion enable these plants to compete more effectively with seedlings of other weed and crop species, which would also apply to roots of *Gypsophila paniculata*, carrots and beetroot.

Stems are telescoped in the form of a **corm** in freesia and cyclamen, or swollen into a **tuber** in potato, or a horizontally growing underground **rhizome** in iris and couch grass. Leaves expanded with food may

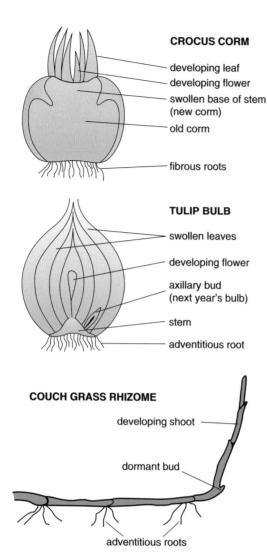

CROCUS CORM
- developing leaf
- developing flower
- swollen base of stem (new corm)
- old corm
- fibrous roots

TULIP BULB
- swollen leaves
- developing flower
- axillary bud (next year's bulb)
- stem
- adventitious root

COUCH GRASS RHIZOME
- developing shoot
- dormant bud
- adventitious roots

Figure 11.6 **Structure of organs** responsible for over-wintering and vegetative propagation

form a large bud or **offset** found in lilies. A **bulb**, as seen in daffodils, tulips, and onions, is largely composed of succulent white leaves enveloping the much reduced stem, found at the base of the bulb (see Figure 11.6).

Other natural means of propagation include lateral stems, which grow horizontally on the soil surface to produce nodal, adventitious roots and subsequently plantlets, e.g. **runners** or **stolons** of strawberries and yarrow. The adventitious nature of stems is exploited when they are deliberately bent to touch the ground, or enclosed in compost, in the method known as **layering**, used in carnations, some apple rootstocks, many deciduous shrubs such as *Forsythia*, and pot plants such as *Ficus* and *Dieffenbachia*. The roots of species, especially in the Rosaceae family, are able to produce underground adventitious buds that grow into aerial stems or **suckers**, e.g. pears, raspberries. By all these methods of runners, layering and suckers, the newly developing plant (**propagule**) will subsequently become detached from the parent plant by the disintegration of the connecting stem or root.

Growth retardation

Stem extension growth is controlled by auxins produced by the plant and also by gibberellins that can dramatically increase stem length, especially when externally applied. Growth retardation may be desirable, especially in the production of compact pot plants from species that would normally have long stems, e.g. chrysanthemums, tulips and *Azaleas*. Therefore, artificial chemicals, such as daminozide (Figure 11.7), which inhibit the action of the growth promoting hormones can retard the development of the main stem and also stimulate the growth of side shoots to produce a bushier, more compact plant. Flower production may be inhibited but this can be countered by the application of flower stimulating chemicals.

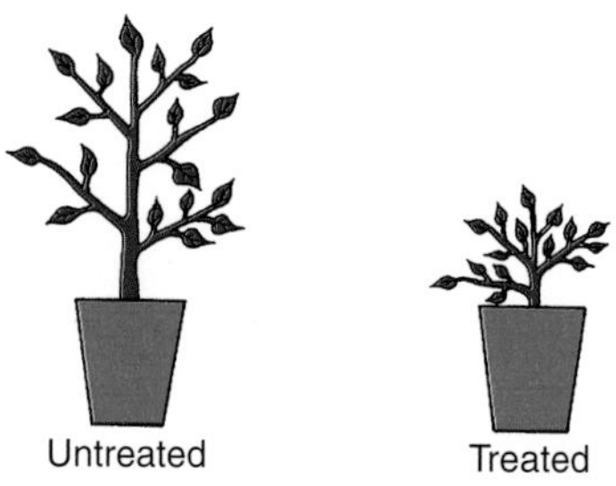

Pruning

Parts of plants can be **pruned** (removed) to reduce the competition within the plant for the available resources. In this way, the plant is encouraged to grow, flower or fruit in a way the horticulturist requires. A reduction in the number of flower buds of, for example, chrysanthemum, will cause the remaining buds to develop into larger flowers; a reduction in fruiting buds of apple trees will produce bigger apples, and the reduction in branches of soft fruit and ornamental shrubs will allow the plants to grow stronger when planted densely. Pruning will also affect the shape of the plant, as meristems previously inhibited by apical dominance will begin to develop. The success of such pruning

Untreated   Treated

Figure 11.7 **Chemical growth retardant** is incorporated into compost used for pot plants such as chrysanthemum

depends very much on the skill of the operator, as a good knowledge of the species habit is required.

A few general principles apply to most **pruning** situations:

- **Young** plants should be trained in a way that will reflect the eventual shape of the more mature plant (formative pruning). For example a young apple tree (called a 'maiden') can be pruned to have one dominant 'leader' shoot, which will give rise to a taller, more slender shape. Alternative pruning strategies will lead to quite different plant shapes. Pruning back all branches in the first few years forms a bush apple. A cordon is a plant where there is a leader shoot, often trained at 45 degrees to the ground, with all side shoots pruned back to one or two buds. Cordon fruit bushes are usually grown against walls or fences. Similarly, fans and espalier forms can be developed.
- The **pruning cut** should be made just above a bud that points in the required direction (usually to the outside of the plant). In this way, the plant is less likely to acquire too dense growth in its centre.
- **Pruning should remove any shoots that are crossing**, as they will lead to dense growth. Some plants, such as roses and gooseberries, are made less susceptible to disease attack by the creation of an open centre to produce a more buoyant (less humid) atmosphere.
- **Weak shoots should be pruned the hardest** where growth within the plant is uneven and strong shoots pruned less, since pruning causes a stimulation of growth.
- **Species that flower on the previous year's growth** of wood (e.g. *Forsythia*) should be pruned soon after flowering has stopped. Conversely, species that flower later in the year on the present year's wood, e.g. *Buddleia davidii*, should be pruned the following spring.

Root pruning was used to restrict over-vigorous cultivars, especially in fruit species, but this technique has been largely superseded by the use of dwarfing rootstock grafted onto commercially grown scions. Root pruning is still seen, however, in the growing of Bonsai plants. Pruning is largely concerned with creating the shape of a plant, and controlling apical dominance, but the removal of dead, damaged and diseased parts is also an important aspect.

The flowering plant

The progression from a vegetative to a flowering plant involves profound physical and chemical changes. The stem apex displays a more complex appearance under the microscope as flower initiation occurs, and is followed, usually irreversibly, by the development of a flower. The stimulus for this change may simply be genetically derived, but often an environmental stimulus is required which links flowering to an appropriate season.

Photoperiodism is a day length stimulated developmental response by the plant.

Photoperiodism

Photoperiodism is a term used to describe the plants various responses to day length, explained here in terms of flowering; other responses include bud dormancy and leaf fall.

Many plant species flower at about the same time each year, e.g. in the UK *Magnolia stellata* in April, *Philadelphus delavayi* in June and chrysanthemum in September. In many cases, flowering is in response to the changing day length, which is the most consistent changing environmental factor, in comparison with above-ground temperature which is more variable.

In a day length sensitive species, the flowering process is 'switched on' by a specific period of daylight (or darkness) called its **critical period**. In the chrysanthemum the critical period is sixteen hours of daylight (or eight hours of darkness), which occurs in September in the UK. If repeated over several weeks, the internal structure of the buds begins to change from a vegetative meristem to a flowering meristem (see p00).

The Phytochrome 'Switch'

Since 1920 much research has attempted to explain the photoperiodic flowering response, including using artificial lighting and investigating genetic and biochemical control. Recently, the following stages have been identified, namely switching on at the leaf, mobilizing leaf genes, moving the message from leaf to bud, and developing a flowering meristem. The first stage represents one of the best examples of the horticulturist manipulating the biology of the plant and an understanding of the science has enabled the grower to control the flowering process with the consequent valuable worldwide industry of year-round flower production. **Phytochrome** is the chemical produced by the plant to operate the switching mechanism.

Phytochrome is a large blue-coloured molecule (molecular weight about 125 000). It is made up of two relatively small colour-sensitive sub-molecules (chromophores) and two very long protein chains. It is thought that the chromophores change their shape in response to light, and that this vital 'day-length message' is passed through the proteins to the next stage in the flowering sequence described. Investigations of phytochrome suggest that, in addition to its involvement in the flowering stimulus, the chemical is used in as many as 24 other light-induced reactions ranging from opening a seed's plumule hook as it emerges from the soil (see p154) to increasing the respiration rate of cells.

A two-way chemical process is involved, requiring a different light colour for each direction. Phytochrome Pr660 is sensitive to red light of wavelength 660 nm, found in daylight from dawn to dusk. Pr660 is changed to a less stable form of phytochrome (Pfr730) after a days' exposure. Pfr730 refers to phytochrome as far red light, is found in shaded conditions and is the form which brings about the plant response.

Day length sensitive plants respond to changing seasons as either **long day plants** or short day plants. In species such as Hosta, sweet pea, Lobelia and radish, long days are essential for flowering, while the flowering of carnations and snapdragons, among others, is improved. In these species, the presence of Pfr730 in a concentration above a critical limit results in the promotion of flowering, because the summer nights are not long enough to allow sufficient Pfr730 to revert back to Pr660. A more accurate term here therefore would be 'short night plant'.

- **Day neutral** species are switched on to flowering by a range of situations involving plant size and development and temperature, e.g. Begonia elatior and tomato.

- **Short day plants**, e.g. chrysanthemum, poinsettia and kalanchoe, respond differently in that the presence of Pfr730 above a critical limit inhibits flowering. In chrysanthemum the critical period of dark is eight hours and this condition over a period of several weeks will induce flowering. To enable all-year-round production as cut flowers and pot plants, the day length manipulation is sophisticated. Immature plants must initially be prevented from flowering, then flower buds must later be induced, often at a time of year when the natural day length would not be suitable.

Artificial control of flowering

A long night may be broken artificially using a technique called **night-break lighting**. Incandescent tungsten bulbs produce a high proportion of red light and are cheap to run. Hung about 1 m above the crop and spaced to give about 150 lux for four hours ensures that the Pfr730 critical level is not reached. **Cyclic lighting** saves electricity and uses a series of brief alternating light and dark cycles to replace one continuous break. High pressure sodium lamps are used where they are installed for supplementary lighting (see p114), this saves expense in providing two systems. Crops such as chrysanthemums can be induced to flower in the summer by imposing a long night regime artificially, using opaque black cloth or plastic curtains to cover the crop (see Figure 11.8). A night of nine to fifteen hours causes the Pfr730 level to drop below the critical limit and the flowering process to be initiated.

Figure 11.8 **Daylength control** provided by blackout curtains to control flowering in chrysanthemum.

Flower initiation

Flower initiation can be stimulated largely by photoperiodic or temperature changes, or a complex interaction between temperature and day length. Cold temperatures experienced during the winter bring about flower initiation (i.e. **vernalization**) in many biennial species such as *Brassica*, lettuce, red beet, *Lunaria* and onion. The period for the response depends on the exact temperature, as with budbreak and seed dormancy (see stratification). The optimum temperature for many of these responses is about 4°C. Hormones are involved in causing the flower apex to be produced. The balance of auxins, gibberellins and cytokinins is important, but some species respond to artificial treatment of one type of chemical; for example, the day length requirement for chrysanthemum plants can be partly replaced by gibberellic acid sprays.

Extended flower life

The flower opens to expose the organs for sexual reproduction. The life of the flower is limited to the time needed for pollination and fertilization, but it is often commercially desirable to extend the life of a cut flower or flowering pot plant. In cut flowers, water uptake must be maintained and dissolved nutrients for opening the flower bud are termed an **opening solution**.

Vase life can be extended by the addition of sterilants and sugar to the water. A **sterilant**, e.g. silver nitrate, in the water can reduce the risk of blockage of xylem by bacterial or fungal growth. **Ethylene** has a considerable effect on flower development, and can bring about premature death (**senescence**) of the flower after it begins to open. Cut flowers should therefore never be stored near to fruit, e.g. apples or bananas, which produce ethylene. Some chemicals, such as sodium thiosulphate, reduce the production of ethylene in carnations and therefore extend their life.

Removal of dead flowers

The removal of dead flowers, an activity called **dead-heading**, is an effective way to help maintain the appearance of a garden border. Examples of species needing this procedure are seen in bedding plants which flower over several months, e.g. African Marigold (*Tagetes erecta*); in herbaceous perennials, e.g. Delphinium and Lupin; in small shrubs, e.g. *Penstemon fruticosus*; and in climbers, e.g. sweet pea and *Rosa* 'Pink Perpetue'. As flowers age, they begin to use up a considerable amount of the plant's energy in the production of fruits. Also, hormones produced by the fruit inhibit flower development. With species such as those mentioned above, the maturation of fruits will considerably reduce the plant's ability to continue producing flowers. The act of dead-heading, therefore, will greatly improve subsequent flowering. An added bonus is that plants that have been dead-headed may continue to flower many weeks longer than those allowed to retain their dead flowers.

Many species such as Wax Begonia (Begonia *x semperflorens-cultorum*), and Busy Lizzie (*Impatiens wallerana*) used as bedding plants have been specially bred as F1 hybrids (see p144) where, in this case, flowers do not produce fruits containing viable seed. In such cases, there is not such a great need to deadhead, but this activity will help prevent unsightly rotting brown petals from spoiling the appearance of foliage and newly-produced flowers.

The ageing plant

At the end of an annual plant's life, or the growing season of perennial plants, a number of changes take place. The changes in colour associated with autumn are due to pigments that develop in the leaves and stems and are revealed as the chlorophyll (green) is broken down and absorbed by the plant.

Pigments are substances that are capable of absorbing light; they also reflect certain wavelengths of light which determine the colour of the pigment. In the actively growing plant, chlorophyll, which reflects mainly green light, is produced in considerable amounts, and therefore the plant, especially the leaves, appears predominantly green. Other pigments are present; e.g. the carotenoids (yellow) and xanthophylls (red), but usually the quantities are so small as to be masked by the chlorophyll. In some species, e.g. copper beech (*Fagus sylvat-ica*) other pigments predominate, masking chlorophyll. These pigments also occur in many species of deciduous plants at the end of the growing season, when chlorophyll synthesis ceases prior to the abscission of the leaves. Many colours are displayed in the leaves at this time in such species as *Acer platanoides*, turning gold and red, *Prunus cerasifera* 'Pissardii' with light purple leaves, European larch with yellow leaves, Virginia creeper (*Parthenocissus* and *Vitus spp.*) with red leaves, beech with brown leaves, *Cotoneaster* and *Pyracantha* with coloured berries, and *Cornus* species, which have coloured stems. These are used in **autumn colour** displays at a time when fewer flowering plants are seen outdoors (see Figure 11.9).

In deciduous woody species the leaves drop in the process of **abscission**, which may be triggered by shortening of the day length. In order to reduce risk of water loss from the remaining leaf scar, a corky layer is formed before the leaf falls. Auxin production in the leaf is reduced, this stimulates the formation of the abscission layer, and abscisic acid is involved in the process. Auxin sprays can be used to achieve a premature leaf fall in nursery stock plants thus enabling the early lifting of bare-root plants. Ethylene inhibits the action of auxin, and can therefore also cause premature leaf fall, for example, in *Hydrangea* prior to cold treatment for flower initiation.

Figure 11.9 **Autumn colour** in (a) Blueberry, (b) *Viburnum* and (c) *Photinia*, showing loss of chlorophyll and emergence of xanthophylls.

Check your learning

1. Describe the factors which affect seed germination.

2. Define the terms epigeal, hypogeal, dormancy and viability.

3. Describe the process of phototropism.

4. Define the term photoperiodism.

5. Describe two plant growth responses to auxin.

Further reading

Bleasdale, J.K.A. (1983). *Plant Physiology in Relation to Horticulture*. Macmillan.

Cushnie, J. (2007). *How to Prune*. Kyle Cathie.

Gardner, R.J. (1993). *The Grafter's Handbook*. Cassell.

Hart, J.W. (1990). *Plant Tropisms and Other Growth Movements*. Unwin Hyman.

Hartman, H.T. *et al.* (1990). *Plant Propagation, Principles and Practice*. Prentice-Hall.

Leopold, A.C. (1977). *Plant Growth and Development*. 2nd edn. McGraw-Hill.

Machin, B. (1983). *Year-round Chrysanthemums*. Grower Publications.

MAFF. *Quality in Seeds*. Advisory Leaflet 247.

MAFF (1979). *Guide to the Seed Regulations*. (HMSO).

Stanley, J. and Toogood, A. (1981). *The Modern Nurseryman*. Faber.

Taiz, L. and Zeiger, E. (2006). *Plant Physiology*. Sinauer Associates.

Thomas, B. and Vince-Prue, D. (1996). *Photoperiodism in Plants*. Academic Press.

Wareing, P.F. and Phillips, I.D.J. (1981). *The Control of Growth and Differentiation in Plants*. Pergamon.

Chapter 12　Plant propagation

Summary

This chapter includes the following topics:

- **Seed propagation**
- **Dormancy**
- **Vegetative propagation**
- **Comparison between seed and vegetative propagation**
- **Budding and grafting**

with additional information on the following:

- **Tissue culture**

Figure 12.1　**Bedding plants**

A plant's life cycle can be seen to end with the process of senescence (p163) and dying. The time taken to get to this point varies enormously from one species to another with many ephemerals living for only a few months, whereas many trees last for hundreds of years. Before it dies the plant has normally ensured continued life by either sexual or asexual reproduction: not many plants employ both methods to produce offspring. Sexual reproduction leads to the formation of seeds in higher plants (see p67). The ability of plants to reproduce asexually is used in horticulture as vegetative propagation.

Seed propagation

Most plants reproduce sexually, which leads to the formation of seeds (see p103). By the nature of this process the seeds produced show a **variation** in characteristics to a greater or lesser extent. Typically plants produced from seed will not be uniform in their growth and will exhibit differences in size, flower colour, etc. This variation can be controlled by skilled plant breeders (see p144) to the extent that a high degree of uniformity can be achieved in bedding, e.g. for flower colour, and vegetable seeds, e.g. for size and 'once over harvesting' (see hybrids p144).

Sexual reproduction is the production of new individuals by the fusion of a nucleus from the male (in pollen) and that of a female (in the ovule) to form a zygote (see Chapters 7 and 10).

Seeds germinate when provided with the right conditions regarding:

- water
- air (oxygen)
- temperature

and, for some, an exposure to light, or, for others, an absence of light, as described in detail in Chapter 11 (see p150). In some cases germination will not take place even if otherwise favourable conditions prevail (consider the seeds that fall into the warm, moist soil in the autumn but do not germinate until the following spring or later). These seeds are exhibiting dormancy, which has to be broken to allow germination to occur (see p150). This is a survival mechanism that helps prevent the seed start germinating just when conditions are about to become unfavourable for growth.

Physical dormancy is a mechanism, such as a hard seed coat, which has to be broken before water and oxygen can get in. Rather than wait, growers can speed up the process by **scarification**, e.g. sand papering or filing the coat; 'chipping' or 'nicking' it with a knife or, as in the trade, by adding acids. Water can then get in quickly through the thin or damaged seed coat and start the germination process. For many seeds simply adding hot water is sufficient to remove the waterproofing qualities of the seed and let water in.

Physiological dormancy includes the effect of abscisic acid in the dry seed which inhibits development of the embryo. Germination cannot begin until its concentration is reduced. In temperate areas an exposure to prolonged cold gradually destroys the inhibitor. Growers can overcome this mechanism by exposing the seed to cold artificially. **Stratification** is the usual method of overcoming this form of dormancy. The seeds are placed in layers of moist sphagnum moss and grit within a polythene bag. The seeds are allowed to take up water in the warm, but once swollen the bag and its contents are chilled but not frozen. For some species, they are ready to germinate after a month, others take much longer. Once the dormancy of most species is broken they do not develop further until all the normal requirements for germination are met. Care needs to be taken because some species start to germinate once their chilling period has been experienced.

Many seeds develop dormancy on storage. It is possible to avoid the problem by sowing 'green' seed. Seed can be collected when it is mature and with adequate food reserves, but before the dormancy mechanisms become established (soft seed coat, low abscisic acid), and sown straight away.

Purchasing seeds, especially vegetable and flower seeds, has the advantage of convenience and the protection of the regulations (see p152). A check of the date should always be made to ensure that the seeds are from the last seed harvest. The seeds are usually supplied in foil packets. Once opened the seeds deteriorate rapidly so should be sown immediately but, so long as they are kept dry and cold in a resealed packet, most seeds will remain viable for a year and some, often the larger seeds, for many years (see p168).

There are difficulties when it comes to seeds from trees or shrubs because there are fewer regulations to protect the buyer. In the preparation of seeds for sale, the drying process used often:

- increases the dormancy effect (harder coats);
- adversely affects the energy reserves;
- damages the embryo;

so reducing seed viability (see p152).

Seeds, especially finer ones, are often coated (with a clay) to make sowing easier and more precise. This **pelleted seed** can help reduce wastage. Likewise, water-soluble seed tapes can be used. A gel (or wallpaper paste) containing seeds can be used for **fluid drilling**; the gel is squeezed out of a plastic bag like icing a cake. Some seeds that are difficult to germinate can be **primed**; the germination process is started but then arrested. The dried seed purchased can be drilled or sown as normal and rapid and reliable germination follows.

Collecting seed can prove to be cheaper. Although there are attractions in keeping their **own seed**, growers need to be aware of difficulties associated with seed variation (see p144) and the risk of disease (see p147). Seed collectors can ensure they take the seed at the ideal time, especially when the intention is to avoid dormancy problems, and care can be taken in drying so as to minimize loss of viability. Where the hardiness of the plant is in doubt then, although a hardy parent does not always produce hardy seed, the chances of success are raised by taking seed from a known hardy specimen. There are other advantages, particularly when it comes to trees and shrubs, because seed can be taken from desirable forms, beneficial even though there will be variation in the offspring.

The majority of seed should be collected as they ripen. Seed in dry fruits should be collected on a dry day. It should be noted that when enclosed in a fruit, the seed is ready to collect before the fruit matures ready for dispersal. A collecting bag, plastic rather than cloth, to keep hands free is an advantage and it is essential to label samples with the name of the plant and from where collected. The seeds should be kept in small batches and kept cool (to prevent the embryo from heating up). The seed should be prepared for stratification and/or sown as quickly as possible for maximum benefit.

Dry seed needs to be prepared from the material collected. Flower stems can be tied lightly then hung upside down in a dry place with a brown paper bag over them; shake from time to time to collect the seed. Large seed heads should be broken up into trays on paper and left to dry. Cones or small seed heads collected when nearly dry should be placed in an open paper bag and left to complete their drying gently. The flesh of fruits, which often contains germination inhibitor, should have the majority of the flesh removed before being squeezed through a sieve with a presser board. The seeds with any remaining flesh should then be put in a jar of warm water and soaked for a few days after which the water is poured off. This is repeated until the flesh has been removed. The remaining skin is then picked off and the seeds dried. Sieves can be used to remove any superfluous pieces before putting the seed into paper packets ready for sowing.

Storing seed which is to be used within a few days requires little more than keeping them at room temperature in a polythene bag to maintain the moisture levels at which they were collected. If they are to be kept for a few weeks then the seeds need to be stored cool, but not frozen. Seed to be stored for longer periods than this, as when commercially produced for sale, is dried, placed in air proof packets (commonly foil), vacuumed to remove air and kept cool. Some of the large fleshy seeds, such as lilies and hellebores, are best left to mature and collected before they are dispersed. Other seed such as anemones is collected and sown 'green', i.e. before maturing.

Sowing and aftercare in protected environments

The ideal conditions for raising plants from seed can be achieved in a protected environment such as a glasshouse or cheaper alternatives such as polythene tunnels or cold frames (see p16).

Most seeds grown in protected culture are sown into **containers** (see Figure 12.2):

- seed trays
- half trays

Figure 12.2 **Range of containers for** growing plants: (a) traditional clay pots (b) standard seed tray and half tray (c) standard plastic pots in range of sizes, compared with (d) 'long toms' and (e) half pots (f) biodegradable pots (g) compressed blocks (h) square or (i) round pots in trays (j) various 'strips' in trays and (k) typical commercial polystyrene bedding plant tray.

- pots (as deep as wide)
- pans ('half pots')
- long toms.

These must have adequate drainage to allow excess water out or for the water in the capillary matting to pass into the compost (see p392). Square shapes utilize space better, but are harder to fill properly in the corners. Although more expensive, rigid plastic is easier to manage. Rims on containers give more rigidity and make them easier to stack. Gardeners can make use of plastic food containers so long as they are given sufficient drainage holes. All containers should be clean before use (see hygiene p269). There are also disposable pots made of compressed organic matter, paper or 'whalehide' through which roots will emerge, which makes them useful for the planting out stage.

For production horticulture there is a wider range of materials, including polystyrene, for once only use; cost and presentation of the plants becomes the main consideration. Too large a container is a waste of compost and space whereas one that is too small can lead to the seedlings having to be spaced out before they are ready; if left they become overcrowded and susceptible to damping off diseases (see p246).

Seed composts are commonly equal parts peat and sand mixes with lime and a source of phosphate. **Potting composts** into which seedlings are transferred and young plants established tend to have a higher proportion of peat with lime and a full range of nutrients. Many advocate the use of sterilized loam which makes the compost easier to manage and increasingly alternatives to peat are being utilized (see compost mixes, p390).

Sowing seeds. The container is generously overfilled with seed compost. Care is taken to ensure that there are no air pockets by tapping on the bench and the corners are fully filled. With a sawing action across the container top, the surplus compost is 'struck off' using a straight edge. The compost is then lightly firmed to just below the rim of the container using an appropriately sized presser board. Seeds are then sown on the surface at the rate recommended. Many advocate that when using trays, half the seed is sown then, to achieve an even distribution, the other half is sown after turning the container through 90 degrees. The seeds are then covered with sieved compost or fine grade vermiculite to their own thickness. Finer seeds are often sown in equal parts of fine dry sand to help distribution, lightly pressed into the surface and then left uncovered. Larger or pelleted seed tends to be 'space or station sown' i.e. placed at recommended distances in a uniform manner.

The seeds are then labelled with name of plant and the sowing date. The compost is then watered either from above gently, with a fine rose, or by standing the container in water. A fungicide can be added to the water to protect against damping off diseases. The moist conditions around the seed must be maintained and this is most easily done by covering with a sheet of glass, clear plastic or kitchen film. The container should be kept in a warm place (approximately 20°C). If necessary, a sheet of paper can be used to shade the seeds from the direct sunlight or to minimize temperature fluctuations. There are advantages in placing the seed containers in a closed propagator (see p176). Covers on the container should be removed as soon as the seedlings appear and they must now be well lit to avoid etiolating (see p153), but not exposed to strong sunlight. Watering must be maintained, but without waterlogging the compost. In production horticulture much of the work is done by machines. Pots are rarely filled by hand and increasingly the whole process is automated, including the seed sowing.

Pricking out. When the seedlings are large enough to be handled, they should be transplanted into potting compost prepared as for the seed tray. Each seedling is eased with a dibber, lifted by the seed leaves, dropped into a hole made in the new compost, gently firmed and watered in. The seedlings are normally planted in rows with space, typically 24 to 40 per seed tray, for them to grow on to the next stage.

Many will be planted out when ready, but others will continue in pots and as they need more space they are '**potted off**' (moved into another container). Those continuing in containers are '**potted on**' when they outgrow their container. As they are moved to the next size container, the compost, especially the nutrient level, is chosen to suit the stage reached (see JIP p386).

Hardening off is required to ensure that the seedlings raised in a protected environment can be put out into the open ground without a check in growth caused by the colder conditions, wind chill and variable water supply. As the pricked off seedlings become established they are moved to a cooler situation, typically a cold frame, which starts the process of hardening off by providing a closed environment without heat. After a few weeks the cold frame is opened up a little by day and closed at night. If tender plants are threatened by cold or frosts they can be given extra protection in the form of easily handled insulation such as bubble wrap or coir matting put over the frame. Watering has to be continued and usually the plants will use up the fertilizer in the compost and need applications of liquid fertilizer. The hardening process then continues with the frame lid ('light') gradually being opened up more to allow air circulation day and night. Ideally the plants have been fully exposed to the outdoor conditions by the time it is ready to plant them out.

The young plants are very susceptible to fungal diseases while in the frame because of the high density of planting and the difficulty with keeping the humidity level right. The need to maintain air circulation is essential as the opportunity arises. Excessive feeding with high nitrogen fertilizers should also be avoided because it can create soft growth which makes them vulnerable to disease and excessively soft and vigorous plants can be checked on planting out.

Bedding plants (see Figure 12.1) are raised as described above with the sequence geared to producing the plants ready to plant out at the right time. For this, the time when required and the growth rates of the selected plants needs to be known. Usually the seeds are sown into seed trays or pans with very lightly firmed peat compost. Care should be taken to avoid waterlogging in shallow containers (see p385) which leads to the seeds rotting off, poor seedling development, attack from sciarid fly (see p218) or fungal diseases (see p246). There are methods of raising plants from seed without containers by using blocks of compressed peat (see p393). Larger seeds can be sown into rockwool modules to create 'plug plants' (see p393).

Sowing in the open

The success of sowing outdoors depends greatly on preparing the seedbed; the tilth needs to be matched to the type of seed, soil texture and the expected weather conditions (see p313). The area to be prepared should be free draining. It is thoroughly dug or ploughed depending on the scale of operation. Weeds are buried and organic matter is incorporated in the process. Ideally this is done in the autumn especially if it is a heavy soil; the raw soil is then exposed to the action of frost and rain (see p312). In the spring the mellow, weathered, soil is knocked down with rake or harrow to form the right tilth that provides water and oxygen for seed germination; broad beans can be sown into a very rough seedbed very early in the spring whereas smaller seed sown into warmer conditions should go into a much finer tilth.

Weeds need to be dealt with by creating a false or stale seedbed (see p191), hoeing or using weedkillers. Nutrients, especially phosphate fertilizer, are worked in and the ground levelled to receive the seed. Seed are usually sown in rows (drills) or broadcast, depending on the circumstances. Some seeds will more appropriately be station sown. On a larger scale, seeds are drilled with appropriate equipment. Seeds should be at the right depth, covered to their own diameter and sown when ground temperatures are suitable for the plants concerned (see p39). The sowing rate will depend on the species and the likely losses, which can be estimated from the field conditions, the germination percentage and the viability of the seed (see p152).

There are advantages to providing protection for the developing plants in the form of windbreaks or floating mulches (see p17). Where residual herbicides are not used there needs to be ongoing control of emerging weeds while they are in competition with the seedlings and young plants. If the seedbed was well watered then there should, normally, be no further need to irrigate; indeed there are advantages in not doing this in terms of water conservation, to encourage deeper rooting and prevent capping of the soil (see p313).

Vegetative propagation

All plants have the potential to reproduce asexually. In plants this practice is known as **vegetative propagation**; pieces of the parent plant are removed and these develop into wholly independent plants (see p157).

Asexual reproduction is the creation of a new individual by the division of the genetic material and cytoplasm of the parent cell.

Adventitious plant tissues or organs are those growing where they are not usually found on the plant i.e. roots that have not arisen from the radicle in the seed and buds that have not developed from the plumule.

All living cells contain a nucleus with a complete set of genetic information (see genetic code, Chapter 10), with the potential to become any specialized cell type (totipotency). Only part of the total information is brought into operation at any one time and for any position in the plant. If parts of the plant are removed, then cells lose their orientation in the whole plant and are able to produce organs in positions not found in the usual organization. These are described as **adventitious** and can, for example, be roots on a stem cutting, buds on a piece of root, or roots and buds on a piece of leaf used for vegetative propagation.

Characteristics of propagation from vegetative parts

Vegetative propagation is used in horticulture to produce numbers of plants from a single parent plant. This group of plants, or **clone**, is an extension of the parent plant and therefore all will have the same genetic characteristics. The greatest advantage for horticulturists is to be able to reproduce a cultivar in which all the resulting plants exhibit consistent characteristics. There are some cultivars that can only be reproduced by vegetative means. Seeds produced without fertilization (i.e. by **apomixis**) found in Alchemilla, Rosaceae, Poaceae and Taraxacum present a special case of natural clonal propagation.

In vegetatively propagated cultivars, changes can occur (see mutations) and differing clonal characteristics within the same cultivar can be distinguished in some species, i.e. 'sports' e.g. the leaf colour and plant habit of × *Cupressocyparis leylandii*.

Natural vegetative propagation (see p157)

Many plant species use their ability for vegetative propagation in their normal pattern of development (see vegetative propagation p172). The production of these vegetative **propagules**, as with the production of seed, is to increase numbers and provide a means by which the plant survives adverse conditions. The energy storage for this purpose makes them attractive as food for us, e.g. potatoes, onions, carrots.

All vegetative propagation is a form of division that has been exploited by mankind for a very long time. In many cases little more than breaking up the plant or taking the natural propagules is involved.

Divisions

Most gardeners will be familiar with dividing **herbaceous perennials**. This usually arises because the shoots become overcrowded and the thick clumps that develop often have woody or bare centres. Borders are rejuvenated by carefully lifting the clump ('crown'), preferably with a ball of soil, teasing off most of the soil carefully from the roots and splitting it with back to back forks for good leverage. Whilst smaller specimens can be pulled apart by hand, some are so tough as to require knives or spades. This division can be undertaken in the autumn as plants die down, but is usually better done in the spring as the new shoots appear. The younger sections with strong shoots can be replanted in prepared ground (normally with many surplus pieces to give away or sell). This should be done before the roots have dried and the plants are then watered in.

Alpines (cushions, carpets, mat formers, rosette types) lend themselves to increase by division which is popular because it is cheap, simple and quick. Commonly the divisions are made in mid-spring as the plants begin to grow and they establish most easily. The only problem for gardeners is that it rather spoils the rock garden if separate stocks are not kept; they can restrict themselves to taking rooted pieces from the edge of the clump. Whilst gardeners can plant these straight into the garden, commercially they are grown on in pots until an attractive plant is produced. This is enhanced by the addition of a suitable grit on the surface of the compost.

Aquatic plants can be dealt with in essentially the same way as herbaceous perennials. **Water lilies** are usually lifted in late spring and the growing shoot (the 'eye') is cut away with a piece of the tuber and some root. These pieces are firmed into a pot or basket containing a minimum of sieved clay loam and powdered charcoal and grown on with the water level at the rim. **Marginals** such as reed mace and sweet flag are divided in late spring. The pieces are cleaned up of excess and dying leaves before replanting. **Submerged plants** (oxygenators) tend to be vigorous and need to be reduced on a regular basis. If more plants are wanted then pieces broken off can be tied in bundles with wire and returned to the water.

House plants that develop clumps such as *Maranta*, spider plants (*Chlorophytum*), African violets (*Saintpaulia*) and mother-in-law's-tongue (*Sansevieria*) can be propagated by divisions. This can usually be done by breaking the clump with the fingers to minimize damage to the roots, with help from a knife only to get started if too tough. The pieces are put into a pot just big enough to take the roots with potting compost. Care needs to be taken to fill without leaving cavities by constantly tapping the pot on the bench as it is filled. The plants need to be given a warm environment, ideally in a propagator unit or polythene bag, which will help reduce water losses until established.

Suckers produced by many trees and shrubs can be a problem as they divert energy from the main purpose of the plant and make a messy area around the base. However, they can be used as a source of new plants in the case of many, such as *Rhus typhina*, raspberries and woody house plants or palms. (Suckers arising from grafted material, e.g. roses or apples, will reproduce the rootstock.) Soil from around the base of the plant is removed to expose the point where the suckers can be removed with a knife or pruning shears, ideally with some root attached. Their relatively large size and lack of root means they need to be kept watered until established. They are normally heeled into a trench and in the nursery protection from sun and wind is provided over the next year until there is a good root system.

Rhizomes

Rhizomes such as border irises can be divided as other herbaceous perennials. Again this becomes necessary to get rid of the developing bare patch where the rhizomes grow away from their starting point. Normally about 10 cm of non-woody rhizome is cut off with each fan of leaves. The leaves are reduced to a third to reduce water loss and wind rocking. Some species, such as *Bergenia*, are flowering in the early spring so it is best to lift the plants in mid-winter and remove the rhizomes. These should be washed and the dormant buds found. Sections with a bud can be taken off and rooted in potting compost in trays by burying them horizontally to half their depth. It is advantageous to provide 'bottom heat' by standing the trays on soil warming cables. The plants are not ready to go out until the fibrous roots have emerged from the bottom of the tray and have been 'hardened off'.

Bulbs

Bulbs can yield several plants if divided in an appropriate way. **Scaly bulbs** (see p158) such as lilies and fritillary are propagated by **scaling** whereby the outer scales are pulled off and put in a polythene bag with a suitable moist material such as vermiculite or pushed to half their depth in open propagating compost and covered with polythene. For **tunicate**

bulbs (see p158), such as tulips, the daughter bulbs within the parent bulb can be removed in late summer and grown on in open compost in a warm environment. Bulbs with a tight structure, such as hyacinths and daffodils, are cut into pairs of scales, **twin scaling**. The outer scales are removed and the remaining bulb is cut vertically into several segments. These are then split with a clean knife into pairs of scales with a piece of the base plate and treated as scales. **Chipping** is used with non-scaly bulbs whereby the bulb is simply divided vertically into many pieces, each with a piece of basal plate. For these methods it is important to maintain hygienic conditions and use a suitable fungicide to minimize the introduction of fungal diseases to the cut surfaces.

Artificial methods of propagation

The artificial methods of vegetative propagation encompass most organs of the plant. Cuttings are parts of plants that have been carefully cut away from the parent plant, and which are then used to produce a new plant. Many species can be propagated in this way. Different methods may be necessary for different species. Only healthy parent plants should be used. Hygienic use of knives, compost and containers is strongly recommended. Cuttings are normally taken from parts of the plant exhibiting juvenile growth. Below is a brief description of the most common methods used for taking cuttings.

Cuttings

Stem cuttings can be taken from stems that have attained different stages of maturity. **Hardwood cuttings** are from pieces of dormant woody stem containing a number of buds, which grow out into shoots when dormancy is broken in spring. The base of the cutting is cut cleanly to expose the cambium tissue from which the adventitious roots will grow (e.g. in rose rootstocks, Forsythia, and many deciduous ornamental shrubs). In Hydrangea and currant the stems show evidence of pre-formed adventitious roots (root-initials), which aid the process of root establishment. Hardwood cuttings are normally taken in late autumn (they are 15–25 cm in length), and are often placed with half their length immersed in a growing medium containing half compost and half sand. A 12-month period is often necessary before the cuttings can be lifted.

Semi-ripe cuttings are taken from stems that are just becoming woody. They are normally taken from mid-summer to early autumn. Most cuttings of this kind are 5–10 cm long. Rooting in a sand/compost mixture may be achieved in cold frames or, more quickly, in a heated structure at about 18°C. *Eleagnus* (Oleaster) will root only if heat is provided. Many shrub and tree species, e.g. holly and conifers, are propagated as 'heeled' cuttings. Here, the semi-ripe cutting is taken in such a way that a one centimetre sliver of last year's wood (the **heel**) is still attached. The heel cambium facilitates root formation and, hence, easier establishment of cuttings.

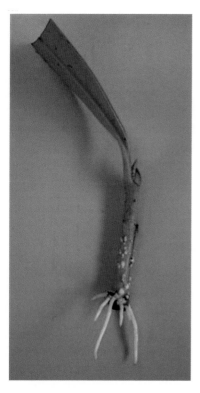

Figure 12.3 **Rooted cuttings**

> **Grafting** involves the union of a **scion** (portion of stem) with a **rootstock** (root system) taken from another plant. **Budding** is a type of graft that involves inserting a single bud into a stock.

Stems without a woody nature are used for the propagation of plants such as *Fuchsia*, *Pelargonium* and *Chrysanthemum*. These are called **softwood cuttings** (see Figure 12.3), and they are most often taken in late spring and early summer. The area of leaf on these cuttings should be kept to a minimum to reduce water loss. Misting (spraying the plants with fine droplets of water to increase humidity and reduce temperature) can further reduce this risk by slowing down the transpiration rate.

Automatic misting employs a switch attached to a sensitive device used for assessing the evaporation rate from the leaves. The cool conditions favouring the survival of the aerial parts of the cutting, however, do not encourage the division of cells in the cambium area of the root initials. The temperature in the rooting medium may be increased with electric cables producing **bottom heat**. These special conditions for the success of cuttings are provided in propagation benches in a greenhouse.

Leaf cuttings are also susceptible to wilting before the essential roots have been formed, and will benefit from mist, provided the wet conditions do not encourage rotting of the plant material. Leaves of plants such as *Begonia*, *Streptocarpus* and *Sansevieria* are divided into pieces from which small plantlets are initiated, while leaves plus petioles are used for *Saintpaulia* propagation. Nursery stock species, e.g. *Camellia* and *Rhododendron*, require a complete leaf and associated axillary bud in a **leaf-bud cutting.**

Root cuttings may be an option when other methods are not seen to succeed. This method is used for species such as *Phlox paniculata* and *Anchusa azurea* (*alkanet*). Roots about a centimetre in thickness are taken in winter and cut into 5 cm lengths. They are inserted vertically into a sand/compost mixture in most species, but thinner-rooted species such as *Phlox* are placed horizontally. It is important that root cuttings are not, inadvertently, placed upside-down, as this will prevent establishment.

Budding and grafting

Grafted plants are commonly used in top-fruit, grapes, roses and amenity shrubs with novel shapes and colours. Rootstocks resistant to soil-borne pests and disease are sometimes used when the desired cultivars would succumb if grown on their own roots, e.g. grapevines, tomatoes and cucumbers grown in border soils. Grafting is not usually attempted in monocotyledons, since they do not produce continuous areas of secondary cambium tissue suitable for successful graft-unions.

In top fruit, grafting is used for several reasons:

- a grafted plant will establish more quickly than a seedling;
- plants derived from seedlings will show different (usually inferior) qualities of fruiting compared with their commercially useful parent plants so a means of vegetative propagation is advantageous; the cultivars are, therefore, clones derived from one original parent;
- to control the size of the tree through the choice of dwarfing rootstock (see Table 12.1), e.g. the M9 apple rootstock, causes the grafted scion

cultivar to be considerably dwarfed. Reduced levels of auxin and cytokinin in the rootstock possibly, bring this about.

Table 12.1   Fruit rootstock

	Vigorous	Semi-vigorous	Semi-dwarfing	Dwarfing
Apples	MM104	MM106	M26	M9
Pears		Quince A	Quince C	
Plums	Brompton	St Julien A		Pixy

There are numerous grafting methods that have been developed for particular plant species. Several principles common to all methods can be briefly mentioned. Firstly, the scion and stock should be genetically very similar. Secondly, the scion and stock will need to have been carefully cut so that their cambial components are able to come in contact. In this way, there will be a higher likelihood of **callus growth** (resulting from cambial contact), which quickly leads to graft establishment. Thirdly, the graft union should be sealed with grafting tape to maintain the graft contact, to prevent drying-out and to keep out disease organisms such as *Botrytis*. Fourthly, the buds on the stem taken as scion material should, ideally, be dormant (leafy material would quickly dry out). The rootstock should be starting active growth, and thus bring water, minerals, and nutrients to the graft area.

Tissue culture

Tissue culture is a method used for vegetative propagation based on the phenomenon that any part of a plant from a single cell to a whole apical meristem can grow into a whole plant (see totipotency). The explant, the piece of the plant taken, is grown in a sterile artificial medium that supplies all vitamins, mineral and organic nutrients. The medium and explant are enclosed in a sterile jar or tube and subjected to precisely controlled environmental conditions. This method has advantages over conventional propagation techniques, since large numbers of propagules can be produced from one original plant. It has particular value with rare or novel plants. An added advantage is the reduced time taken for bulking up plant stocks. Some species that traditionally propagate only by seed, e.g. orchids and asparagus, can now be grown by this means.

One of the problems of conventional vegetative propagation is that diseases and pests are passed on to the propagules. Disease levels, particularly virus, in their growing tips can be greatly reduced by exposing stock plants to high temperatures. Following this heat-treatment, a **meristem-tip** can be dissected out of the stem and grown in a tissue culture medium, to produce stock that is free from disease (e.g. chrysanthemum stunt viroid, see Chapter 15). This method of propagation is now used for species including *Begonia, Alstroemeria, Ficus, Malus, Pelargonium*, Boston fern (*Nephrolepsis exaltata*), roses and many others.

Figure 12.4 Tissue Culture

In all the methods described, **cell division** (see mitosis) must be stimulated in order to produce the new tissues and organs. The correct balance of hormones produced by the cells triggers this **initiation**. Auxins are found to stimulate the initiation of adventitious roots of cuttings. In the propagation of cuttings, the bases may be dipped in powder or liquid formulations of auxin-like chemicals such as naphthalene acetic acid to achieve this result. The number of roots is increased and production time reduced. The precise concentration of chemical in the cells is critical in producing the desired growth response.

A large amount of hormone can bring about an inhibition of growth rather than promotion. For this reason, manufacturers of **hormone powders** and dips produce several distinct formulations with differing hormone concentrations, relevant to the hardwood, the semi-ripe, and the softwood cutting situations. Also different organs respond to different concentration ranges; e.g. the amount of auxin needed to increase stem growth would inhibit the production of roots. The same principle applies to another group of chemicals important in cell division, the cyto-kinins, which can be applied to increase the incidence of plantlet formation.

Both auxin and cytokinin must be included in a tissue culture medium, at concentrations appropriate to the species and the type of growth required; the proportions of each determines whether it is roots or stems that are promoted. Short initiation is promoted by a high cytokinin to auxin ratio whereas high auxin to cytokinin ratios favour root initiation. The subsequent weaning of plantlets from their protected environment in tissue culture conditions requires care and usually conditions of high relative humidity, shade and warmth.

Check your learning

1. Compare the production of plants by seed and vegetative propagation with regard to (a) advantages and (b) disadvantages.

2. State the ideal conditions for storing seed for long periods and explain how storage conditions affect the seed.

3. State two types of dormancy in seeds and for each describe how dormancy may be broken.

4. Explain why fruit trees are usually propagated by grafting.

5. State the principles that underlie the practice of tissue culture.

Further reading

Cassells, A.C. and Graham, P.B. (2006). *Dictionary of Plant Tissue Culture*. Haworth Press.

Dirr, M.A. and Heuser, C.W. (2006). *Reference Manual of Woody Plant Propagation from Seed to Tissue Culture*. 2nd edn pb. Varsity Press.

Donnely, D.J. and Vidaver, W.E. (1995). *Glossary of Plant Tissue Culture*. Cassell.

Gardner, R.J. (1993). *The Grafter's Handbook*. Cassell.

Hartman, H.T. *et al.* (1990). *Plant Propagation, Principles and Practice*. Prentice-Hall.

Kyle, L. and Kleyn, J.G. (1996). *Plants from Test Tubes*. 3rd edn. Timber Press.

Macdonald, B. (2006). *Practical Woody Plant Propagation for Nursery Growers*. Timber Press.

McMillan Browse, P. (1979). *Hardy Woody Plants from Seed*. Grower Books.

McMillan Browse, P. (1999). *Plant Propagation*. 3rd edn. Mitchell Beasley.

MAFF. *Quality in Seeds*, Advisory Leaflet 247.

MAFF (1979). *Guide to the Seed Regulations*. (HMSO).

Stanley, J. and Toogood, A. (1981). *The Modern Nurseryman*. Faber.

Chapter 13   Weeds

Summary

This chapter includes the following topics:

- Definition of a weed
- Damage caused by weeds
- Weeds as alternate hosts of pests and diseases
- Definitions of ephemeral, annual and perennial weeds
- Characteristics of ephemeral, annual and perennial weeds
- Spread of weeds
- Physical methods of control
- Chemical methods of control
- Mode of action of herbicides

with additional information on the following:

- Identification of weeds
- Weed biology
- Mosses and Liverworts

Figure 13.1 **Chickweed and sowthistle** crowding out cabbages

Damage

A **weed** is a plant of any kind that is growing in the wrong place.

Figure 13.2 **Ragwort,** a poisonous weed

Problems caused by weeds may be categorized into seven main areas:

Competition between the weed and the plant for water, nutrients and light may prove favourable to the weed if it is able to establish itself quickly. A large cleaver plant (*Galium aparine*), for example, may compete for a square metre of soil. The cultivated plants are therefore deprived of their major requirement and poor growth results. The extent of this competition is largely unpredict able, varying with climatic factors such as temperature and rainfall, soil factors such as soil type, and cultural factors such as cultivation method, plant spacing and quality of weed control in previous seasons. Large numbers of weed seeds may be introduced into a plot in poor quality composts or farmyard manure. The uncontrolled proliferation of weeds will inevitably produce serious plant losses.

Drainage (see Chapter 19) depends on a free flow of water along ditches. Dense growth of weeds such as chickweed may seriously reduce this flow and increase waterlogging of horticultural land.

Machinery such as mowing machines and harvesting equipment may be fouled by weeds, such as knotgrass, that have stringy stems

Poisonous plants. Ragwort (see Figure 13.2), sorrel and buttercups are eaten by herbivorous animals when more desirable food is scarce. Also, poisonous fruits of plants such as black nightshade may be attractive to children and also contaminate mechanically harvested crops such as blackcurrants and peas for freezing.

Seed quality is lowered by the presence of weed seeds. For example fat hen can contaminate batches of carrot seed.

Tidiness is important for a well-maintained garden. The amenity horticulturist may consider that any plant spoiling the appearance of plants in pots, borders, paths or lawns should be removed, even though the garden plants themselves are not affected.

Alternate hosts of pests and diseases. Pests and diseases are commonly harboured on weeds. Chickweed supports whitefly, red spider mite and cucumber mosaic virus in greenhouses. Sowthistles are commonly attacked by chrysanthemum leaf miner. Groundsel is everywhere infected by a rust which attacks cinerarias (see Figure15.9). Charlock may support levels of club root, a serious disease of brassica crops. Fat hen and docks allow early infestations of black bean aphid to build-up. Speedwells may be infested with stem and bulb nematodes.

Weed identification

As with any problem in horticulture, recognition and identification are essential before any reliable control measures can be attempted. The weed **seedling** causes little damage to a crop, but will quickly grow to be the damaging adult plant bearing seeds that will spread. The seedling stage is relatively **easy to control**, whether by physical or by chemical methods. Identification of this stage is therefore important and with a little practice the gardener or grower may learn to recognize the important weeds using such features as cotyledon and leaf shape, colour and hairiness of the cotyledons and first true leaves (see Figure 13.3).

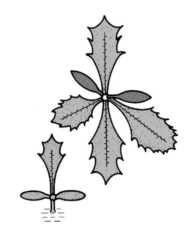

Chickweed (×1.5)
Bright green. Cotyledons have a light-coloured tip and a prominent mid-vein. True leaves have long hairs on their petioles

Groundsel (×1.5)
Cotyledons are narrow and purple underneath. True leaves have step-like teeth

Large field speedwell (×1.5)
Cotyledons like the 'spade' on playing cards. True leaves hairy, notched and opposite

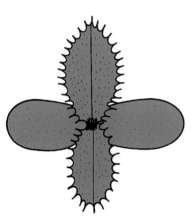

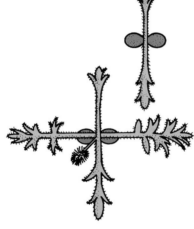

Creeping thistle (×1.5)
Cotyledons large and fleshy. True leaves have prickly margins

Yarrow (×1.5)
Small broad cotyledons. True leaves hairy and with pointed lateral lobes

Broad-leaved dock (×1.5)
Cotyledons narrow. First leaves often crimson, rounded with small lobes at the bottom

Figure 13.3 Seedlings of common weeds. Notice the difference between cotyledons and true leaves. (Reproduced by permission of Blackwell Scientific Publications)

Within any crop or bedding display, a range of different weed species will be observed. Changes in the weed flora may occur because of environmental factors such as reduced pH, because of new crops that may encourage different weeds to develop, or because repeated use of one herbicide selectively encourages certain weeds, e.g. groundsel in lettuce crops or annual meadow grass in turf. Horticulturists must watch carefully for these changes so that their chemical control may be adjusted. The mature weeds may be identified using an **illustrated flora** book, which shows details of leaf and flower characters.

Weed biology

Three types of weed

An **ephemeral weed** is a weed that has several life cycles in a growing season.

An **annual weed** is a weed that completes its life cycle in a growing season.

A **perennial weed** is a weed that lives through several growing seasons.

The range of weed species includes algae, mosses, liverworts, ferns and flowering plants. These species display one or more special features of their life cycle which enable them to compete as **successful weeds** against the crop, and cause problems for the horticulturist.

- **Ephemeral weeds**, such as groundsel and chickweed, produce seeds through much of the year. Weed seeds often germinate more quickly than crop seeds and thus emerge from the soil to crowd out the developing plants. Their seeds germinate throughout the year. Their roots are often quite shallow.
- **Annual weeds**, such as speedwells, annual meadow grass and fat hen, are similar to the ephemerals in their all-year round seed production. Their seeds take longer to ripen those of ephemerals. They may develop deeper roots than ephemerals.
- **Perennial weeds**, such as creeping thistle, couch-grass, yarrow and docks, have long-lived root system. Each species has an underground organ that is difficult to control. The creeping thistle has long lateral roots; couch has long lateral rhizomes; yarrow has long lateral roots and docks have deep, swollen roots.

Whilst seed production may be high, especially in the last three of the four above-mentioned species, it is the spreading underground organs that present the main problems to horticulturalists. The large quantities of food stored in their vegetative organs enable these species to emerge quickly from the soil in spring, often from considerable depths if they have been ploughed in. The fragmentation of underground organs by cultivation machinery often enables these species to propagate vegetatively and increase their numbers in disturbed soils.

Spread of weeds

Weeds may be spread in a number of ways:

- fruits such as those in Himalayan balsam discharge seeds **explosively** to a considerable distance;
- seeds of species from the Asteraceae family such as groundsel, thistles, and dandelion, are carried along in the **wind** by a seed 'parachute';
- seeds of chickweed and dandelion may be spread by the moving **water** in ditches;
- fruits of the cleavers weed (see Figure 13.4) **stick** to clothes and hair of humans and animals in a manner similar to 'Velcro'. Chickweed seed is held in a similar way;
- groundsel and annual meadow grass seeds **become sticky** in damp conditions and are able to stick to boots and machinery wheels;
- a proportion of the seeds of groundsel, annual meadow grass, yarrow and dock survive digestion in the guts of **birds**;

Figure 13.4 *Young cleavers*. Seeds on older plants stick to the fur of animals

- chickweed and annual meadow grass seed is also able to survive **mammal** digestive systems;
- cut stems of slender speedwell are moved by **grass mowers**;
- **ants** carry around the seeds of speedwell;
- underground horizontal **roots**, **stolons** and **rhizomes** of perennial weeds such as thistle, yarrow and couch respectively slowly spread the weed from its point of origin;
- ploughs and rotavators move around **cut underground fragments** of thistles, yarrow, dandelion, and couch;
- commercial **seed stocks** can be contaminated with seeds of weeds such as speedwells and couch.

Other aspects of weed biology

Particular **soil conditions** may favour certain weeds. Sheep's sorrel (*Rumex acetosella*) prefers acid conditions. Mosses are found in badly drained soils. Knapweed (*Centaurea scabiosa*) competes well in dry soils. Common sorrel (*Rumex acetosa*) survives well on phosphate-deficient land. Yorkshire fog grass (*Holcus lanatus*) invades poorly fertilized turf. Nettle and chickweed prefer highly fertile soils.

The **growth habit** of a weed may influence its success. Chickweed and slender speedwell produce horizontal (prostrate) stems bearing numerous leaves that prevent light reaching emergent crop seedlings. Groundsel and fat hen have an upright habit that competes less for light in the early period of weed growth. Perennial weeds such as bindweed, cleavers and nightshades are able to grow alongside and climb up woody plants, such as cane fruit and border shrubs, making control difficult.

Annual **seed production** may be high in certain species. A scentless may-weed plant (perennial) may produce 300 000 seeds, fat hen (annual) 70 000 and groundsel (annual) 1000. A dormancy period is seen in many weed species. In this way, seed germination commonly continues over a period of 4 or 5 years after seed dispersal, presenting the grower with a continual problem. Groundsel is something of an exception, since many of its seeds germinate in the first year.

Perennial weeds with swollen underground organs provide the greatest problems to the horticulturist in long-term crops such as soft fruit and turf because foliage-acting and residual herbicides may have little effect.

Fragmentation of above-ground parts may be important. A lawnmower used on turf containing the slender speedwell weed cuts and spreads the delicate stems that, under damp conditions, establish (like cuttings) in other parts of the lawn.

Greenhouse production generally suffers less from weed problems because composts and border soils are regularly sterilized.

Some important horticultural weeds

Specific descriptions of identification, damage, biology and control measures are given for each weed species. Detailed discussion of **weed control** measures (cultural, chemical and legislative) is presented in Chapter 16.

Ephemeral weeds

Chickweed (*Stellaria media*). Plant family – Caryophyllaceae

Damage. This species is found in many horticultural situations as a weed of flowerbeds, vegetables, soft fruit and greenhouse plantings. It has a wide distribution throughout Britain, grows on land up to altitudes of 700 m, and is most important on rich, heavy soils.

Life cycle. The seedling cotyledons are pointed with a light-coloured tip while its true leaves have hairy petioles (see Figure 13.5). The adult plant has a characteristic lush appearance and grows in a prostrate manner over the surface of the soil; in some cases it covers an area of 0.1 square metres, its leafy stems crowding out young plants as it increases in size. Small white, five-petalled flowers are produced throughout the year, the flowering response being indifferent to day length. The flowers are self-fertile.

An average of 2500 disc-like seeds (1 mm in diameter) may result from the oblong fruit capsules produced by one plant. Since the first seed may be dispersed within 6 weeks of the plant germinating and the plant continues to produce seed for several months, it can be seen just how

Figure 13.5 (a) Chickweed seedling (b) Chickweed plant

prolific the species is. The large numbers of seed (up to 14 million/ha) are most commonly found in the top 7 cm of the soil where, under conditions of light, fluctuating temperatures and nitrate ions, they may overcome the dormancy mechanism and germinate to form the seedling. Many seeds, however, survive up to the second, third and occasionally fourth years. Figure 13.6 shows that germination can occur at any time of the year, with April and September as peak periods. Chickweed is an alternate host for many aphid transmitted viruses (e.g. cucumber mosaic), and the stem and bulb nematode.

Spread. The seeds are normally released as the fruit capsule opens during dry weather; they survive digestion by animals and birds and may thus be dispersed over large distances. Irrigation water may carry them into channels and ditches.

Control. This weed is controlled by a combination of methods. Physical controls include partial sterilization of soil in greenhouses while hoeing in the spring and autumn periods prevents the seedling from developing and flowering. Mulching is effective against germinating weeds.

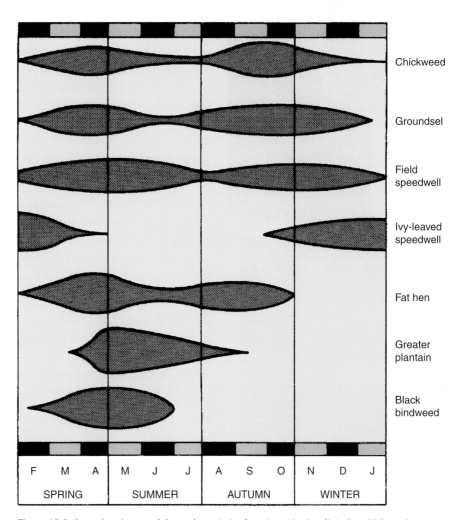

Figure 13.6 Annual and perennial weeds: periods of seed germination. Note that chickweed, groundsel and field speedwell seeds germinate throughout the year. Many other species are more limited. (Reproduced by permission of Blackwell Scientific Publications)

Amateur gardeners can use the non-selective, non-persistent herbicide, **glufosinate-ammonium plus fatty acids** for control in such situations as ornamental beds containing woody perennials, and in cane fruit.

Professional horticulturalists use pre-emergent contact sprays such as **paraquat** applied before a crop emerges. A soil-applied root-acting herbicide such as **propachlor** is used on a crop such as strawberries before weeds germinate. A foliage-acting herbicide such as **linuron** is applied for chickweed control in potatoes.

Groundsel (*Senecio vulgaris*). Plant family – Asteraceae (Compositae)

Damage. This is a very common and important weed, particularly on heavy soil. Its high level of seed production and the ability of its seed to germinate soon after release lead to dense mats of the weed. It grows on both rich and poor soils up to almost 600 m in altitude.

Life cycle. The seedling cotyledons are narrow, purple underneath, and the first true leaves have step-like teeth (see Figure 13.7). The adult plant has an upright habit, and produces as many as 25 yellow, small-petalled flower heads. Flowering occurs in all seasons of the year. Produces about 45 column-shaped seeds, 2 mm in length densely packed in the fruit head. As can be seen in Figure 13.5, the seeds may germinate at any time of the year, with early May and September as peak periods. Since there may be more than three generations of groundsel per year (the autumn generation surviving the winter), and each generation may give rise to a thousand seeds, it is clear why groundsel is one of the most successful colonizers of cultivated ground. Its role as a symptomless carrier of the wilt fungus, *Verticillium*, increases its importance in certain crops, e.g. tomatoes and hops.

Spread. The seeds bear a mass of fine hairs. These hairs, in dry weather, can parachute seeds along on air currents for many metres. In wet weather the seeds become sticky and may be carried on the feet of animals, including humans. The seeds survive digestion by birds, and thus can be transported in this way.

Control. A combination of control methods may be necessary for successful control. Physical control is by hoeing or by mechanical cultivation, particularly in spring and autumn to prevent developing seedlings from flowering. Care should be taken not to allow uprooted flowering groundsel plants to release viable seed.

The *amateur* gardener can use the herbicide, **glufosinate-ammonium plus fatty acids**, for control in such situations as ornamental beds containing woody perennials, and in cane fruit.

The *professional* horticulturalist can use contact herbicides, such as **paraquat** to control the weed in all stages of growth on paths or in fallow soil (but never on garden plants or turf). Soil-acting chemicals such as **propachlor** on brassicas kill off the germinating seedling. An established groundsel population, especially in a crop such as lettuce (a fellow member of the Asteraceae family)

Figure 13.7 (a) Groundsel seedling (b) Groundsel plant

requires careful choice of herbicide to avoid damage to the crop (see also **propyzamide**, p279).

Annual weeds

While there are at least 50 successful annual weed species in horticulture, this book can cover only a few examples that illustrate the main points of life cycles and control. Two species, annual meadow grass and speedwell, are described.

Annual meadow grass (*Poa annua*). Plant family – Poaceae (Graminae)

Damage. This species is a quite small annual (or short-term perennial) found on a range of ornamental and sports grass surfaces, on paths and in vegetable plots (see Figure 13.8). It is able to establish quickly on bare ground. It does not thrive on acid soils or those low in phosphates. Despite its relatively small size, it often emerges in sufficient quantities to smother crop seedlings. Its seed may be present as an impurity in commercial grass seed. Special selections of this species are used in seed mixtures for lawns.

Life cycle. Flowers can occur at any time of year and are usually self-pollinated. About 2000 seeds per plant are produced from April to September. Plants will flower and seed even when mown regularly. Seeds germinate from February to November with the main peaks in early spring and autumn. Some seed will germinate soon after their release; others can remain viable in soil for at least 4 years. This weed species can be the host of a number of nematode species that also attack important crops.

Spread. There is no obvious dispersal mechanism. Most seeds fall around the parent plant and become incorporated into the soil. Seeds may be carried around on boots and wheels of machinery. Worms may bring seeds to the soil surface in worm casts.

Figure 13.8 Annual meadow grass plant

Control is achieved by a variety of methods. The physical action of hoeing normally controls the weed especially when it is in the young stage. Deep digging-in of seedlings and young plants is also usually effective. Mulching is effective against germinating weeds in flower beds and fruit areas. The *amateur* gardener can use the non-selective, non-residual herbicide, **glufosi-nate-ammonium plus fatty acids**, for control in situations such as ornamental beds containing woody perennials and in cane fruit. The *professional* horticulturalist may use **paraquat** or **glyphosate** for total chemical control, but these two chemicals should never be sprayed in the vicinity of growing crops. Care should be taken not to walk on grass after application of these chemicals. **Chlorpropham** may be used as a soil applied chemical on crops such as currants, onions and chrysanthemum.

Figure 13.9 (a) **Field speedwell seedling** (b) **Field speedwell plants**

Speedwells (*Veronica persica* and *V. filiformis*). Plant family – Scrophulariaceae

Damage. The first species, the large field speedwell (*V. persica*) is an important weed in vegetable production, crowding out young crop plants and reducing growth of more mature stages. The second species, the slender or round-leaved speedwell (*V. filiformis*), once considered a desirable rock garden plant introduction from Turkey, has become a serious turf problem.

Life cycle. The seedling cotyledons are spade shaped, while the true leaves are opposite, notched and hairy (see Figure 13.9) in both species. The adult plants have erect, hairy stems and rather similar broad-toothed leaves. *V. persica* produces up to 300 bright blue flowers, 1 cm wide, per plant. The flowers are self-fertile and occur throughout the year, but mainly between February and November. The adult plant produces an average of 2000 light brown boat-shaped seeds 2 mm across. The seeds of this species germinate below soil level all year round, but most commonly from March to May (see Figure 13.5), the winter period being necessary to break dormancy. Seeds may remain viable for more than 2 years. *V. filiformis* produces self-sterile purplish-blue flowers between March and May, and spreads by means of prostrate stems which root at their nodes to invade fine and coarse turf, especially in damp areas. Segments of this weed cut by lawnmowers easily root and further increase the species. Seeds are not important in its spread.

Spread. Seeds of *V. persica* falling to the ground may be dispersed by ants. Seed of this species can be spread as contaminants of crop seed. *V. filiformis* does not produce seed. Its slow spread is mainly by means of grass-cutting machinery.

Control. Field speedwell (*V. persica*) is controlled by a combination of methods. The physical action of hoeing or mechanical cultivation, particularly in spring, prevents developing seedlings from growing to mature plants and producing their many seeds.

The *amateur* gardener can use the herbicide, **glufosinate-ammonium plus fatty acids** for control in situations such as ornamental beds containing woody perennials and in cane fruit.

For the *professional* grower, total contact herbicides, such as **paraquat**, may be sprayed to control the weed on paths or in fallow soils. Soil-acting chemicals such as **chlorpropham** on crops such as lettuce, onions and chrysanthemums kill off the germinating weed seedling. A contact, foliar chemical such as **clopyralid**, may be sprayed on to brassicas, young onions, and strawberries to control emerging seedlings.

The slender speedwell (*V. filiformis*) represents a different problem for control. Physical controls such as regular close mowing and spiking of turf removes the high humidity necessary for this weed's establishment and development.

The *amateur* gardener has difficulty controlling this weed with the turf herbicides available. The *amateur* or *professional* grower can control the weed in a turf seedbed using a total, contact chemical such as **glufosinate-ammonium plus fatty acids**, a few weeks before sowing the turf seed. This '**stale seed bed**' method leaves the turf to establish relatively undisturbed by weeds. Organic growers may remove stale seedbed weeds by hoeing. Professional growers can use a selective contact chemical such as **chlorthal dimethyl** which is effective against this weed in established grass.

Perennial weeds

Five species (creeping thistle; couch; yarrow; dandelion and broad-leaved dock) are described below to demonstrate the different features of their biology (particularly the perennating organs) that make them successful weeds. The flowering period of these weeds is mainly between June and October (see Figure 13.10), but the main problem for gardeners and growers is the plant's ability to survive and reproduce vegetatively.

Creeping thistle (*Cirsium arvense*). Plant family – Asteraceae (Compositae)

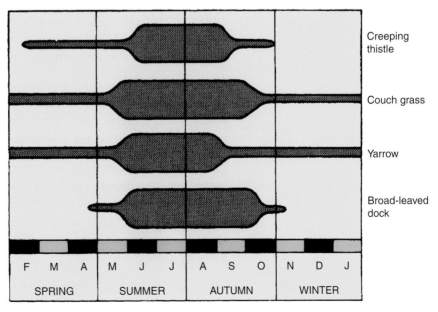

Figure 13.10 **Perennial weeds**: periods of flowering. Most flowers and seeds are produced between June and October. Annual weeds commonly flower throughout the year. However the slender speedwell flowers only between March and May

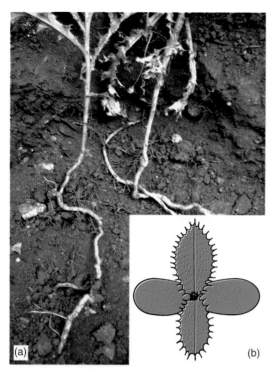

(a)   (b)

Figure 13.11 **Creeping thistle plant** showing (a) lateral roots, (b) seedling

Damage. This species is a common weed in grass and perennial crops, e.g. apples, where it forms dense clumps of foliage, often several metres across.

Life cycle. The seedling cotyledons are broad and smooth, the true leaves are spiky (see Figure 13.11). The mature plant is readily recognizable by its dark green spiny foliage growing up to 1 m in height. It is found in all areas, even at altitudes of 750 m, and on saline soil. The species is dioecious, the male plant producing spherical and the female slightly elongated purple flower heads from July to September. Only when both sexes of the plant are within about 100 m of each other does fertilization occur in sufficient quantities to produce large numbers of brown, shiny fruit, 4 mm long. The seeds may germinate beneath the soil surface in the same year as their production, or in the following spring, particularly when soil temperatures reach 20°C. The resulting seedlings develop into a plant with a taproot which commonly reaches 3 m down into the soil.

Spread. Seeds are wind-borne using a parachute of long hairs. The mature plant produces lateral roots which grow out horizontally about 0.3 m below the soil surface and may spread the plant as much as 6 m in one season. Along their length adventitious buds are produced that, each spring, grow up as stems. Under permanent grassland, the roots may remain dormant for many years. Soil disturbance, such as ploughing, breaks up the roots and may result in a worse thistle problem.

Control. The seedling stage of this weed is not normally targeted by the gardener/grower. The main control strategy is primarily against the perennial root system. For both *amateur* and *professional* horticulturalist, cutting down plants at the flower bud stage when sugars are being transferred from the roots upwards is a physical control measure that partly achieves this objective.

The *amateur* gardener can use another physical control, removing roots by deep digging. The amateur gardener is also able to use herbicide products which contain a mixture of dicamba, MCPA and mecoprop-P (all of these are translocated down to the roots).

The *professional* grower can use the technique of deep ploughing to expose and dry off roots. Products containing the three active ingredients mentioned in the last paragraph are also available to the professional. Effectiveness of herbicide translocation to the roots is greatest when applied in autumn, at a time when plant sugars are similarly moving down the plant.

Couch grass (*Agropyron repens*). Plant family – Poaceae (Graminae)

Damage. This grass, sometimes called 'twitch', is a widely distributed and important weed found at altitudes up to 500 m. It is able to quite rapidly take over plots growing ornamentals, vegetables or fruit.

Life cycle. The dull-green plant is often confused, in the vegetative stage, with the creeping bent (*Agrostis stolonifera*). However, the small 'ears' (ligules) at the leaf base characterize couch. The plant may reach a metre in height and often grows in clumps. Flowering heads produced from May to October resemble perennial ryegrass, but, unlike ryegrass, the flat flower spikelets are positioned at right angles to the main stem in couch. Seeds (9 mm long) are produced only after cross-fertilization between different strains of the species, and the importance of the seed stage, therefore, varies from field to field. The seed may survive deep in the soil for up to 10 years.

Spread. Couch seeds may be carried in grass seed batches over long distances. From May to October, stimulated by high light intensity, overwintered plants produce horizontal rhizomes (see Figure 13.12) just under the soil; these white rhizomes may spread 15 cm per year in heavy soils, 30 cm in sandy soils. They bear scale leaves on nodes that, under apical dominance, remain suppressed during the growing period. In the autumn, rhizomes attached to the mother plant often grow above ground to produce new plants that survive the winter. If the rhizome is cut by cultivations such as digging or ploughing, fragments containing a node and several centimetres of rhizome are able to grow into new plants. The rapid growth and extension of couch plants provides severe competition for light, water and nutrients in any infested crop.

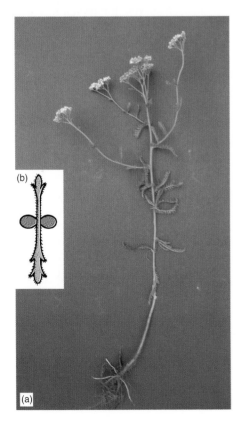

Figure 13.12 **Couch grass plant** showing rhizomes

Control is achieved by a combination of physical and chemical methods. In fallow soil, deep digging or ploughing (especially in heavy land) exposes the rhizomes to drying. Further control by rotavating the weed when it reaches the one or two leaf stage disturbs the plant at its weakest point, and repeated rotavating will eventually cut up couch rhizomes into such small fragments that nodes are unable to propagate.

The *amateur* gardener can use the herbicide, **glufosinate-ammonium plus fatty acids** for control in such situations as ornamental beds containing woody perennials and in cane fruit.

For the *professional* horticulturalist, a translocated herbicide such as **glyphosate**, sprayed onto couch in fallow soils during active weed vegetative growth, kill most of the underground rhizomes. In established fruit, **glufosinate-ammonium** is recommended.

Yarrow (*Achillea millifolium*). Plant family – Asteraceae (Compositae)

Damage. This strongly scented perennial, with its spreading flowering head (Figure 13.13), is a common hedgerow plant found on most soils at altitudes up to 1200 m. Its persistence, together with its resistance to herbicides and drought in grassland, makes it a serious turf weed.

Figure 13.13 **Yarrow plant** showing (a) stolons, (b) seedling

Figure 13.14 **Dandelion plant** growing in turf

Life cycle. The seedling leaves are hairy and elongated, with sharp teeth (Figure 13.3). The mature plant has dissected pinnate leaves produced throughout the year on wiry, woolly stems, which commonly reach 45 cm in height, and which from May to September produce flat-topped white-to-pink flower heads. Each plant may produce 3000 small, flat seeds annually. The seeds germinate on arrival at the soil surface.

Spread. Seeds are dispersed by birds. When not in flower, this species produces below-ground and above-ground stolons which can grow up to 20 cm long per year. In autumn, rooting from the nodes occurs.

Control. Control of this weed may prove difficult. Routine scarification of turf does not easily remove the roots. For the *amateur* and *professional*, products containing **2,4-D** and **mecoprop** are used against yarrow.

Dandelion (*Taraxacum officinale*). Plant family – Asteraceae (Compositae)

Damage. This species is a perennial with a stout taproot. It is a weed in lawns (see Figure 13.14), orchards and on paths. Several similar species such as mouse-ear hawkweed (*Hieracium pilosella*) and smooth hawk's beard (*Crepis capillaris*) present problems similar to dandelion in turf.

Life cycle. Seedlings emerge mainly in March and April. Flowers are produced from May to October. An average of 6000 seeds is produced by each plant. Most seeds survive only one year in the soil, but a few may survive for five years. Mature plants can survive for 10 years.

Spread. Seeds are wind dispersed by means of tiny 'parachutes' and may travel several hundred metres. They are also able to spread in the moving water found in ditches and by animals through their digestive systems. The plant may regenerate from roots, after being chopped up by spades or rotavators.

Control. Physical removal of the deep root by a sharp trowel is recommended, but this leaves bare gaps in turf for invasion by other weeds. For the *amateur* gardener and *professional* groundsman, products containing the two translocated ingredients, **2,4-D** and **dicamba**, are able to kill the stout penetrating root of the dandelion.

Broad-leaved dock (*Rumex obtusifolius*). Plant family – Polygonaceae

Damage. This is a common perennial weed of arable land, grassland and fallow soil.

Life cycle. The seedling cotyledons are narrow (see Figure 13.3). Seedling true leaves are often crimson coloured. The mature plant is readily identified by its long (up to 25 cm) shiny green leaves (see Figure 13.15), known to many as an antidote to 'nettle rash'. The plant may grow 1 m tall, producing a conspicuous branched inflorescence of small green flowers from June to October. The seed represents an important stage in this perennial weed's life cycle, surviving many years in the soil, and most commonly germinating in spring. Like most *Rumex* spp, the seedling develops a stout, branched taproot (see Figure 13.15), which may penetrate the soil down to 1 m in the mature plant, but most commonly reaches 25 cm. Segments of the taproot, chopped by cultivation implements are capable of producing new plants.

Spread. The numerous plate-like fruits (3 mm long) may fall to the ground or be dispersed by seed-eating birds such as finches. They are sometimes found in batches of seed stocks.

Control. High levels of seed production, a tough taproot and a resistance to most herbicides present a problem in the control of this weed. For the **amateur** gardener, in turf, physical removal of the deep root by a sharp trowel is recommended, but this leaves bare gaps in turf for invasion by other weeds. A product containing **dicamba, MCPA**, and **mecoprop-P** is effective against young plants.

(b)

(a)

Figure 13.15 **Broad-leaved dock** showing (a) swollen taproot, (b) seedling

For the **professional** horticulturalist, attempts in fallow soils to exhaust the root system by repeated ploughing and rotavating have proved useful. Young seedlings are easily controlled by translocated chemicals such as **2,4-D**, but the mature plant is resistant to all but a few translocated chemicals, e.g. **asulam**, which may be used on grassland (not fine turf), soft fruit, top fruit and amenity areas, during periods of active vegetative weed growth when the chemical is moved most rapidly towards the roots.

Mixed weed populations

In the field a wide variety of both annual and perennial weeds may occur together. The horticulturalist must recognize the most important weeds in their holding or garden, so that a decision on the precise use of chemical control with the correct herbicide is achieved. Particular care is required to match the concentration of the herbicide to the weed species present. Also, the grower must be aware that continued use of one chemical may induce a change in weed species, some of which may be tolerant to that chemical.

Mosses and liverworts

These simple plants may become weeds in wet growing conditions. The small cushion-forming moss (*Bryum* spp.) grows on sand capillary benches and thin, acid turf that has been closely mown. Feathery moss

Figure 13.16 **Pot plant compost** covered with moss and liverwort

(*Hypnum* spp.) is common on less closely mown, unscarified turf. A third type (*Polytrichum* spp.), erect and with a rosette of leaves, is found in dry acid conditions around golf greens. Liverworts (*Pellia* spp.) are recognized by their flat (thallus) leaves growing on the surface of pot plant compost (see Figure 13.16).

These organisms increase only when the soil and compost surface is excessively wet, or when nutrients are so low as to limit plant growth. Cultural methods such as improved drainage, aeration, liming, application of fertilizer and removal of shade usually achieve good results in turf. Control with contact scorching chemicals, e.g. alkaline ferrous sulphate, may give temporary results. Moss on sand benches becomes less of a problem if the sand is regularly washed.

Check your learning

1. Define the term 'weed'.

2. Describe five ways in which weeds reduce plant productivity.

3. Describe how weeds may harbour pests and diseases.

4. Define the term 'ephemeral weed'. Give an example.

5. Describe the characteristics of ephemeral weeds.

6. Describe a physical control appropriate to this type of weed.

7. Describe a named chemical used for control of the weed example given.

8. Describe the mode of action of this herbicide, stating in which horticultural situation(s) it is used.

9. Describe four methods by which weeds spread.

10. Describe three ways of reducing moss on lawns.

Further reading

Hance, R.J. and Holly, K. (1990). *Weed Control Handbook, Vol. 1 Principles*. Blackwell Scientific Publications.

Hill, A.H. (1977). *The Biology of Weeds*. Edward Arnold.

Hope, F. (1990). *Turf Culture: A Manual for Groundsmen*. Cassell.

Roth, S. (2001). *Weeds: Friend or foe?* Carroll and Brown UK.

The UK Pesticide Guide (2007). British Crop Protection Council.

Weed Guide (1997). Hoechst Schering.

Williams, J.B. and Morrison, J.R. (2003). *Colour Atlas of Weed Seedlings*. Manson Publishing.

Chapter 14
Horticultural pests

Summary

This chapter includes the following topics:

- **Definition of the term 'pest'**
- **Life cycles of pests**
- **Damage caused by pests**
- **Relationship between life cycle and control**
- **Methods of limiting the effects of pests on plant growth**

Figure 14.1 **Red spider mite** symptoms on palm

Mammal pests

A small selection of important mammal pests is included here.

The rabbit (*Oryctolagus cuniculus*)

The rabbit is common in most countries of central and southern Europe. It came to Britain around the eleventh century with the Normans, and became an established pest in the nineteenth century.

Damage. The rabbit may consume 0.5 kg of plant food per day. Young turf and cereal crops are the worst affected, particularly winter varieties that, in the seedling stage, may be almost completely destroyed. Rabbits may move from cereal crops to horticultural holdings. Stems of top fruit may be **ring barked** by rabbits, particularly in early spring when other food is scarce. Vegetables and recently planted garden-border plants are a common target for the pest, and fine turf on golf courses may be damaged, thus allowing lawn weeds, e.g. yarrow, to become established.

Life cycle. The rabbit's high reproductive ability enables it to maintain high populations even when continued control methods are in operation. The doe, weighing about 1 kg, can reproduce within a year of its birth, and may have three to five litters of three to six young ones in 1 year, commonly in the months of February to July. The young are blind and naked at birth, but emerge from the underground 'maternal' nest after only a few weeks to find their own food. Large burrow systems (**warrens**), penetrating as deep as 3 m in sandy soils, may contain as many as 100 rabbits. Escape or **bolt holes** running off from the main burrow system allow the rabbit to escape from predators.

Control. Rabbit control is, by law, the responsibility of the land owner. **Preventative** measures, available to both *amateur* and *professional* horticulturists are effective. **Wire fencing**, with the base 30 cm **underground** and facing outwards, represents an effective barrier to the pest, while thick plastic sheet **guards** are commonly coiled round the base of exposed young trees (see Figure 16.2). Repellant chemicals, e.g. **aluminium ammonium sulphate**, may be sprayed on bedding displays and young trees.

Small spring traps placed in the rabbit hole, winter **ferreting** or **long nets** placed at the corner of a field to catch herded rabbits are methods used as **curative** control. **Shotguns** are used on large horticultural holdings by holders of gun licences.

Gassing is an effective method, but must be applied only by trained operators. Crystals of powdered **sodium cyanide** are introduced into the holes of warrens by means of long-handled spoons or by power operated machines. On contact with moisture hydrocyanic acid is released as a gas and, in well-blocked warrens, the rabbits are quickly killed. Care is required in the storage and use of powdered cyanide, where an **antidote**, amyl nitrite, should be readily available.

Myxamatosis, a flea-borne virus disease of the rabbit, causing a swollen head and eyes, was introduced into Britain in 1953, and within a few years greatly reduced the rabbit population. The development of weaker virus strains, and the increase in rabbit resistance, has combined to reduce this disease's effectiveness in control, although its importance in any one area is constantly fluctuating.

The brown rat (*Rattus norvegicus*)

The brown rat, also called the common rat, is well known by its dark-brown colour, blunt nose, short ears and long, scaly tail.

Damage. Its diet is varied; it will eat **seeds**, **succulent stems**, **bulbs and tubers**, and may grind its teeth down to size by the unlikely act of gnawing at plastic **piping** and electric cables. A rat's average annual food intake may reach 50 kg, a large amount for an animal weighing only about 300 g.

Life cycle. This species has considerable reproductive powers. The female may begin to breed at 8 weeks of age, producing an average of six litters of six young ones per year. Its unpopular image is further increased by its habit of fouling the food it eats, and by the potentially lethal human bacterium causing **Weil's disease**, which it transmits through its urine.

Control. *Amateur* gardeners are able to use products containing **aluminium ammonium sulphate** to deter rats. **Sonic** devices are sometimes used to disturb the animal and provide a round-the-clock deterrent.

Baiting with rat poison should be performed by trained operatives. This method is best achieved by a preliminary survey of rat numbers in buildings and fields of the horticultural holding, and by the identification of the 'rat-runs' along which the animals travel. Baits containing a mixture of **anticoagulant** poison and food material such as oatmeal are placed near the runs, inside a container that, while attracting the rat, prevents access by children and pets. Drainage tiles or oil drums drilled with a small rat-sized hole often serve this purpose. The poison, e.g. **difenacoum**, takes about three days to kill the rat and, since the other rats do not associate the rat's death with the chemical, the whole family may be controlled. The bait should be placed wherever there are signs of rat activity, and repeated applications every three days for a period of three weeks should be effective.

Strains of rat resistant to some anticoagulants are found in some areas, and a range of chemicals may need to be tried before successful control is achieved. The poison, when not in use, should be safely stored away from children and pets. Dead rats should be burnt to avoid poisoning of other animals.

The grey squirrel (*Sciurus carolinensis*)

This attractive-looking 45 cm long creature was introduced into Britain in the late nineteenth century, at a time when the red squirrel population was suffering from disease. The grey squirrel became dominant in most areas, with the red squirrel surviving in isolated areas such as the Isle of Wight.

Damage. The horticultural damage caused by grey squirrels varies with each season. In spring, germinating **bulbs** may be eaten, and the **bark** of many tree species stripped off (see **ring barking**, p95). In summer, **pears**, **plums** and **peas** may suffer. Autumn provides a large wild food source, although **apples** and **potatoes** may be damaged. In winter, little damage is done. Fields next to wooded areas are clearly prone to squirrel damage.

Life cycle. Squirrels most commonly produce two litters of three young ones from March to June, in twig platforms (**dreys**) high in the trees. The female may become pregnant at an early age (6 months). As the squirrels have few natural enemies, and this species lives high above ground, control is difficult.

Control. *Amateur* and *professional* horticulturists may need to reduce grey squirrel numbers. During the months of April–July, when most damage is seen, **cage traps** containing desirable food, e.g. maize seed, reduce the squirrel population to less damaging levels. **Spring traps** placed in natural or artificial tunnels achieve rapid results at this time of year if placed where the squirrel moves. *Professional* horticulturists are able to use **poisoned bait** containing a formulation of anticoagulant chemical, e.g. **warfarin**, when placed in a well-designed ground-level **hopper** (one hopper per 3 ha). This can achieve successful squirrel control without affecting other small wild mammal numbers. In winter and early spring, the destruction of squirrel nests by means of long poles may achieve some success.

Figure 14.2 **Mole hill** in grass

The mole (*Talpa europea*)

The mole is found in all parts of the British Isles except Ireland.

Damage. This dark-grey, 15 cm long mammal, weighing about 90 g, uses its shovel-shaped feet to create an underground system 5–20 cm deep and up to 0.25 ha in extent. The tunnel contents are excavated into mole hills (see Figure 14.2). The resulting **root disturbance** to grassland and other crops causes wilting, and may result in serious losses.

Life cycle. In its dark environment, the solitary mole moves, actively searching for earthworms, slugs, millipedes and insects. About 5 hours of activity is followed by about 3 hours of rest. Only in spring do males and females meet. In June, one litter of two to seven young ones are born in a grass-lined underground nest, often located underneath a dense thicket. Young moles often move above ground, reach maturity at about 4 months, and live for about 4 years.

Control. Natural predators of the mole include tawny owls, weasels and foxes. The main control methods are **trapping** and **poison baiting**, usually carried out between October and April, when tunnelling is closer to the surface. *Amateur* or *professional* horticulturalists can use pincer or half barrel **traps** placed in fresh tunnels and inserted carefully so as not to greatly change the tunnel diameter. The soil must be replaced so that the mole sees no light from its position in the tunnel. The mole enters the trap, is caught and starves to death. In serious mole infestations, trained operators use **strychnine** salts mixed with earthworms at the rate of 2 g ingredient per 100 worms. Single worms are carefully inserted into inhabited tunnels at the rate of 25 worms per hectare. DEFRA authority is required before purchasing strychnine, a highly dangerous chemical, which must be stored with care.

Deer

Deer may become pests in land adjoining woodland where they hide. Muntjac and roe deer **ring-bark** (see p95) trees and eat succulent crops. High fences and regular shooting may be used in their control.

Bird pests

A couple of important bird pests are included here.

The wood-pigeon (*Columba palumbus*)

Damage. This attractive-looking, 40 cm long, blue-grey pigeon with white underwing bars is known to horticulturists as a serious pest on most outdoor edible crops. In **spring**, seeds and seedlings of crops such as brassicas, beans and germinating turf may be systematically eaten. In **summer**, cereals and clover receive its attention; in **autumn**, tree fruits may be taken in large quantities, while in **winter**, cereals and brassicas are often seriously attacked, the latter when snowfall prevents the consumption of other food. The wood-pigeon is invariably attracted to **high protein** foods such as seeds when they are available.

Life cycle. Wood-pigeons lay several clutches of two eggs per year from March to September. The August/September clutches show highest survival. The eggs, laid on a nest of twigs situated deep inside the tree, hatch after about 18 days, and the young ones remain in the nest for 20–30 days. Predators such as jays and magpies eat many eggs, but the main population-control factor is the availability of food in winter. Numbers in the British Isles are boosted a little by migrating Scandinavian pigeons in April, but the large majority of this species is resident and non-migratory.

Control. The wood-pigeon spends much of its time feeding on wild plants, and only a small proportion of its time on crops. Control of the whole population, therefore, seems ethically unsound and is both costly and impracticable. Physical control involves the protection of particular fields by means of **scaring devices** which include scarecrows, bangers (firecrackers or gas guns), artificial hawks on wires, or rotating orange and black vanes, all which disturb the pigeons. Changing the **type** and **location** of the device every few days helps prevent the pigeons from becoming indifferent. The use of the **shotgun** by licenced operators from hidden positions such as hides and ditches (particularly when plastic decoy pigeons are placed in the field) is an important additional method of scaring birds and thus protecting crops.

The bullfinch (*Pyrrhula pyrrhula*)

This is a strikingly-coloured, 14 cm long bird, characterized by its sturdy appearance and broad bill. The male has a rose-red breast, blue-grey back and black headcap. The female has a less striking pink breast and yellowish-brown back.

Damage. From April to September the bird progressively feeds on seeds of wild plants, e.g. chickweed, buttercup, dock, fat hen and blackberry. From September to April, the species forms small flocks that, in addition to feeding on buds and seeds of wild species, e.g. docks, willow, oak and hawthorn, turn their attention to **buds of soft and top fruit**. Gooseberries are attacked from November to January, apples from February to April and blackcurrants from March to April. The birds are shy, preferring to forage on the edges of orchards, but as winter advances they become bolder, moving towards the more central trees and bushes. The birds nip buds out at the rate of about 30 per minute, eating the central meristem tissues. Leaf, flower and fruit development may thus be seriously reduced, and since in some plums and gooseberries there is no regeneration of fruiting points, damage may be seen several years after attack.

Life cycle. The bullfinch produces a platform nest of twigs in birch or hazel trees, and between May and September lays two to three clutches of four to five pale blue eggs, with purple-brown streaks. It can thus quickly re-establish numbers reduced by lack of food or human attempts at control. They are mainly resident, only rarely migrating.

Control. *Amateur* and *professional* horticulturists can use fine mesh **netting** or **cotton** or synthetic **thread** draped over trees. *Professional* horticulturists spray bitter chemicals such as **ziram** on to shrubs and trees at the time of expected attack to limit bird numbers. Some additional control is achieved by catching birds (usually immature individuals) in specially designed **traps**, which close when the bird lands on a perch to eat seeds. Trapped birds are then taken to a non-horticultural location. Trapping may be started as early as September. Large-scale reinvasion by the same birds in the same season is unlikely as they are territorial, rarely moving more than two miles throughout their lives. Bullfinch trapping is permitted only in scheduled areas of wide-scale fruit production, e.g. Kent.

Mollusc pests

Slugs

This group of serious pests belongs to the phylum Mollusca, a group including the octopus and whelk and the slug's close relatives, the snails, which cause some damage to plants in greenhouses and private gardens.

Damage. The slug lacks a shell and this permits movement into the soil in search of its food source: seedlings, roots, tubers and bulbs. It feeds

by means of a file-like tongue (**radula**), which cuts through plant tissue held by the soft mouth, and scoops out cavities in affected plants (see Figure 14.3). In moist, warm weather it may cause above-ground damage to leaves of plants such as border plants, establishing turf, lettuce and Brussels sprouts. Slugs move slowly by means of an undulating foot, the slime trails from which may indicate the slug's presence. Horticultural areas commonly support populations of 50 000 slugs per hectare.

Life cycle. Slugs are **hermaphrodite** (bearing in their bodies both male and female organs), mate in spring and summer, and lay clusters of up to 50 round, white eggs in rotting vegetation, the warmth from which protects this sensitive stage during cold

Figure 14.3 **Slug damage on carrot**

periods. Slugs range in size from the black keeled slug (*Milax*) 3 cm long, to the garden slug (*Arion*) which reaches 10 cm in length. The mottled carnivorous slug (*Testacella*) is occasionally found feeding on earthworms.

Control. *Private* gardeners use many non-chemical forms of control, ranging from baits of grapefruit skins and stale beer to soot sprinkled around larger plants. A nematode (*Phasmarhabditis hermaphrodita*) is increasingly being used to limit slug numbers. The most effective methods available to both amateur and professional horticulturists at the moment are three chemicals, **aluminium sulphate** (an irritant), **metaldehyde** (which dehydrates the slug) and **methiocarb** (which acts as a stomach poison). The chemicals are most commonly used as small-coloured **pellets** (which include food attractants such as bran and sugar), but metaldehyde may also be applied as a drench. Some growers estimate the slug population using small heaps of pellets covered with a tile or flat stone (to prevent bird poisoning) before deciding on general control. Use of **metaldehyde** and **methiocarb** in gardens has recently been claimed as a major contribution to the decline of the thrush numbers. A simple device such as that seen in Figure 16.2, using a modified plastic milk carton (containing slug pellets) prevents the entry of mammals and birds.

Insect pests

Belonging to the large group of **Arthropods**, which include also the **woodlice**, **mites**, **millipedes** and **symphilids** (see Table 14.1), insects are horticulturally the most important arthropod group, both as pests, and also as beneficial soil animals.

Structure and biology

The body of the adult insect is made up of segments, and is divided into three main parts: the head, thorax and abdomen (see Figure 14.4). The **head** bears three pairs of moving **mouthparts**. The first, the mandibles in insects (such as in caterpillars and beetles) have a **biting** action (see Figure 14.5). The second and third pairs, the maxillae and labia in these insects help in pushing food into the mouth. In the aphids, the mandibles and maxillae are fused to form a delicate tubular stylet, which **sucks** up liquids from the plant phloem tissues. Insects remain aware of their environment by means of compound eyes which are sensitive to movement (of predators) and to colour (of flowers). Their antennae may have a touching and smelling function.

The **thorax** bears three pairs of legs and, in most insects, two pairs of wings. The **abdomen** bears breathing holes (spiracles) along its length, which lead to a respiratory system of tracheae. The blood is colourless, circulates digested food and has no breathing function. The digestive system, in addition to its food absorbing role, removes waste cell products from the body by means of fine, hair-like growths (malpighian tubules) located near the end of the gut.

Since the animal has an **external skeleton** made of tough chitin, it must shed and replace its 'skin' (**cuticle**, see Figure 14.4) periodically by a process called **ecdysis**, in order to increase in size.

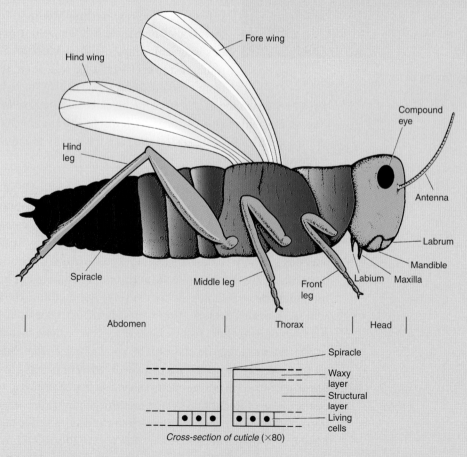

Figure 14.4 **External appearance of an insect**. Note the mouthparts, spiracles and cuticle, the three main entry points for insecticides

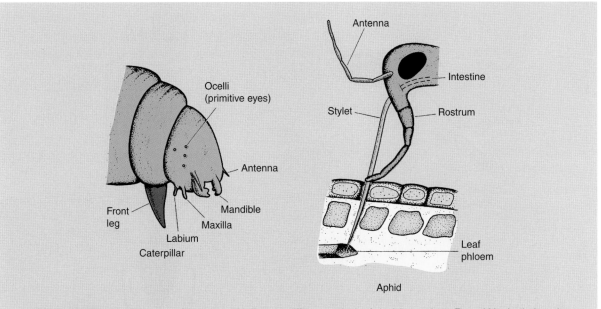

Figure 14.5 Mouthparts of the caterpillar and aphid. Note the different methods of obtaining nutrients. The aphid selectively sucks up dilute sugar solution from the phloem tissue

The two main groups of insect develop from egg to adult in different ways. In the first group (Endopterygeta), typified by the aphids, thrips and earwigs, the egg hatches to form a first stage, or instar, called a **nymph**, which resembles the adult in all but size, wing development and possession of sexual organs. Successive nymph instars more closely resemble the adult. Two to seven instars (growth stages) occur before the adult emerges (see Figure 14.6). This development method is called **incomplete metamorphosis**.

In contrast, the second group of insects 'Endopterygota' including the moths, butterflies, flies, beetles and sawflies undergo a **complete metamorphosis**. The egg hatches to form a first **instar**, called a **larva** which usually differs greatly in shape from the adult. For example the larva (caterpillar) of the cabbage white bears little resemblance to the adult butterfly. Some other dam aging larval stages are shown in Figure 14.7 and these can be compared with the often more familiar adult stage. The great change (**metamorphosis**) necessary to achieve this transformation occurs inside the pupa stage (see Figure 14.6).

The method of **overwintering** differs between insect groups. The aphids survive mainly as the eggs, while most moths, butterflies and flies survive as the pupa. The **speed** of increase of insects varies greatly between groups. Aphids may take as little as 20 days to complete a life cycle in summer, often resulting in vast numbers in the period June–September. On the other hand, the wireworm, the larva of the click beetle, usually takes four years to complete its life cycle.

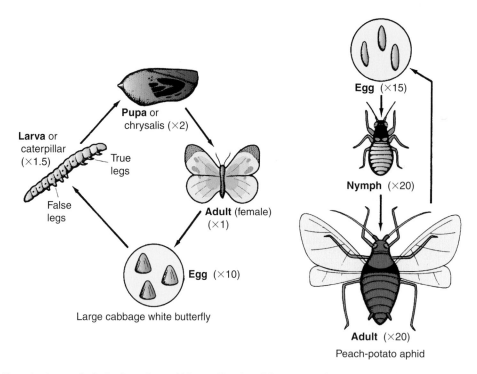

Figure 14.6 Life cycle stages of a butterfly and an aphid pest. Note that all four stages of the butterfly's life cycle are very different in appearance. The nymph and adult of the aphid are similar

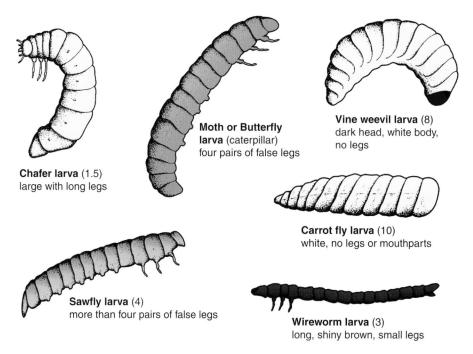

Figure 14.7 Insect larvae that damage crops. Identification into the groups above can be achieved by observing the features of colour, shape, legs and mouthparts

Insect groups are classified into their appropriate order (Table 14.1) according to their general appearance and life cycle stages. There follows now a selection of insect pests in which each species' particular features of life cycle are given. Whilst comments on control are mentioned here, the reader should also refer to Chapter 16 for details

Table 14.1 Arthropod groups found in horticulture

Group	Key features of group	Habitat	Damage
Woodlice (*Crustacea*)	Grey, seven pairs of legs, up to 12 mm in length	Damp organic soils	Eat roots and lower leaves
Millipedes (*Diplopoda*)	Brown, many pairs of legs, slow moving	Most soils	Occasionally eat underground tubers and seed
Centipedes (*Chilopoda*)	Brown, many pairs of legs, very active with strong jaws	Most soils	Beneficial
Symphilids (*Symphyla*)	White, 12 pairs of legs, up to 8 mm in length	Glasshouse soils	Eat fine roots
Mites (*Acarina*)	Variable colour, usually four pairs of legs (e.g. red spider mites)	Soils and plant tissues	Mottle or distort leaves, buds, flowers and bulbs; soil species are beneficial
Insects (*Insecta*)	Usually six pairs of legs, two pairs of wings		
Springtails (*Collembola*)	White to brown, 3–10 mm in length	Soils and decaying humus	Eat fine roots; some beneficial
Aphid group (*Hemiptera*)	Variable colour, sucking mouthparts, produce honeydew (e.g. greenfly)	All habitats	Discolour leaves and stems; prevent flower pollination; transmit viruses
Moths and butterflies (*Lepidoptera*)	Large wings; larva with three pairs of legs, four pairs of false legs and biting mouthparts (e.g. cabbage-white butterfly)	Mainly leaves and flowers	Defoliate leaves (stems and roots)
Flies (*Diptera*)	One pair of wings, larvae legless (e.g. leatherjacket)	All habitats	Leaf mining, eat roots
Beetles (*Coleoptera*)	Horny front pair of wings which meet down centre; well-developed mouthparts in adult and larva (e.g. wireworm)	Mainly in the soil	Eat roots and tubers (and fruit)
Sawflies (*Hymenoptera*)	Adult like a queen ant; larvae have three pairs of legs, and more than four pairs of false legs (e.g. rose-leaf curling sawfly)	Mainly leaves and flowers	Defoliation
Thrips (*Thysanoptera*)	Yellow and brown, very small, wriggle their bodies (e.g. onion thrips)	Leaves and flowers	Cause spotting of leaves and petals
Earwigs (*Dermaptera*)	Brown, with pincers at rear of body	Flowers and soil	Eat flowers

of specific types of control (cultivations, chemicals, etc.) and for explanations of terms used.

Aphids and their relatives (*Order Hemiptera*)

This important group of insects has the egg–nymph–adult life cycle and sucking mouthparts.

Peach-potato aphid (Myzus persicae)

This and similar species are often referred to by the name 'greenfly' (Figure 14.8).

Damage. It is common in market gardens and greenhouses. The nymph and adult of this aphid may cause three types of damage. Using its sucking stylet, it may inject a digestive juice into the plant phloem, which in young organs may cause severe **distortion**. Having sucked up sugary phloem contents, the aphid excretes a sticky substance called **honey-dew**, which may block up leaf stomata and reduce photosynthesis, particularly when dark-coloured fungi (sooty moulds) grow over the honeydew. Thirdly, the aphid stylets may **transmit viruses** such as virus Y on potatoes and tomato aspermy virus on chrysanthemums.

Life cycle. The aphid varies in colour from light green to orange, measures 3 mm in length (see Figure 14.9) and has a complex life cycle, shown in Figure 14.8, alternating between the winter host (peach) and the many summer hosts such as potato and bedding plants. In spring and summer, the females give birth to nymphs directly without any egg stage (a process called vivipary), and without fertilization by a male (a process called parthenogenesis). Spread is by the summer flighted females. Only in autumn, in response to decreasing daylight length and outdoor temperatures, are both sexes produced, which having wings, fly to the winter host, the peach. Here, the female is fertilized and lays

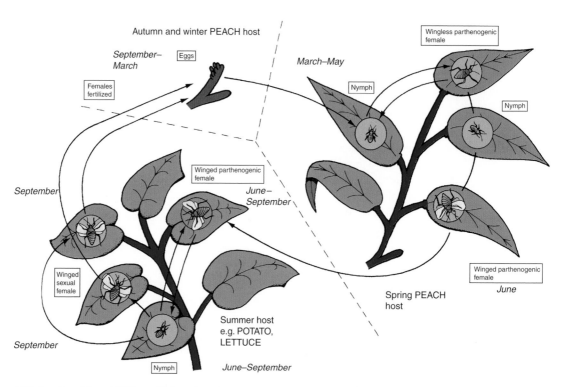

Figure 14.8 Peach-potato aphid life cycle throughout the year. Female aphids produce nymphs on both the peach and summer host. Winged females develop from June to September. Males are produced only in autumn. Eggs survive the winter. In greenhouses the life cycle may continue throughout the year

Figure 14.9 **Rose aphid**

thick-walled black eggs. In glasshouses, the aphid may survive the winter as the nymph and adult female on plants such as begonias and chrysanthemums, or on weeds such as fat hen.

Spread occurs in early summer and autumn by winged females.

Control. The peach-potato aphid can be controlled in several ways. In outdoor crops, several organisms, e.g. ladybirds, lacewings, hoverflies and parasitic fungi (see **biological control**, Chapter 16), naturally found in the environment, may reduce the pest's importance in favourable seasons. In the greenhouse, an introduced parasitic wasp (*Aphidius matricariae*) is available to amateurs and professionals.

The *amateur* gardener uses an aphicide containing **pyrethrins.** Two other chemicals are used by both amateur and professional horticulturists: a contact chemical **bifenthrin** and a soap concentrate containing **fatty acids**.

For the *professional*, there is a biological control using the fungus *Verticillium lecanii*, and many chemical controls, including **pirimicarb** used outdoors and in glasshouses as a spray, and **nicotine** used as a smoke in glasshouses.

There are many other horticulturally important aphid species. The black bean aphid (*Aphis fabae*), which overwinters on *Euonymus* bushes, may seriously damage broad beans, runner beans and red beet. The rose aphid (*Macrosiphum rosae*) attacks young shoots of rose.

Spruce-larch adelgid (Adelges viridis)

This relative of the aphid may cause serious damage on spruce grown for Christmas trees.

Damage. Although nursery trees less than four years of age are rarely badly damaged, early infestation in the young plant may result in serious damage as it gets older. In June–September the adults move to **larch**, acquire woolly white hairs and may cause defoliation of the leaves.

Life cycle. The green female adult develops from overwintering nymphs on spruce, and in May (year 1) lays about 50 eggs on the dwarf shoots. The emerging nymphs, injecting poisons into the shoots, cause abnormal growth into pineapple galls, which spoil the tree's appearance. After a further year (year 2) on this host, the female adelges returns to the spruce, where it lives for another year (year 3) before the gall-inducing stages are produced.

Spread is by flighted females.

Control. In Christmas tree production, the adelges may be controlled by sprays of **deltamethrin** in May, when the gall-inducing nymphs are developing.

Glasshouse whitefly (Trialeurodes vaporariorum)

This small insect, looking like a tiny moth, was originally introduced from the tropics, but now causes serious problems on a range of glasshouse food and flower crops. It should not be confused with the very similar, but slightly larger cabbage whitefly on brassicas.

Damage. All stages after the egg have sucking stylets, which extract a sugary liquid from the phloem, often causing large amounts of honeydew and sooty moulds on the leaf surface. Plants that are seriously attacked include fuchsias, cucumbers, chrysanthemums and pelargoniums. Chickweed, a common greenhouse weed, may harbour the pest over winter in all stages of the pest's life cycle.

Figure 14.10 Adult glasshouse whitefly, actual size is about 1 mm long

Life cycle. The adult glasshouse whitefly (Figure 14.10) is about 1 mm long and is able to fly from plant to plant. The fertilized female lays about 200 minute, white, elongated oval eggs in a circular pattern on the **lower leaf surface** over a period of several weeks. After turning black, the eggs hatch to produce nymphs (crawlers), which soon become flat immobile scales. The last scale instar is thick-walled and called a ' pupa' from which the male or female adult emerges. Three days later, the female starts to lay eggs again. The whole life cycle takes about 32 days in spring, and about 23 days in the summer.

Spread is mainly by introduced plants, or more rarely by chance arrivals of adult through doors or vents.

Control of glasshouse whitefly is achieved in several ways.

Amateur and *professional* horticulturalists should remove weeds (such as chickweed or sowthistle) harbouring the pest from crop to crop. Careful inspection of the lower leaves of introduced plants achieves a similar aim.

There is a reliable form of biological control. This involves a minute wasp (*Encarsia formosa*) which lays an egg inside the last scale stage of the white-fly. The developing whitefly is eaten away by the wasp grub and the scale turns black and soon releases the next generation of wasps (see Chapter 16 for more details). Control is usually most effective when whitefly numbers are low.

Amateur gardeners can use spray products containing specially formulated **fatty acids** to control young and adult pest stages. *Professional horticulturists* can choose between an insecticide that physically blocks the insect's breathing holes (see p204) such as **alginate/polysaccharide**, or a contact insecticide such as deltamethrin, or the above-mentioned **fatty acids**.

It is suggested that serious infestations of this pest receive a regular weekly chemical spray to catch the more sensitive scale and adult stages.

Greenhouse mealy bug (Planococcus citri)

Damage. This pest, a distant relative of the aphid, spoils the appearance of glasshouse crops, particularly orchids, *Coleus* species, cacti and *Solanum* species. All of the stages except the egg suck phloem juices by means of a tubular mouthpart (**stylet**), and when this pest is present in dense masses it produces **honey dew** and may cause leaf drop (see Figure 14.11).

 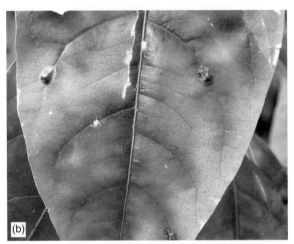

Figure 14.11 (a) **Mealy bug** (b) **Brown scale**

Life cycle. Being a tropical species, it develops most quickly in high temperatures and humidities, and at 30°C completes a life cycle within about 22 days. The adult measures about 3 mm in length and produces fine waxy threads.

Spread. Adults have wings, but the most important spread is by introduction of plant material newly infested with small nymphs, or from plant to plant when leaves are touching.

Control. Mealy bugs are difficult pests to control, as the thick cuticle resists chemical sprays, and the droplets fall off the waxy threads. An introduced tropical ladybird, *Cryptolaemus montrouzieri*, is commercially available for controlling the pest and is most effective at temperatures above 20°C. Spray products containing pure **fatty acids** are effective against this pest and are available to both *amateur* and *professional* horticulturists.

A recently reported related problem on beech has been the woolly beech aphid (*Phyllapis fagi*).

Brown scale (Parthenolecanium corni)

The female scale, measuring up to 6 mm, is tortoise shaped (see Figure 14.11) and has a very **thick cuticle**. It may be a serious pest

Figure 14.12 (a) **Bud blast on rhododendron** transmitted by a leafhopper (b) **Capsid damage on apple.**(c) **Capsid damage on potato leaf**

outdoors on vines, currants, top fruit and cotoneasters, sucking phloem sugars out of the plants. It is also found in greenhouses on vines, peaches and Amaryllis, causing stunted growth and leaf defoliation. The nymphs are mobile, but do not spread far. Transport of infested plant material is the main cause of outbreaks in greenhouses. As with mealy-bug, control is made more difficult because of the thick cuticle, but products containing **fatty acids** give good control. A recently reported scale problem is soft brown scale (*Coccus hesperidum*) in greenhouses.

Leaf hoppers (Graphocephala fennahi)

These slender, light green insects, about 3 mm long, well known in their nymph stage as 'cuckoo spit', are found on a wide variety of crops, e.g. potato, rose, *Primula* and Calceolaria. The adults can fly from plant to plant. They live on the **undersurface** of leaves, causing a mottling of the upper surface. In strawberries, they are vectors of the **green-petal** disease, while in rhododendron they carry the serious **bud blast** disease that kills off the flower buds (see Figure 14.12). August and September sprays of products containing **fatty acids** prevent egg laying inside buds of rhododendron, and thus reduce the entry points for the fungus disease.

Common green capsid (Lygocoris pabulinus)

This very active, light green pest measuring 5 mm in length and resembling a large aphid, occurs on fruit trees and shrubs and flower crops, most commonly outdoors. Owing to the poisonous nature of its salivary juices, young foliage shows distorted growth with small holes, even when relatively low insect numbers are present and fruit is scarred (see Figure 14.12). The adult flies from plant to plant. The chemicals used against aphids control this pest.

Thrips (*Order Thysanoptera*)

Whilst classified under the broad grouping of the Exopterygota (like the aphids), thrips also produce a survival stage which is often soil borne, and is loosely called a pupa (a life cycle stage normally reserved for members of the other main insect group, the Endopterygota). Due to their increased activity during warm humid weather, thrips are sometimes called ' thunder flies' (and are known for their ability to get into human's hair in sultry summer weather and cause itching).

Onion thrips (Thrips tabaci)

Damage. Thrips' mouthparts are modified for piercing and sucking, and the toxic salivary juices cause silvering in onion leaves, straw-brown spots several mm in diameter on cucumber leaves, and white streaks on carnation petals.

Life cycle. The 1 mm long, narrow-bodied insect has feather-like wings. The last instar of the life cycle, called the **pupa**, descends to the soil, and it is this stage which overwinters. In greenhouses there may be seven generations per year, while outdoors one life cycle is common.

Spread. Adults may be blown considerable distances from nursery to nursery in the wind. (The occurrence in Britain of Western flower thrip (*Frankliniella occidentalis*) on both greenhouse and outdoor flower and vegetable crops has created serious problems for the industry, particularly because it transmits the serious tomato spotted wilt virus, and is able to pupate on the plant, deep within dense flowers such as carnation.)

Control. Thrips infestations may be reduced in greenhouses by the use of fine screens over vents, and by a double door system.

Amateur gardeners are able to use a recently introduced product for thrip control containing **natural plant extracts** which block the insect's breathing holes.

Professional growers use the predatory mite *Neoseiulus cucumeris*, and the predatory bug *Orius laevigatus*. Chemicals used include the ingredient **abamectin** and the above named **natural plant extract**. Western flower thrip has shown greater resistance to chemical control than the other thrips and a careful rotation of chemical groups has proved necessary.

Earwigs (Forficula auricularia)

These pests belong to the order Dermaptera, and bear characteristic 'pincers' (cerci) at the rear of the 15 mm long body. They gnaw away at leaves and petals of crops such as beans, beet, chrysanthemums and dahlias, usually from July to September, when the nymphs emerge from the parental underground nest. They usually **spread** by crawling on the surface of the soil, but they can also fly. Upturned flower pots containing straw are sometimes used in greenhouses for trapping these shy nocturnal insects. The professional grower may use **pirimiphos-methyl** as a spray or smoke.

Moths and butterflies

This group of insects belongs to the large grouping, the Endopterygota (see p205) which has different life cycle details from the aphid group. The order (Lepidoptera) characteristically contains adults with four large wings and curled feeding tubes. The larva (**caterpillar**), with six small legs and eight false legs, is modified for a leaf-eating habit (see Figures 14.6 and 14.7). Some species are specialized for feeding inside fruit (codling moth on apple, see Figure 14.15), underground (cutworms), inside leaves (oak leaf miner), or inside stems (leopard moth). The gardener may find large webbed caterpillar colonies of the lackey moth (*Malacosoma neustria*) on fruit trees and hawthorns, or the juniper webber (*Dichomeris marginella*) causing webs and defoliating junipers.

Large cabbage-white butterfly (Pieris brassicae)

Damage. Leaves of cabbage, cauliflower, Brussels sprouts and other hosts such as wallflowers and the shepherd's-purse weed are progressively eaten away. The defoliating damage of the larva may result in skeletonized leaves.

Figure 14.13 (a) **Cabbage-white adult**. (b) **Yellow underwing moth** with cutworm larva and brown

Life cycle. This well-known pest on cruciferous plants emerges from the overwintering pupa (chrysalis) in April and May and, after mating, the females (see Figure 14.13) lay batches of 20 to a 100 yellow eggs on the underside of leaves. Within a fortnight, groups of first instar larvae emerge and soon moult to produce the later instars, which reach 25 mm in length and are yellow or green in colour, with clear black markings. They have well-developed mandibles. Pupation occurs usually in June, in a crevice or woody stem, the pupa (chrysalis) being held to its host by silk threads. A second generation of the adult emerges in July, giving rise to more damaging larval infestation than the first. The second pupa stage overwinters.

Spread. The species is spread by the adults. (Care should be taken not to confuse the cabbage-white larva with the large smooth green or brown larva of the cabbage moth, or the smaller light green larva of the diamond-backed moth, both of which may enter the hearts of cabbages and cauliflowers, presenting greater problems for control.)

Control. There are several forms of control against the cabbage-white butterfly. A naturally occurring small wasp (*Apantales glomeratus*) lays its eggs inside the pest larva (see p273). A virus disease may infect the pest, causing the larva to go grey and die. Birds such as starlings eat the plump larvae. When damage becomes severe, amateur gardeners and professional growers can use spray products containing **bifenthrin**.

Winter moths (Operophthera brumata)

Damage. These are pests which may be serious on top fruit and ornamentals, especially woody members of the Rosaceae family. The caterpillars eat away leaves in spring and early summer and often form other leaves into loose webs, reducing the plant's photosynthesis. They occasionally scar young apple fruit.

Life cycle. This pest's timing of life cycle stages is unusual. The pest emerges as the adult form from a soil-borne pupa in November and December. The male is a greyish-brown moth, 2.5 cm across its wings, while the female is wingless. The female crawls up the tree to lay 100–200 light-green eggs around the buds. The eggs hatch in spring at bud burst to produce green larvae with faint white stripes. These larvae move in a characteristic looping fashion and when fully grown, descend on silk threads at the end of May before pupating in the soil until winter.

Spread is slow because the females do not have wings.

Control. A common control is a **grease band** wound around the main trunk of the tree in October which is effective in preventing the flightless female moth's progress up the tree. In large orchards, professional growers use springtime sprays of an insecticide such as **deltamethrin** to kill the young caterpillars.

Cutworm (e.g. Noctua pronuba)

Damage. The larvae of the **yellow underwing moth**, unlike most other moth larvae, live in the soil, nipping off the stems of young plants and eating holes in succulent crops, e.g. bedding plants, lawns, potatoes, celery, turnips and conifer seedlings. The damage resembles that caused by slugs.

Life cycle (see Figure 14.13). The adult moth, 2 cm across, with brown fore-wings and yellow or orange hind wings, emerges from the shiny soil-borne chestnut brown pupa from June to July, and lays about 1000 eggs on the stems of a wide variety of weeds. The first instar caterpillars, having fed on weeds, descend to the soil and in the later instars cause the damage described above, eventually reaching about 3.5 cm in length. They are grey to grey-brown in colour, with black spots along the sides. Several other cutworm species such as **heart and dart moth** (*Agrotis exclamationis*) and **turnip moth** (*Agrotis segetum*) may cause damage similar to that of the yellow underwing. In all three species, their typical caterpillar-shaped larvae should not be confused with the legless leatherjacket which is also a common underground larva. The cutworm species normally have two life cycles per year, but in hot summers this may increase to three.

Spread. The larvae are able to crawl from plant to plant, but most spread is by the actively flying adults.

Control. *Amateur* gardeners remove a good proportion of the cutworm larvae as they dig plots over. Good weed control reduces cutworm damage. A soil-directed spray of *bifenthrin* will achieve some control of cutworm.

For the *professional* horticulturist, soil drenches of residual insecticides such as **chlorpyrifos** have proved successful against the larva stage of this pest.

Leopard moth (Zeuzera pyrina)

Damage. The caterpillar of this species tunnels into the branches and trunk of a wide range of tree species, such as apple, ash, birch, and lilac. The tunnelling may weaken the branches of trees which in high winds commonly break.

Life cycle. The Leopard moth has an unusual life cycle. The moth is large, 5–6 cm across, and is white with black spots. In early summer the female lays dark-yellow eggs on the bark of the tree. The emerging caterpillar (see Figure 14.14) enters the stem by a bud, and then

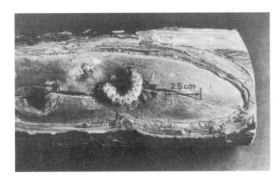

Figure 14.14 **Leopard moth larva** emerging from an apple stem

tunnels for 2–3 years in the heartwood. It has bacteria in its gut which help to digest the xylem tissue that it eats. It eventually reaches 5 cm in length, pupating in the tunnel, and finally emerging from the branch the following summer as the adult. Spread is only by adults.

Control. Where tunnels are observed, a piece of wire may be pushed along the tunnel to kill the larva.

Other moths worthy of mention here are the fruit-invading species such as codling moth (*Cydia pomonella*) on apple, plum moth (*Cydia funebrana*) and pea moth (*Cydia nigricana*) each of which needs accurately timed insecticidal control to avoid fruit damage by the pest (see also **pheremone traps**), because insecticidal control inside the fruit is not possible.

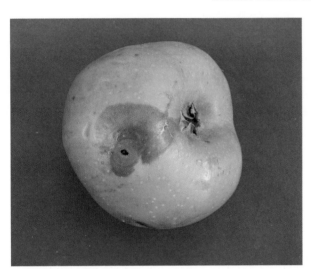

Figure 14.15 Codling moth damage

Recently reported moth pests are the Holm oak leaf-mining moth (*Phyllonorycter messaniella*), the horse chestnut leaf miner (*Cameraria obridella*) and the Leek moth (*Acrolepiosis assectella*).

Flies

This group of insects, of the order Diptera, are characterized by having only a single pair of functioning wings. The hind wings are modified into little stubs which act as balancing organs. The larvae are legless, elongated, and their mouthparts, where present, are simple hooks. The larvae are the only stage causing crop damage.

Carrot fly (Psila rosae)

Damage. This is a widespread and serious pest on umbelliferous crops (carrots, celery and parsnips). The grubs emerging from the eggs eat fine roots and then enter the mature root using fine hooks in their mouths. Damage is similar in all crops. In carrots, seedlings may be killed, while in older plants the foliage may become red, and wilt in dry weather. Stunting is often seen, and affected roots, when lifted, are riddled with **small tunnels** (see Figure 14.16) that make the carrots unsaleable. Damage should not be confused in carrots with **cavity spot**, a condition associated with a *Pythium* species of fungi which produces elongated sunken spots partly circling the root.

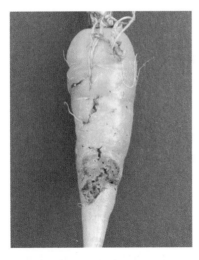

Figure 14.16 **Carrot fly damage**

Life cycle. The adult fly is 8 mm long, shiny black and with a red head. It emerges from the rice-grain-sized overwintering pupa in the soil from late May to early June. The small eggs laid on the soil near the host plant soon hatch to give white larvae which damage the plant (see Figure 14.16). The fully grown larva leaves the host to turn into a cylindrical pale yellow pupa when a month old. A second generation of adults emerges in late July, while in October a third emergence is seen in some areas.

Spread. The adult is the stage which spreads the pest.

Control. The *amateur* gardener can use a variety of controls. Planting carrots after the May emergence has occurred reduces infestation. Covering carrot plots with horticultural **fleece** (see Chapter 16, Figure 16.2), prevents the adut from laying eggs next to carrots. Inter-planting carrots with onions is said to prevent carrot fly from homing in on the carrot crop.

For the *professional* grower, high levels of carrot fly can be prevented by keeping hedges and nettle beds trimmed, thus reducing sheltering sites for the flies. A seed treatment containing **tefluthrin** reduces early larval infestation. A ground-directed spray at a high volume rate of **lambda-cyhalothrin** reaches the larva in the soil.

Chrysanthemum leaf miner (Phytomyza syngenesiae)

Damage. The leaf miners are a group of small flies, the larvae of which can do serious damage to horticultural crops by tunnelling through the leaf. This species is found on members of the plant family Asteraceae. Plants attacked include chrysanthemum, cineraria and lettuce.

Life cycle. The flies emerge at any time of the year in greenhouses, but normally only between July and October outdoors. These adults, which measure about 2 mm in length and are grey-black with yellow underparts, fly around with short hopping movements. The female lays about 75 minute eggs singly inside the leaves, causing white spot symptoms to appear on the upper leaf surface. The **larva** stage is greenish white in colour, and tunnels into the pal-lisade mesophyll of the leaf, leaving behind the characteristic mines seen in Figure 14.17. On reaching its final instar, the 3.5 mm long larva develops within the mine into a brown pupa, from which the adult emerges. The total life cycle period takes about three weeks during the summer months.

Figure 14.17 **Chrysanthemum leaf** miner damage

Spread. This pest is spread by the adult stage.

Control. Weed hosts such as groundsel and sowthistle should be controlled. Yellow sticky traps remove many of the adult flies. Certain chrysanthemum cultivars show some resistance.

Amateur gardeners have no effective insecticide product to control the larva inside the leaf.

Professional growers use tiny wasps such as *Diglyphus isaea* and *Dacnusa sibirica* to parasitize the tunnelling leaf-miner larvae. Products containing **abamectin** may be used outdoors and in greenhouses. (The occurrence of South American leaf miner (*Liriomyza huidobrensis*) and American serpentine leaf miner (*Liriomyza trifolii*) which are able to damage a wide variety of greenhouse plants has, in recent years, created many problems for horticulture).

Leatherjacket (Tipula paludosa)

Damage. This is an underground pest which is a natural inhabitant of grassland and causes most problems on golf greens. After ploughing up of grassland, leatherjackets may also cause damage to the crops such as potatoes, cabbages, lettuce and strawberries. This pest is particularly damaging in prolonged wet periods when the roots of young or succulent crops may be killed off. Occasionally lower leaves may be eaten.

Life cycle. The adult of this species is the crane fly, or 'daddy-long-legs', commonly seen in late August. The females lay up to 300 small eggs on the surface of the soil at this period, and the emerging larvae feed on plant roots during the autumn, winter and spring months, reaching a length of 4 cm by June. They are cylindrical, grey-brown in colour, **legless** and possess hooks in their mouths for feeding. During the summer months, they survive as a thick-walled pupa.

Spread is achieved by the adults.

Control. *Amateur* gardeners remove this pest as they dig in autumn and spring. Crops sown in autumn are rarely affected, as the larvae are very small at this time. There are no pesticide products recommended to control this pest.

Professional growers and groundsmen use products containing the residual **chlorpyrifos**, which is drenched into soil to reduce the larval numbers.

Sciarid fly (Bradysia spp.)

Damage. The larvae of this pest (sometimes called **fungus gnat**) feed on fine roots of greenhouse pot plants such as cyclamen, orchid and freesia, causing the plants to wilt. Fungal strands of mushrooms in commercial houses may be attacked in the compost.

Life cycle. The slender black females, which are about 3 mm long, fly to suitable sites (freshly steamed compost, moss on sand benches and well-fertilized compost containing growing plants), where about 100 minute eggs are laid. The emerging legless larvae are translucent-white with a black head, and during the next month grow to a length of 3 mm before briefly pupating and starting the next life cycle.

Spread is achieved by the adults.

Control. *Amateur* gardeners and *professional* horticulturists use yellow sticky traps to catch the flying adults in greenhouses. The pest can be reduced by avoiding overwatering of plants. Biological control by the tiny nematode *Steinernema feltiae* is now available. *Professional* growers in mushroom houses attempt to exclude the flies from mushroom houses by means of fine mesh screens placed next to ventilator fans. A predatory mite (*Hypoaspis miles*) is used to control the larvae. The larvae also may be controlled by the insecticide, **diflubenzuron**, incorporated into composts.

Beetles

This group of insects in the order Coleoptera is characterized in the adult by hard, horny forewings (elytrae) which, when folded, cover the delicate hind wings used for flight. The meeting point of these hard wing cases produces the characteristic straight line down the beetles back over its abdomen. Most beetles are beneficial, helping in the breakdown of humus, e.g. dung beetles, or feeding on pest species (see **ground beetle**). A few beetles, e.g. wireworm, raspberry beetle and vine weevil, causes crop damage.

Vine weevil (Otiorhyncus sulcatus)

This species belongs to the beetle group but, as with all weevils, possesses a longer snout on their heads than other beetles.

Damage. The larva stage is the most damaging, eating away roots of crops such as cyclamen and begonias in greenhouses, primulas, strawberries, young conifers and vines outdoors, causing above-ground symptoms similar to root diseases such as vascular wilt. Close inspection of the plant's root zone will, however, quickly show the unmistakable white grubs (see Figure 14.18). The adults may eat out neat holes or leaf edges of the foliage of hosts such as rhododendron, raspberry and grapes, and many herbaceous perennials (see Figure 14.18). Several related species, e.g. the clay-coloured weevil (*Otiorhyncus singularis*) cause similar damage to that of the vine weevil.

Figure 14.18 (a) **Vine weevil larva** and pupa (b) **Adult Vine Weevil** damage on *Tellima*

Life cycle. The adult is 9 mm long, black in colour, with a rough textured cuticle (see Figure 14.19). The forewings are fused together, the pest being incapable of flight. No males are known. The female lays eggs (mainly in August and September) in soil or compost, next to the roots of a preferred plant species. Over a period of a few years, she may lay a thousand eggs as she visits many plants. The emerging larvae are white, legless and with a characteristic chestnut-brown head. They reach 1 cm in length in December when they pupate in the soil before developing into the adult.

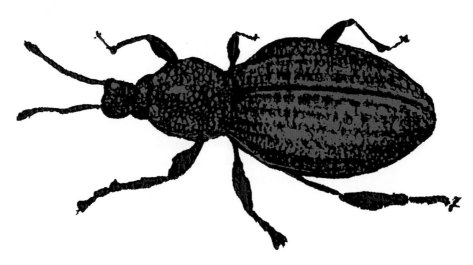

Figure 14.19 **Vine weevil adult**-9 mm in length

Spread is achieved by the female adult crawling around at night, or by the movement of pots containing grubs.

Control. *Amateur* gardeners sometimes use traps of corrugated paper placed near infested crops. Inspection of plants at night by torchlight may reveal the feeding adult.

Amateurs and professionals can use the nematode (*Steinemena carpocapsae*) by incorporating it into compost or soil.

Professional growers have residual chemicals, **imidacloprid** or **chlorpyrifos** incorporated or drenched into compost or soil.

Wireworm (Agriotes lineatus)

Damage. This beetle species is commonly found in grassland, but will attack most crops. Turf grass may be eaten away by the larvae (**wireworms**) resulting in dry areas of grass. The pest also bores through potatoes to produce characteristic narrow tunnels, while in onions, brassicas and strawberries the roots are eaten. In tomatoes, the larvae bore into the hollow stem.

Life cycle. The 1 cm long adult (click beetle) is brown-black and has the unusual ability of flicking itself in the air when placed on its back. The female lays eggs in weedy ground in May and June and the larvae, after hatching, develop over a four year period. Fully grown wireworms are about 2.5 cm long, shiny golden-brown in colour, and possess short legs (see Figure 14.7). After a three week pupation period in the soil, usually in summer, the adult emerges and in this stage survives the winter.

Spread is by means of the flighted adults.

Control. Some *amateur* gardeners dig in green manure crops to lure wireworms away from underground roots and tubers. *Professional* growers may reduce serious damage to young crops by using a seed dressing containing **tefluthrin**.

Garden Chafer (Phyllopertha horticola)

Damage. This pest is increasingly proving a problem on turf where the large white grubs eat the roots. Small yellow patches appear in the lawn or sports area, notably in summer when the grubs are becoming fully grown. Further damage can occur when starlings, crows, moles, foxes and even badgers dig up parts of the lawn to reach the succulent prey.

Life cycle. The adult is a broad, 1 cm long, light-brown beetle with a bottle-green upper thorax and head. Adults emerge from soil-borne pupae in May and June. They feed on leaves of a variety of plants, sometimes badly damaging fruit tree foliage. Eggs are laid near grasses in early summer and emerging larvae develop during the next 10 months into the characteristic stout white grub with a light-brown head. Their body, which may reach nearly 2 cm in length, is curved and bears well-developed legs (see Figure 14.20). The larvae are the winter- survival stage. Pupation occurs in early spring.

Spread is by means of the flighted adults.

Figure 14.20 **Chafer grub**-2 cm in length

Control. For the *amateur* gardener, maintenance of a healthy, well-fertilized lawn will lessen chafer attack.

Groundsmen may use a biological control, involving a nematode, *Heterorhabditis bacteriophora* available for application during the summer. The nematode enters the chafer grub through its spiracles, and proceeds to release bacteria which digest the grub's body. A soil-applied insecticide containing **imadocloprid** is used against this pest.

Raspberry beetle (Byturus tomentosus)

The developing fruit of raspberry, loganberry and blackberry may be eaten away by the 8 mm long, golden-brown larvae of this pest. Only one life cycle per year occurs, the larva descending to the soil in July and August, pupating in a cell from which the golden-brown adult emerges to spend the winter in the soil. The adult female lays eggs in the host flower the following June. **Spread** is by means of the flighted adult.

Control. Since the destructive larval stage may enter the host fruit and thus escape insecticidal control, the timing of the spray is vital. In raspberries, a contact chemical such as **bifenthrin**, applied when the fruit is pink, achieves good control.

Other beetle pests

Springtime attack of flea beetle (*Phyllotreta* species) on leaves of young cruciferous plants (e.g. cabbages and stocks) is a serious problem to amateur and professional horticulturist alike (see Figure 14.21). In recent years, four other increasingly common beetle problems have been reported. These are viburnum beetle on *Viburnum opulus*, *V. tinus* and *V. lantana*, rosemary leaf beetle (*Chrysolina americana)* on lavender, rosemary and thyme, red lily beetle (*Lilioceris lilii*) on lilies, and asparagus beetle (*Crioceris asparagi*).

Figure 14.21 **Flea beetle** damage

Figure 14.22 **Viburnum sawfly** damage

Sawflies

This group, together with bees (see Chapter 10), wasps and ants, are classified in the order Hymenoptera, characterized by adults with two pairs of translucent wings and with the fore- and hind-wings being locked together by fine hooks. The slender waist-like first segments of the abdomen give these insects a characteristic appearance. A sawfly larva is shown in Figure 14.22.

Gooseberry sawfly (Nematus ribesii)

Damage. This is an important pest on gooseberries, redcurrants and whitecurrants, but not on blackcurrants. Extensive damage to foliage may be caused by the caterpillars. In some cases, the leaves of the whole bush may be skeletonized.

Life cycle. The adults emerge from their overwintering soil-borne cocoons in late April. Adults measure about 1 cm in length and resemble flying ants. The male has a black abdomen, the female a yellow abdomen. The female, in early May, lays elongate 1 mm long, light-green eggs in rows along the veins of the leaves situated low down in the host bush. Emerging young caterpillars eat small holes in these leaves. Maturing caterpillars (reaching up to 2 cm in length and identifiable by their **green appearance with numerous black spots** and with a black head) consume whole leaves and move outwards and upwards from their original positions. After 3–4 weeks of feeding, the mature caterpillars drop to the soil and pupate. There are 2–3 more lifecycle generations of the sawfly from the second emergence in early June to last soil pupation in late September.

Spread is by the flighted adults.

Control. *Private* gardeners often pick off the first few young caterpillars found on the base leaves of the bush at egg-germination times (in late April, early June, early July and late August). In autumn and winter periods, the removal of any mulch from around the bushes and disturbance of soil in the area encourage birds to seek out overwintering pupae. Insecticide products containing **pyrethrins** control this pest.

Professional growers use two ingredients derived from plants. The insecticides, **derris** or **nicotine**, are effective against the young caterpillars if directed on the under-leaf surface, and with special attention being paid to the lower, central leaves.

No biological control is currently available for gooseberry sawfly, and the insecticide formulated from the bacterium, *Bacillus thuringiensis*, whilst effective on many caterpillars of moths, is ineffective against this sawfly species.

Rose leaf-rolling sawfly (Blennocampa pusilla)

The black shiny adults, resembling winged queen ants, emerge from the soil-borne pupa in May and early June. Eggs are inserted into the leaf lamina of roses, which, in responding to the pest, rolls up tightly. The emerging larva, which is pale green with a white or brown head, feeds on the rolled foliage. It reaches a length of 1 cm by August, when it descends to the ground and forms an underground cocoon to survive the winter until it pupates in March. All types of roses are affected, although climbing roses are preferred. Damage caused by leaf-rolling tortrix caterpillars, e.g. *Cacoecia oparana*, may be confused with the sawfly, although the leaves are less curled.

Spread of the sawfly is achieved by means of the flighted adults.

Amateur gardeners may need to control this pest with an insecticide containing **pyrethrins**. Nursery stock *growers* may use products containing **pirimiphos-methyl**.

Springtails (order Collembola)

This group of primitive wingless insects, about 2 mm in length (Figure 14.23) has a spring-like appendage at the base of the abdomen. They are very common in soils, and normally aid in the breakdown of soil **organic matter**. Two genera, *Bourletiella* and *Collembola*, however, may do serious damage to conifer seedlings and cucumber roots respectively.

Spread is slow since all stages are wingless.

Figure 14.23 **Springtail** – about 2 mm in size – can jump by means of spring at the end of its body

Mite pests

The mites (Acarina) are classified with spiders and scorpions in the Arachnida. Although similar to insects in many respects they are distinguished from them by the **possession of four pairs of legs, a fused body structure and by the absence of wings** (see Figure 14.24). Many of the tiny soil-inhabiting mites serve a useful purpose in breaking down plant debris. Several above-ground species are serious pests on plants. The life cycle is composed of egg, larva, nymph and adult stages.

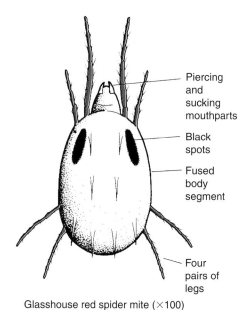

Piercing and sucking mouthparts

Black spots

Fused body segment

Four pairs of legs

Glasshouse red spider mite (×100)

Figure 14.24 Glasshouse **red spider mite**. Note that the red spider mite may be light green or red in colour. Its extremely small size (0.8 mm) enables it to escape attention

Glasshouse red spider mite (Tetranychus urticae and T. cinnabarinus)

Damage. The piercing mouthparts of the mites inject poisonous secretions which cause localized death of leaf mesophyll cells. This results in a fine mottling symptom on the leaf (see Figure 14.1), not to be confused with the larger spots caused by **thrips**. In large numbers the mites can kill off leaves and eventually whole plants. Fine silk strands are produced in severe infestations, appearing as 'ropes' (see Figure 14.25) on which the mites move down the plant. On flowering crops such as chrysanthemums, these ropes make the plant unsaleable.

Life cycle. This pest is of tropical origin and thrives best in high greenhouse temperatures. Both species are 1 mm in length. The **first** species (*T. urticae*) is yellowish in colour, with two black spots (see Figure 14.25). The female lays about 100 tiny spherical eggs on the underside of the leaf, and after a period of three days the tiny six-legged larva moults to produce the nymph stage that resembles the adult. The life cycle length varies markedly from 62 days at 10°C, to 6 days at 35°C when the pest's multiplication potential is extremely high. In autumn, when the daylight period decreases to 14 h and temperatures fall, egg production ceases and the fertilized females, which are now red in colour, move into the greenhouse structures to hibernate (diapause), representing foci for the next spring's infestation.

Figure 14.25 (a) Glasshouse **red spider** in the centre (b) webbing

Spread occurs when adults and nymphs crawl from plant to plant when leaves are touching. Wind currents can move mites attached to their silk strands. The small size and under-leaf habitat of the pest combine to keep its existence on introduced plants away from the gaze of growers, especially older growers.

The **second** species (*T. cinnabarinus*), which is dark reddish-brown, has a similar life cycle to *T. urticae*, but does not hibernate. The hibernation habit of *T. urticae* leads to it being a common pest on annual crops such as tomatoes, cucumbers and chrysanthemums, while *T. cinnabarinus* is found more commonly on the perennial crops such as carnations, arums and hothouse pot plants. The two species often occur together on summer hosts.

Control may be achieved in several ways. ***Amateur*** gardeners and ***professional*** growers should carefully check incoming plants for the presence of the mite, using a hand lens if necessary. A predatory mite, *Phytoseiulus persimilis* is commonly introduced into cucumber, chrysanthemum and tomato crops in spring. For the ***amateur*** there are products containing **fatty acids** that control the mite. For the professional, winter fumigation of greenhouse structures with chemicals such as **formalin** or burning **sulphur** kills off many of the hibernating females. A pesticide containing **abamectin** is commonly used.

Gall mite of blackcurrant (Cecidophyopsis ribis)

Damage. Mites living inside the blackcurrant bud damage the meristem and induce the bud to produce many scale leaves, which gives the bud its unusual swollen appearance (see Figure 14.26). These buds either fail to open or produce distorted leaves. In addition to the mechanical damage, the mite carries the virus responsible for **reversion disease** (see Chapter 15), which stunts the plant and reduces fruit production.

Figure 14.26 (a) Big bud symptoms on **blackcurrant** (b) Erineum mite damage on **grape leaf**

Life cycle. Unlike red spider mite, this species, sometimes called **big-bud mite**, is elongated in shape and is minute (0.25 mm) in size. It spends most of the year living inside the buds of blackcurrants and, to a lesser extent, other *Ribes* species. Breeding takes place inside the buds from June to September, and January to April.

Spread. In May the mites emerge and are spread on silk threads and on the bodies of aphids to infest newly emerging buds and plants.

Control. The mite is controlled in three ways. **Clean planting material** is essential for the establishment of a healthy crop. **Pruning** out of stems with big bud and destruction of reversion-infected plants slows down the progress of the pest. *Amateur* gardeners and *professional* growers can spray a fine formulation of **sulphur** during the May–June period when the mites are migrating. There are no chemical ways to control the mite in the bud. Recently there has been a reported problem on hazel in UK caused by the hazel big bud mite (*Phytopus avellanae*).

Tarsonemid mite (Tarsonemus pallidus)

Damage. Distortion of developing leaves and flowers resulting from small feeding holes and injected toxins are the main symptom of this pest. This may happen to such an extent that leaves and petals are stunted and misshapen, and flowers may not open properly. Plant species affected are *Amaranthus*, *Fuchsia*, pelargonium and cyclamen (the pest is sometimes referred to as 'cyclamen mite') A closely related but distinct strain is found on strawberries.

Life cycle. This spherical mite, only 0.25 mm in length, lives in the unex-panded buds of a wide variety of pot plants. In greenhouses, the adults may lay eggs all the year round, and the **2 weeks life cycle** period can cause a rapid increase in its numbers.

Spread occurs mainly on transported bulbs, corms and plants.

Control. Care should be taken to prevent introduction of infested plants and propagation material into greenhouses. For the *amateur*, there is no recommended chemical product. For the professional grower, the contact acaricide, **abamectin**, is effective against the mite. Addition of a **wetter/spreader** may help the spray penetrate the tight-knit scale leaves of buds.

Other mites

Four other horticulturally important mites require a mention. The fruit tree red spider mite (*Panonychus ulmi*) causes serious leaf mottling of ornamental Malus and apple. Conifer spinning mite (*Oligonychus ununguis*) causes spruce leaves to yellow, and the mite spins a web of silk threads. Bulb-scale mite (*Steneotarsonemus laticeps*) causes internal discoloration of forced narcissus bulbs. Bryobia mite (e.g. *Bryobia rubrioculus*) attacks fruit trees, and may cause damage to greenhouse crops, e.g. cucumbers, if blown in from neighbouring trees.

Other arthropods

In addition to insects and mites, the phylum Arthropoda contains three other horticulturally relevant classes, the Crustacea (woodlice), Symphyla (symphilids) and Diplopoda (millipedes). The Chilopoda (centipedes) superficially resemble millipedes, but are unrelated and are useful general predators.

Woodlouse (Armadillidium nasutum)

The damage is confined mainly to stems and lower leaves of succulent glasshouse crops such as cucumbers, but occasionally young transplants may be nipped. A relative of marine crabs and lobsters, the woodlouse has adapted for terrestrial life, but still requires damp conditions to survive. In damp soils it may number over a million per hectare, and greatly helps the breakdown of plant debris, as do earthworms. In greenhouses, where plants are grown in hot, humid conditions, this species may multiply rapidly, producing two batches of 50 eggs per year. The adults roll into a ball when disturbed. Partial soil sterilization by steam effectively controls woodlice.

Symphilid (Scutigerella immaculata)

In greenhouse crops, **root hairs** are removed and may cause lettuce to mature without a heart. Infectious fungi, e.g. *Botrytis*, may enter the roots after symphilid damage. These delicate white creatures, with 12 pairs of legs, resemble small millipedes. The adult female, 6 mm long, lays eggs in the soil all the year round, and the development through larvae to the adult takes about 3 months. Symphilids may migrate 2 m down into soil during hot, dry weather. The recognition of this pest is made easier by dipping a suspect root and its surrounding soil into a bucket of water and searching for symphilids which float on the water surface.

Millipedes

These elongated, slow-moving creatures are characterized by a thick cuticle and the possession of many legs, two pairs to each body segment. Many species are useful in breaking down soil organic matter, but two pest species, the flat millipede (*Brachydesmus superus*) and a tropical species (*Oxidus gracilus*), can cause damage to roots of strawberries and cucumbers respectively.

Centipedes

These animals resemble millipedes, but are much more active. They help control soil pests by searching for insects, mites and nematodes in the soil.

Nematode pests

This group of organisms, also called **eelworms**, is found in almost every part of the terrestrial environment, and range in size from the large animal parasites, e.g. *Ascaris* (about 20 cm long) in livestock, to the tiny soil-inhabiting species (about 0.5 mm long). Non-parasitic species may be beneficial, feeding on plant remains and soil bacteria, and helping in the formation of **humus** (see p321). The general structure of the nematode body is shown in Figure 14.28. A feature of the plant parasitic species is the **spear** in the mouth region, which is thrust into plant cells. Salivary enzymes are then injected into the plant and the plant juices sucked into the nematode (see Figure 14.27). Nematodes are very active animals, moving in a wriggling fashion in soil moisture films, most actively when the soil is at **field capacity**, and more slowly as the soil either waterlogs or dries out. Five horticulturally important types are described below.

Potato cyst nematode (*Globodera rostochiensis* and *G. pallida*)

Damage. This serious pest is found in most soils that have grown potatoes. Leaves become yellow and plants become stunted (see Figure 14.29) and occasionally die. The distribution of damage in the field is characteristically in patches. Tomatoes grown in greenhouses and outdoors may be similarly affected. The pest may be diagnosed in the field by the tiny, mature white or yellow females seen on the potato roots (a hand-lens is useful for this observation).

Life cycle. A proportion of the eggs in the soil hatch in spring, stimulated by chemicals produced in potato roots. The larvae invade the roots, disturbing **translocation** in xylem and phloem tissues, and sucking up plant cell contents. When the adult male and female nematodes are fully developed, they migrate to the outside of the root, and the now swollen female leaves only her head inserted in the plant tissues (see Figure 14.28). After fertilization, the white female

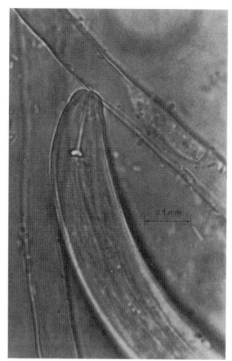

Figure 14.27 **Nematode feeding**: note the spear inside the mouth, used to penetrate plant tissues

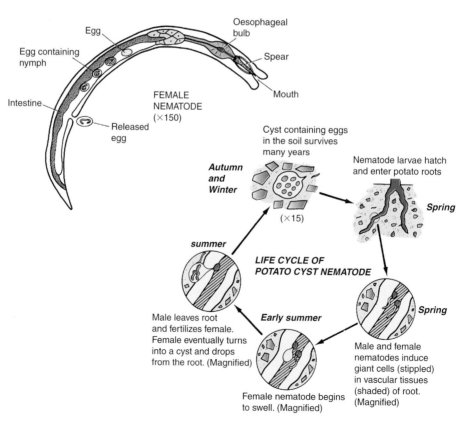

Figure 14.28 The generalized structure of a **nematode** and life cycle of **potato cyst nematode**

Figure 14.29 Potatoes stunted by **potato cyst nematode**

swells and becomes almost spherical, about 0.5 mm in size, and contains 200–600 eggs. As the potato crop reaches harvest, the female changes colour. In *G. rostochiensis* (the golden nematode), the change is from white to **yellow** and then to dark brown, while in the other species, *G. pallida*, no yellow phase is seen. The significance of the species difference is seen in its control. Eventually the dark-brown female dies and falls into the soil. This stage, which looks like a minute brown onion, is called the **cyst**, and the eggs inside this protective shell may survive for 10 years or more.

Spread. This nematode spreads with the movement of infested soil. In peat-soil areas, it often travels along with the wind-blown soil.

Control. Several forms of control are available against this pest. Since it attacks only potatoes and tomatoes, **rotation** is a reliable way of overcoming the problem. It has been found that an average soil population of 10 cysts per 100 gm of soil results in about 3 t/ha decline in yield. Thus a soil count for cysts can indicate to a grower (or gardener) whether a field or plot should be used for a potato crop.

Early cultivars of potatoes are lifted before most nematodes have reached the cyst stage, and thus escape serious damage. Some potato cultivars, such as 'Pentland Javelin' and 'Maris Piper', are **resistant** to golden nematode strains found in Great Britain, but not to *G. pallida*. Since the golden nematode is dominant in the south of England, use of resistant cultivars has proved effective in this region.

There is no chemical control for the *amateur* gardener, but the *professional* grower may use a residual chemical such as **oxamyl**, incorporated as granules into the soil at planting time. This provides economical control when the nematode levels are moderate to fairly high, but is not recommended at low levels because it is uneconomic, or at high levels because the chemical kills insufficient nematodes.

Stem and bulb eelworm (*Ditylenchus dipsaci*)

Damage. The damage caused by this species varies with the crop attacked. Onions show a loose puffy appearance (called bloat); carrots have a dry mealy rot; the stems of beans are swollen and distorted. Narcissus bulbs show brown rings when cut across and their leaves show raised yellow streaks.

Life cycle. This species attacks many plants, e.g. narcissus, onions, beans and strawberries. Several strains are known, but their host ranges are not fully defined. The 1 mm long nematodes enter plant material and breed continuously, often with thousands of individuals in one plant. When an infected plant matures, the nematodes dry out in large numbers, appearing as white fluffy **eelworm wool** that may survive for several years in the soil. Weeds, such as bindweed, chickweed and speedwells, act as alternate hosts to the pest.

Spread. This pest spreads mainly in infested planting material.

Control is achieved in several ways. Control of **weeds** (see chickweed); **rotation** with resistant crops, e.g. lettuce, brassicas (and cereals in commercial bulb growing areas); use of clean, nematode-free **seed** in onions; **warm-water treatment** (see Chapter 16) onions and narcissus at precisely controlled temperatures. All these methods help reduce this serious pest.

Chrysanthemum eelworm (*Aphelenchoides ritzemabosi*)

Damage. The first symptom is blotching and purpling of the leaves, which spreads and becomes a dead brown, **V-shaped** area between the veins. The lower leaves are worst affected. When buds are infested, the resulting leaves may be misshapen. In addition to chrysanthemum, this nematode also attacks *Saintpaulia* and strawberries.

Life cycle. Most species of nematode live in the soil. This 1 mm long nematode spends most of its life cycle inside young leaves of the species mentioned above. The adults move along films of water on the surface of the plant, and enter the leaf through the stomata. They breed rapidly, the females laying about 30 eggs, which complete a life cycle in 14 days. During the winter they live as adults in stem tissues, but a few overwinter in the soil.

Spread. This pest spreads mainly in infested stools.

Control. Greenhouse-grown chrysanthemums are rarely affected, as they are raised from pest-free cuttings. Warm-water treatment of dormant chrysanthemum stools, e.g. at 46°C for 5 min, is very effective for outdoor grown plants. Dispose of all plant debris.

Root knot eelworm (*Meloidogyne spp.*)

Damage. This nematode, a very serious pest in tropical areas of the world, can be important in UK glasshouse production. It causes large root galls, up to 4 cm in size on the roots of plants such as chrysanthemum, *Begonia*, cucumber and tomato, resulting in wilting and poor plant growth.

Life cycle. The swollen female lays 300–1000 eggs inside the root and on the root surface. These eggs can survive in root debris for over a year, and are an important source of subsequent infestations. The larvae hatch from the eggs and search for roots, reaching soil depths of 40 cm and surviving in damp soil for several months. On entering the plant, the nematode larvae stimulate the adjoining root cells to enlarge. These cells block movement of water to the root stele (see **root structure**), which results in wilting symptom so commonly seen with this pest.

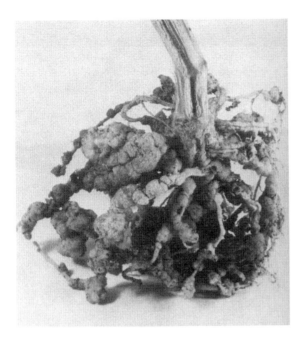

Figure 14.30 **Root knot nematode** damage on **cucumber**

Spread is mainly caused by the movement of infested soil.

Control. For the *amateur,* **resistant** tomato rootstocks, e.g. KVNF, may be used on grafted plants. For the *professional*, nutrient film and soilless methods of growing reduce the pest's likely importance in a crop. **Partial steam sterilization** effectively controls the nematode only if the soil temperature reaches 99°C to a depth of 45 cm. Less stringent sterilization often results in a severe infestation in the next crop. Chemical sterilization of soil with the chemical fumigant **metam-sodium** is effective if a damp seedbed tilth is first prepared. A 2.5 cm layer of clean soil or compost placed around roots of infested plants allows some new root growth. Care should be taken not to transfer infested soil (together with transplants) from one greenhouse to another.

Migratory plant nematodes

The species of nematodes described above spend most of their life cycle inside plant tissues (**endoparasites**). Some species, however, feed only from the outside of the root (**ectoparasites**). The dagger nematodes (e.g. *Xiphinema diversicaudatum*) and needle nematodes (e.g. *Longidorus elongatus*), which reach lengths of 0.4 and 1.0 cm respectively, attack the young roots of crops such as rose, raspberry and strawberry, and cause stunted growth. In addition, these species transmit the important **viruses**, arabis mosaic on strawberry and tomato black ring on ornamental cherries. The nematodes may survive on the roots of a wide variety of weeds.

Control is achieved by professional growers by the injection of a fumigant chemical, e.g. **dichloropropene** or incorporation of **dazomet** in fallow soils.

Check your learning

1. Define the term 'pest'.

2. Describe the damage caused by a chosen mammal pest.

3. Describe the life cycle of this pest.

4. Describe how the life cycle of this pest is related to its control.

5. Describe the available control measures for this pest.

6. Describe the damage caused by a chosen invertebrate (slug, insect, mite, or nematode) pest.

7. Describe the life cycle of this pest.

8. Describe how the life cycle of this pest is related to its control.

9. Describe the available control measures for this pest.

Further reading

Alford, D.V. (1984). *A Colour Atlas of Fruit Pests*. Wolfe Science.

Alford, D.V. (2002). *Pests of Ornamental trees, shrubs and flowers*. Wolfe Publishing.

British Crop Protection Council (2007). *The UK Pesticide Guide*.

Brown, L.V. (2008). *Applied Principles of Horticultural Science*. 3rd edn. Butterworth–Heinemann.

Buckle, A.P. and Smith, R.H. (1996). *Rodent Pests and Their Control*. Oxford University Press.

Buczacki, S. and Harris, K. (1998). *Guide to Pests, Diseases and Disorders of Garden Plants*. Collins.

Dent, P. (1995). *Integrated Pest Management*. Chapman & Hall.

Gratwick, M. (1992). *Crop Pests in the UK*. Chapman & Hall.

Greenwood, P. and Halstead, A. (2007). *Pests and Diseases*. Dorling Kindersley.

Heinz, K. (2004). *Biocontrol in protected culture*. Ball Publishing.

Hope, F. (1990). *Turf Culture: A Manual for Groundsmen*. Cassell.

Ingram, D.S. *et al.* (2002). *Science and the Garden*. Blackwell Science Ltd.

Savigear, E. (1992). *Garden Pests and Predators*. Blandford.

Scopes, N. and Stables, L. (1992). *Pests and Disease Control Handbook*. British Crop Protection Council.

Chapter 15 Horticultural diseases and disorders

Summary

This chapter includes the following items:

- Definition of term 'disease'
- Damage caused by fungal diseases
- Life cycles of fungal diseases
- Relationship between life cycle and control
- Control measures for disease
- Damage caused by bacteria
- Minimizing bacterial problems
- Damage caused by viruses
- Minimizing virus spread
- Definition of term 'physiological disorder'
- Symptoms of physiological disorders
- Methods of avoiding physiological disorders

Figure 15.1 **Powdery mildew** on **courgette**

A disease is an unhealthy condition in a plant caused by a fungus, bacterium or virus.

Below are described some of the most important horticultural diseases caused by fungi, bacteria and viruses.

Structure and biology of fungi, bacteria and virus

Fungi, commonly called moulds, cause serious losses in all areas of horticulture. They are thought to have common ancestors with the filamentous algae, a group including the present-day green slime in ponds. Some details of their classification are given in Chapter 4.

A fungus is composed, in most species, of microscopic strands (**hyphae**) which may occur together in a loose structure (**mycelium**), form dense resting bodies (**sclerotia**, see Figure 15.2) or produce complex underground rootlike strands (see **rhizomorphs**). The club root group of fungi is quite different, producing a jelly-like structure (**plasmodium**) inside the cells of the host plant.

The hyphae in most fungal species are capable of producing spores. Wind-borne spores are generally very small (about 0.01 mm), not sticky and often borne by hyphae protruding above the leaf surface, e.g. grey mould, so that they catch turbulent wind currents. Water or rain-borne spores are often sticky, e.g. damping off. Minute asexual spores produced without fusion of two hyphae commonly occur in seasons favourable for disease increase, e.g. humid weather for downy mildews and dry, hot weather for powdery mildews. Sexual spores, produced after hyphal fusion, commonly develop in unfavourable conditions, e.g. a cold, damp autumn. They may be produced singly, as in the downy mildews, or in groups within a protective hyphal spore case, often observable to the naked eye, as in the powdery mildews. Different genera and species are identified by microscopic measurement of the shape and size of the spores or of the spore-bearing spore cases.

Horticulturists without microscopes must use symptoms as a guide to the cause of the disease. While disease-causing or parasitic fungi are the main concern of this chapter, in many parts of the environment there are useful saprophytic fungi that break down organic material such as dead roots, leaves, stems and sometimes decaying tree stumps (see Chapter 3) and useful **symbiotic** fungi that may live in close association with the plant, e.g. **mycorrhizal** fungi in fine roots of conifers (see Chapter 18).

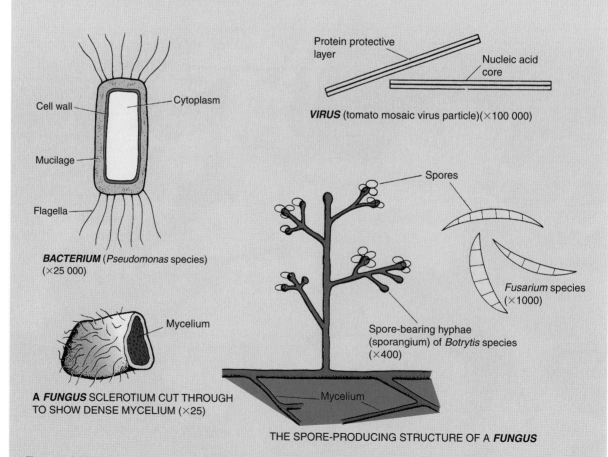

Cell wall — Cytoplasm

Mucilage

Flagella

BACTERIUM (*Pseudomonas* species) (×25 000)

Mycelium

A FUNGUS SCLEROTIUM CUT THROUGH TO SHOW DENSE MYCELIUM (×25)

Protein protective layer

Nucleic acid core

VIRUS (tomato mosaic virus particle)(×100 000)

Spores

Fusarium species (×1000)

Spore-bearing hyphae (sporangium) of *Botrytis* species (×400)

Mycelium

THE SPORE-PRODUCING STRUCTURE OF A **FUNGUS**

Figure 15.2 Microscopic details of a **virus**, **bacterium** and **three fungi**. Note the relative sizes of the organisms.

The spore of a leaf-infecting fungal parasite, after landing on the leaf in damp conditions, produces a germination tube which, being delicate and easily dried out, must enter through the **cuticle** or **stomata** within a few hours before dry, unfavourable conditions recur. Within the leaf, the hyphae grow, absorbing food until, within a period of a few weeks they produce a further crop of spores (see Figure 15.4). Leaf diseases such as potato blight often increase very rapidly when conditions are favourable. Roots may be infected by spores, e.g. in damping off; hyphae, e.g. wilt diseases; sclerotia, e.g. Sclerotinia rot; or rhizomorphs, e.g. honey fungus. Root diseases are generally less affected by short periods of unfavourable conditions and often increase at a slower, more constant rate.

Phyllosphere

On the surface of leaves and stems a population of micro-organisms (mainly bacteria) lives which occupy a microhabitat commonly called the **phyllosphere** (see also **rhizosphere** p322). These bacteria may be 'casual' or 'resident'. Casual organisms such as *Bacillus* spp. mainly arrive from soil, roots and water, and are more common on leaves closer to the ground. These species are capable of rapid increase under favourable conditions, but then may decline. Resident organisms such as *Pseudomonas* spp. may be weakly parasitic on plants, but more commonly persist (often for considerable periods) without causing damage on a wide variety of plants.

There is increasing evidence that phyllosphere bacteria may reduce the infection of diseases such as powdery mildews, *Botrytis* diseases on lettuce and onion, and turf grass diseases. Practical disease control strategies by phyllosphere organisms have not been developed, but there remains the general principle that a healthy, well-nourished plant will be more likely to have organisms on the leaf surface available to reduce fungal infection.

Figure 15.3 White rot on onion. Note the black sclerotia which enable this disease to survive long periods in the soil

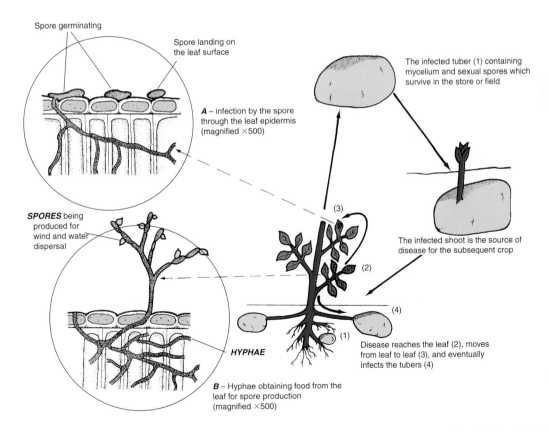

Figure 15.4 Infection and life cycle of potato blight fungus. The left side illustrates microscopic infection of the leaf. The right side shows how the disease survives and spreads

Fungi

The **classification** of fungi referred to in Chapter 4 has practical implications in understanding fungal disease life cycles and control. Species within a fungal division often have similar methods of spread, and of survival. Their similarity of spore and hypha structure and of biochemistry also means that they are often controlled by a common fungicide active ingredient. For this reason, the appropriate fungal division is given for each disease as it appears (e.g. Zygomycota is mentioned against potato blight).

Leaf and flower diseases

Potato blight (*Phytophthora infestans*)

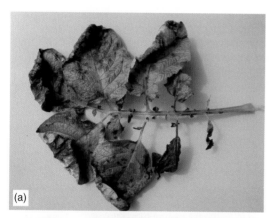

Figure 15.5 **Potato blight**: (a) Leaf symptoms of potato blight on potato, (b) Fruit symptoms of blight on tomato

This fungus is a member of the Zygomycota division of fungi (see p74).

Damage. This important disease is a constant threat to potato production; it caused the Irish potato famine in the nineteenth century. The first symptoms seen in the field are yellowing of the foliage, which quickly goes **black** and then produces a white bloom on the under surface of the leaf in damp weather. The stems may then go black, killing off the whole plant. The tubers may show dark surface spots that, internally, appear as a **deep dry red-brown rot**. This fungus may attack tomatoes, the most notable symptom being the **dark-brown blisters** on the fruit (Figure 15.5).

Life cycle. The fungus survives the winter as mycelium and sexual spores in the tubers (see Figure 15.4, p235). The spring emergence of infected shoots results in the production of asexual spores.

Spread. The spores are spread by wind and land on potato leaves or stems. They can, after infection, result in a further crop of spores within a few days under warm, wet weather conditions. The disease can spread very quickly. Later in the crop, badly infected plants may have tuber infection as rainfall washes spores down into the soil.

Control. *The amateur* gardener should use clean seed, choose resistant varieties, and apply a protective fungicide such as copper sulphate (Bordeaux mixture) before damp conditions appear. *The professional* grower similarly needs to use clean seed. Removal and herbicidal destruction, (e.g. using **dichlobenil**), of diseased tubers from stores or 'clamps' prevents disease spread. Knowledge of the disease's moisture requirement leads to better control. DEFRA are able to measure both the duration of atmospheric relative humidity greater than 92 per cent, and mean daily temperatures greater than 10°C (together known as **critical periods**), and to issue forecasts of potato blight outbreaks. In this way, protectant sprays of chemicals such as **mancozeb** can be applied before infection can take place.

Resistant potato cultivars prevent rapid build-up of disease, although resistance may be overcome by newly occurring fungal strains. Early potato cultivars usually complete tuber production before serious blight attacks, while main-crop top growth may deliberately be killed off with foliage-acting herbicides, such as **diquat**, to prevent disease spread to tubers.

A curative, systemic fungicide such as **benalaxyl** penetrates the leaf and kills the infecting mycelium. This ingredient has within its formulation a protectant ingredient (in this case **mancozeb**) to kill off most germinating spores and thus reduce the development of fungus resistance to the systemic fungicide.

Downy mildew of cabbage and related plants (*Perenospora brassicae*)

This fungus is a member of the Zygomycota group of fungi.

Damage. This serious disease causes a white bloom mainly on the under surface of leaves (see Figure 15.6) which present a more favourable humid microclimate for infection and spore production than the upper leaf surface. Ornamental cruciferous plants such as stocks and wallflowers, brassicas such as cabbage and occasionally weeds such as shepherd's purse are attacked by this fungus. The disease is most damaging when seedlings are germinating, particularly in spring when the young infectable tissues of the host plant and favourable damp conditions may combine to kill off a large proportion of the developing plants.

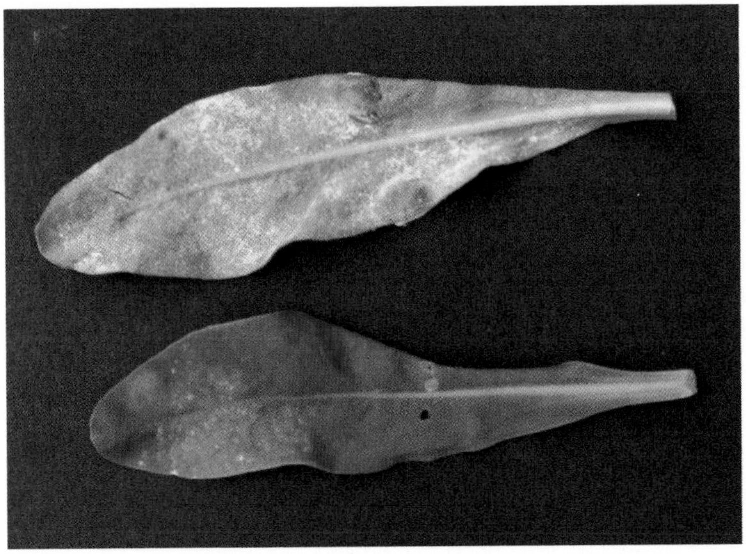

Figure 15.6 **Leaf symptoms of cabbage downy mildew** on stock. Lower leaf symptoms (top), upper leaf symptoms (bottom)

Life cycle and spread. Asexual spores (zoospores) are produced on microscopic structures on the lower leaf surface, mainly in spring and summer, and are spread by wind currents. Thick-walled sexual spores (**oospores**) produced within the leaf tissues fall to the ground with the death of the leaf and survive the winter to initiate the spring infections, when rain splash carries the spores up to the lower leaf surface of seedlings and young plants.

Control. It is not advisable to grow successive brassicas in the same field, and particularly not to sow in spring next to overwintered crops.

The *amateur* grower can use a product containing the protectant fungicide, **mancozeb**. The *professional* horticulturist uses a protective chemical such as chlorothalinal at the seedling stage to kill off spores on the leaf. A combination of a systemic fungicide **metalaxyl** plus chlorothalinal gives better control, while reducing the development of fungus strains resistant to the systemic ingredient.

Other crops such as lettuce and onions are attacked by different downy mildews (*Bremia lactucae* and *Perenospora destructor* respectively) and no cross-infection is seen between crops belonging to different plant families.

Powdery mildew of ornamental Malus and apple (*Podosphaera leucotricha*)

This fungus belongs to the Ascomycota group of fungi.

Damage. Powdery mildews should not be confused with downy mildews. Powdery mildew is distinguished by its dry powdery appearance, most commonly found on the **upper** surface of the leaf, and by its preference for **hot, dry weather** conditions (see examples of powdery mildew in Figure 15.7).

Life cycle and spread. The disease survives the winter as mycelium within the buds which often appear small and shrivelled on twigs which often have a dried, **silvery** appearance. The emergence of the mycelium with the germinating buds in spring results in a white bloom over the young leaves (**primary mildew**). As the spring progresses, asexual spores produced in chains on the upper surface of the leaf are spread as individual spores by wind and cause the destructive **secondary** mildew. This stage of the fungal life cycle involves an **external** mycelium covering the leaf surface, but not entering into the internal leaf tissues other than to sink small peg-like structures (which suck out the leaf's moisture and may cause premature leaf drop). Flowering normally occurs before the secondary infection stage, but infection in young apple fruit may produce a rough skin (**russeting**). This organism may affect other species of fruit such as pears, quinces, medlars and ornamental *Malus*.

Powdery mildews in autumn may produce sexual, dark-coloured spore cases (**cleistothecia**) on the leaf, about 1 mm in size. (Although not important in most horticultural crops as an overwintering stage, they may assume a vital role in powdery mildew of cereals, the sexual spores having the potential to form new strains resistant to fungicides, and new strains capable of overcoming the plant's genetic resistance.)

Control is achieved by two main 'methods'. Pruning of silvered shoots can eliminate a good proportion of the primary mildew. The amateur and professional horticulturist can use a systemic fungicide ingredient such as **myclobutanil** for spring and summer sprays, whilst the professional also has a protective fungicide option such as **dinocap**.

Figure 15.7 **Powdery mildew diseases**: (a) Primary apple powdery mildew on apple shoot. Affected shoot (left), healthy shoot (right), (b) Rose powdery mildew, (c) Gooseberry powdery mildew on fruit.

Other species of powdery mildew commonly occur in horticulture, e.g. *Sphaerotheca pannosa* on rose (Figure 15.7) and *S. fuliginea* on cucumber. Cross-infection between these crops does not occur.

Black spot of roses (*Diplocarpon rosae*)

This fungus belongs to the Ascomycota group of fungi.

Damage. This common disease in garden and greenhouse roses is first seen as dark-brown leaf spots which may be followed by general leaf yellowing and then leaf drop (see Figure 15.8). The infection of young shoots has a slow weakening effect on the whole plant.

Figure 15.8 **Black spot of rose**. Note the leaf-yellowing symptom that accompanies the black spot

Life cycle and spread. Asexual spores (produced within spore cases embedded in the leaves) are released in wet and mainly warm weather conditions, and are then spread a few metres by rain drops or irrigation water before beginning the cycle of infection again. No overwintering sexual stage is seen in Britain, and it is probable that asexual spores surviving in autumn-produced wood or in fallen leaves begin the infection process again the following spring.

Control. Removal of fallen leaves is a very important aspect of control. Resistance is not common in rose cultivars. Both *amateur* and *professional* horticulturists can use a systemic fungicide ingredient such as **myclobutanil**.

The *professional* has also a protectant ingredient, **captan** which, while very effective against this disease, is limited in its action by the vigorous leaf expansion in late spring and summer which leaves unprotected areas of foliage. The addition of a **wetter/spreader** may improve control by spreading the active ingredient more effectively over the leaf surface. In industrial areas the sulphur dioxide in the air may be at a sufficient concentration to help reduce black spot.

Carnation rust (*Uromyces dianthi*)

Rust fungus belongs to the Basidiomycota group of fungi. (The rusts are a distinctive group of fungi which may have very complex life cycles involving five spore-forms within the same fungal species. When these different spore-forms occur on more than one host, e.g. blackcurrant rust (*Cronartium ribicola*), which attacks both blackcurrants and five-needle pines, the close planting of the two crops may give rise to high rust levels. For this reason, in some North American forested regions, blackcurrants may not be planted. Most horticultural rust species are found on only one host species.)

Damage. Carnation rust first appears as an indistinct yellowing of the leaf and stem, soon turning into an elongated raised brown spot which yields brown dust (spores) when rubbed (see Figure 15.9).

Life cycle and spread. The more common thin-walled spores (uredospores) are spread by wind currents and infect the leaf by way of the stomata in damp conditions. The less common thick-walled black spores (teleutospores) may survive and overwinter in the soil. The resistance of carnation cultivars varies, while related species, e.g. pinks or sweet williams, are rarely affected.

Control. Preventative control includes the use of rust-free cuttings, sterilization of border soils and careful maintenance of greenhouse ventilators to prevent damp patches occurring in the crop. *Amateur* and *professional* horticulturist are able to spray products containing a systemic ingredient, **myclobutanil**, and products containing the protectant **mancozeb**.

Figure 15.9 **Rust diseases**: (a) Rose rust, (b) Bean rust, (c) Leek rust, (d) Groundsel rust which affects cinerarias

The occurrence of white rust (*Puccinia horiana*) on chrysanthemums has created serious problems for the horticultural industry and for gardeners, because of the ease with which the disease is carried in cuttings and its speed of increase and spread. Other common rusts are found on antirrhinum, hollyhock, rose and leek, and more recently European pear rust, with an alternate host of juniper, has become more common in the UK.

Stem diseases

Grey mould (*Botrytis cinerea*)

This fungus is classified in the Deuteromycota group of fungi.

Damage. This disease is most commonly recognized by the fluffy, light-grey fungal mass which follows its infection. In lettuce, the whole plant rots off at the base. The plant turns yellow and dies. In tomatoes, infection in damaged side shoots, and yellow spots (**ghost spots**) on the unripe and ripe fruit are found. In many flower crops, e.g. chrysanthemums, infected petals show purple spots which, in very damp conditions, lead to mummified flower heads. This disease may affect many crops.

Life cycle and spread. Grey mould normally requires wounded tissue for infection, which explains its importance in crops which are de-leafed, e.g. tomatoes, or disbudded, e.g. chrysanthemums. Damp conditions are essential for infection and spore production. The millions of spores are carried by wind to the next wounded surface. Black sclerotia, about 2 mm across, produced in badly infected plants, often act as the overwintering stage of the disease after falling to the ground, and are particularly infective in unsterilized soils on young seedlings and delicate plants, e.g. lettuce.

Control. Preventative control may involve soil sterilization. Strict attention to greenhouse humidity control (particularly overnight) reduces the dew formation which is so important in the organism's infection. Cutting out of infected tissue is possible in sturdy stems, e.g. in tomatoes. *Amateur* gardeners at present have no effective chemical control against this disease. The *professional* grower has two protectant fungicide ingredients, **iprodione** and **chlorothalinal**. As well as being sprayed, **iprodione** can also be applied as a paste to cut plant surfaces.

Apple canker (*Nectria galligena*)

This fungus belongs to the Ascomycota group of fungi.

Damage. This fungus causes sunken areas in bark of both young and old branches of ornamental Malus, apples or pears (see Figure 15.10). Poor shoot growth is seen, and the wood may fracture in high winds.

Life cycle and spread. The fungus enters through leaf scars in autumn or through pruning wounds during winter. Care is therefore necessary to prevent infection, particularly in susceptible apple cultivars, e.g. 'Cox's Orange Pippin', by avoiding pruning in damp conditions. Spread is by rain splash.

Control. Removal of cankered shoots may be necessary to prevent further infection, while in cankers of large branches cutting out of brown infected tissue may allow continued use of the branch. Removed tissue should be burnt. *Amateur* and *professional* horticulturists may apply a spray of **copper** (Bordeaux solution) at bud burst (spring) and leaf fall (autumn) to prevent entry of germinating spores.

Dutch elm disease (*Ophiostoma novo-ulmi*)

This fungus belongs to the Ascomycota group of fungi.

Damage. The first symptom of this disease is a yellowing of foliage in one part of the tree in early summer. The foliage then dies off progressively from this area of the tree, often resulting in death within three months. Trees that survive 1 year's infection may fully recover in the following year. All common species and hybrids of elm growing in Great Britain are susceptible to the disease (see Figure 15.11).

Life cycle and spread. The causative fungus lives in the xylem tissues of the stem, and produces a poison that results in a blockage of xylem

Figure 15.10 Apple canker

Figure 15.11 (a) **Dutch Elm Disease**. A young elm tree showing typical yellowing and wilting of leaves, (b) Tunnelling inside elm bark caused by Scolytus beetle larvae

vessels, causing the wilt that is observed. Associated with this disease are two black and red wood-boring species of beetle, *Scolytus scolytus* (5 mm long) and *Scolytus multistriatus* (3 mm long). These beetles bore into elm stems leaving characteristic 'shot holes'. Eggs are then laid, and a fan-shaped pattern of galleries is produced under the bark by the larvae. Later, as adults, they emerge from the wood, carrying sticky asexual and sexual spores of Dutch elm disease to continue its spread to other uninfected elms. Graft transmission of the disease from tree to tree by roots commonly occurs in hedge-grown elms.

Control. For the professional horticulturist, the cost of preventative control on a large number of uninfected trees is uneconomical. However, high pressure injection of a systemic fungicide such as **thiabendazole** which travels upwards through the xylem tissues has proved successful in some cases. Selections of hybrid elms have proved to have high levels of resistance. Examples are the Dutch/French cultivar 'Lutéce' with a complex parentage; the Italian cultivar 'Planio' with resistant Siberian Elm male parentage; and the American 'Princeton Elm' which has shown resistance since the 1920s. The European White Elm (*Ulmus laevis*) exhibits a resistance to the vector beetles, not the fungus itself.

After an estimated 80 per cent removal of the elm population in the UK since the 1970s by Dutch Elm Disease there has been an understandable lack of confidence in planting elms of any species or cultivar around the countryside, especially the English elm. However, there is a growing realization that the native elm contributes considerably to plant and animal biodiversity in the UK. For example, the elm maintains one lichen species (Orange-fruited Elm lichen) and one butterfly (the White letter Hairstreak butterfly) which are both wholly dependant on the elm.

Bleeding Cankers (*Phytophthora species*)

In recent years the following species have been reported on a range of trees: horse chestnut and sweet chestnut affected by *P. cambivora*, and Oak, beech and tulip tree affected by *P. ramorum and P. kernovii*. Symptoms of the problem involve a **dark ooze** coming from the bark. After peeling away the outer bark, deep red, dead patches of inner bark are seen, often with a sharp distinction between dead and living bark. Oaks may be suddenly killed by *P. ramorum*. Control is at the moment limited to scraping away the outer bark in hot weather to allow drying and healing of tissues.

Root diseases

Club root (*Plasmodiophora brassicae*)

This fungus is classified into a quite separate group of fungi, the Plasmodiophorales.

Damage. It causes serious damage to most members of the Cruciferae family, which includes cabbage, cauliflowers, Brussels sprouts, stocks and Alyssum. Infected plants show signs of wilting and yellowing of older leaves, and often severe stunting. On examination, the roots appear stubby and swollen (see Figure 15.12), and may show a wet rot.

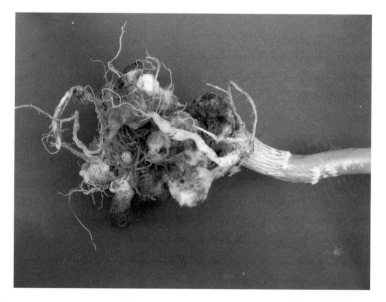

Figure 15.12 Club root on cabbage

Life cycle and spread. The club root organism survives in the soil for more than five years as minute spores which germinate to infect the root hairs of susceptible plants. The fungus is unusual in forming a jelly-like mass (**plasmodium**), not hyphae, within the plant's root tissues. The plasmodium stimulates root cell division and causes cell enlargement, which produces swollen roots. The flow of food and nutrients in phloem and xylem is disturbed, with consequent poor growth of the plant. With plant maturity the spores produced by the plasmodium within the root are released as the root rots.

The disease is favoured by high soil moisture, high soil temperatures and acid soils. Although this fungus does not spread much in undisturbed soils it can be easily carried on infected plants, or on tools and wheels of machinery. In peat soil-growing areas, high winds may carry the disease a considerable distance.

Control. Several preventative control measures may be used by the *amateur* gardener and *professional* grower. Rotation greatly helps by keeping cruciferous crops away from high spore levels in the soil. Liming of soil greatly inhibits spore activity (see **soil fertility**). Recently released cultivars of late-summer cabbage are claimed to have strong resistance to club root. Autumn-sown plants establish in soil temperatures unfavourable to the disease and are normally less infected. Compost made from infected brassica plants should be avoided. The *professional* grower using transplants can prepare a seedbed previously sterilized with a granular product containing **dazomet**, which would ensure healthy transplants.

Damping off (*Pythium* and *Phytophthora species*)

These two fungi belong to the Zygomycota.

Damage. These two similar genera of fungi cause considerable losses to the delicate **seedling** stage. The infection may occur below the soil surface, but most commonly the emerging seedling plumule is infected at the soil surface, causing it to topple (see Figure 15.13). Occasionally the roots of mature plants, e.g. cucumbers, are infected, turn brown and soggy, and the plants die. Rose plants often have high levels of *Pythium* around their roots as they age. Although the mature plant is not seriously affected, it is a common experience that on removal of the plant and replacement with a young rose plant, there is a quite rapid decline in its vigour, called **rose-sickness**.

Life cycle and spread. Both *Pythium* and *Phytophthora* occur naturally in soils as saprophytes, but under damp conditions they produce the asexual spores that cause infection. These spores are spread by water. Sexual spores (oospores) are produced in infected roots (mostly in autumn) and may survive several months of dry or cold soil conditions.

Control. Prevention control is best achieved (both for the *amateur* gardener and the *professional* grower) against

Figure 15.13 Damping off on seedlings. Note the shrivelled, papery appearance of the leaves on the infected plants

these diseases by providing a disease-free growing medium. This may be produced by using fresh compost, by partial sterilization of soil with heat, or (for the ***professional*** grower) by a sterilant such as **dazomet**. Seed producers often coat crop seed with a protective seed dressing (see also p284) such as **thiram** to prevent early infection.

Water tanks with open tops, harbouring rotting leaves, are a common source of infected water and should be cleaned out regularly. Sand and capillary matting on benches in greenhouses should be regularly washed with hot water. The use of door mats soaked in a sterilant such as dilute **formalin** may prevent foot spread of the organisms from one greenhouse to another. Waterlogged soils should be avoided, as these fungi increase most rapidly under these conditions. The ***amateur*** gardener may use a **copper** formulation (known as Cheshunt mixture) as a drench to slow down the increase of damping off. ***Professional*** growers use a product containing **etridiazole**, which may be mixed in with composts, or drenched on to seed trays, pots or border soil growing young plants.

Conifer root rot (*Phytophthora cinnamomi*)

This fungus belongs to the Zygomycota group of fungi.

Damage. This soil-inhabiting fungus is most commonly a problem in nursery stock production nurseries. It causes the foliage of plants to turn grey-green, then brown and eventually to die off completely (see Figure 15.14). Sliced roots show a chestnut brown rot, with a clear line between infected and non-infected tissues. Two hundred plant species, including *Chaemaecyparis*, *Erica* and *Rhododendron* species may be badly attacked.

Figure 15.14 **Conifer root rot**. Note the different shades of colour in individual trees, representing different stages of infection

Life cycle and spread. The disease is commonly introduced on infected stock plants or contaminated footwear. It multiplies most rapidly under wet conditions, within a temperature range of 20°C and 30°C, infecting the root tissues and producing numerous asexual spores, which may be spread by water currents to adjacent plants. Sexual oospores produced further inside the root are released on decay and allow the fungus to survive in the soil for several months without a host.

Control. Preventative control (see **hygienic growing**, p269) is important. Reliable stock plants should be used. Water supply should be checked to avoid contamination. The stock plant area should be elevated slightly higher than the production area to prevent infection by drainage water. Rooting trays, compost and equipment, e.g. knives and spades, should be sterilized (e.g. with formalin) before use. Placing container plants on gravel reduces infection through the base of the pot. The chemical, **etridiazole**, incorporated in compost protects the roots, but does not kill the fungus. Some species, such as *Juniperus horizontalis*, have some tolerance to this disease.

Honey fungus (*Armillaria mellea*)

This fungus belongs to the Basidiomycota group of fungi.

Damage. This fungus primarily attacks trees and shrubs, e.g. apple, lilac and privet. In spring the foliage wilts and turns yellow. Death of the plant may take a few weeks or several years in large trees. Confirming symptoms are the white mycelium, rhizomorphs and toadstools mentioned below (see Figure 15.15).

Figure 15.15 **Honey fungus**. Note the dense clump of honey-coloured toadstools, and also the fungal strands (rhizomorphs) spreading out from the clump.

Life cycle and spread. The infection process involves **rhizomorphs** (sometimes referred to as 'bootlaces'), which radiate out underground from infected trees or stumps for a distance of 7 m, to a depth of 0.7 m. The infected stump may remain a serious source of infection for twenty years or more. The rhizomorphs are the only means of spread for this disease. The nutrients they are able to conduct provide the considerable energy required for the infection of the tough, woody roots. **Mycelium**, moves up the stem **beneath** the bark to a height of several metres and is visible (when the bark is pulled away) as white sheets, smelling of mushrooms. In autumn, clumps of light-brown **toadstools** may be produced, often at the base of the stem. The millions of spores produced by the toadstools are not considered to be important in the infection process. Honey fungus often establishes itself in newly planted trees and shrubs that have been planted too deeply. Deep planting produces less vigorous plants that are more vulnerable to infection. Vigour is reduced because feeding roots which ideally should be growing near the surface of the soil have been located in the subsoil.

Control is difficult. Some genera of plants are less likely to be infected (see Table 15.1). Removal of the disease source, the infected stump, is strongly recommended. In large stumps which are hard to remove a surrounding trench is sometimes dug to a depth of 0.7 m to prevent the

Table 15.1 Levels of resistance to Honey fungus in garden shrubs and trees

Plant species	Latin name	Resistance level
Maple	Acer spp.	susceptible
Box elder	Acer negundo	very resistant
Birch	Betula spp.	susceptible
Box	Buxus sempervivans	resistant
Cedar	Cedrus spp.	susceptible
Cypress	Chamaecyparis lawsoniana	susceptible
	X Cupressocyparis leylandii	susceptible
Eleagnus	Eleagnus spp.	resistant
Holly	Ilex aquifolium	resistant
Privet	Ligustrum spp.	susceptible
Lonicera	Lonicera nitida	resistant
Mahonia	Mahonia spp.	resistant
Apple	Malus spp.	susceptible
Pine	Pinus	susceptible
Cherry and plum	Prunus spp.	susceptible
Laurel	Prunus laurocerasus	resistant
Rhododendron	Rhododendron	susceptible
Sumach	Rhus typhina	resistant
Lilac	Syringa spp.	susceptible
Tamarisk	Tamarix spp.	resistant
Yew	Taxus baccata	very resistant

progress of rhizomorphs. Loosening soil with a fork and then applying a sterilant, e.g. **formalin** in a diluted state, may be applied in situations where there are no crops.

Fusarium patch on turf (now called *Microdochium nivale*)

This belongs to the Deuteromycota group of fungi.

Damage. This disease appears as irregular circular patches of yellow then dead brown grass up to 30 cm in diameter on fine turf. These patches eventually merge (see Figure 15.16). Under extreme damp conditions, dead leaves become slimy and then are covered with a light pink bloom, most evident between May and September.

Figure 15.16 **Fusarium patch on turf**. Note the area of dying turf. In the earlier stages of the disease, distinct circular patches about 30 cm across are seen.

Life cycle and spread. Infection of the leaves by spores and hyphae occurs most seriously between 0°C and 8°C, conditions that are found under a layer of snow (hence its other name, **snow mould**). However, conditions of high humidity at temperatures up to 18°C may result in typical patch symptoms. Spread is by means of water-borne asexual spores under conditions such as autumn dew with no wind. The fungus can survive in frosty or dry summer conditions as dormant mycelium in dead leaf matter or newly infected leaves.

Control. Preventative control measures are important. Avoid high soil nitrogen levels in autumn, as this promotes lush, susceptible growth in autumn and winter. Avoid thatchy growth of the turf, as this encourages high humidity and thus favours the disease organism. The groundsman can drench preventative fungicide such as **iprodione** in autumn to slow down infection of the fungus. Summer-applied systemic fungicide such as **thiophanate methyl** is able during the actively growing period of the year to move within the plants and achieve curative control.

Vascular wilt diseases (*Fusarium oxysporum* and *Verticillium dahliae*)

These fungi belong to the Deuteromycota group of fungi.

Damage. These two organisms infect the **xylem** tissues of horticultural plants, causing the leaves to wilt in hot conditions, a symptom which can also be caused by other factors, e.g. lack of soil moisture (see wilt) and nematode infestation (see root knot nematode p230). The wilt diseases can be recognized by yellowing and eventual browning of the **lower** leaves (see Figure 15.17) and by brown staining of the xylem tissue when it is exposed with a knife. *Verticillium* may attack a wide range of plants, e.g. dahlia, strawberry, lilac, tomato and potato, so that rotation is not a feasible control measure. *Fusarium oxysporum*, however, exists in many

Figure 15.17 **Fusarium wilt on beans**. Note the leaf yellowing

distinct forms, each specializing in crops in different plant families, e.g. tomato, cucumber, bean or carnation.

Life cycle and spread. Both organisms may live as **saprophytes** in the soil. *Fusarium* survives unfavourable conditions as thick-walled asexual spores, while *Verticillium* forms small sclerotia. Infection by both genera occurs through young roots or after nematode attack in older roots. The fungal hyphae enter the root xylem tissue and then move up the stem, sometimes reaching the flowers and seeds. The diseases are spread by water-borne asexual spores. The two fungi have different temperature preferences. *Verticillium* more commonly attacks in springtime, having an optimum infection temperature of 20°C, while *Fusarium* is more common in summer, with an optimum temperature of 28°C.

Control. Control is often necessary in greenhouse crops. Infected crop residues should be carefully removed from the soil at the end of the growing season. The *amateur* gardener or *professional* grower may choose to use peat bags instead of soil. The professional may use partial soil sterilization by steam, or a chemical sterilant such as **metam-sodium**.

In unsterilized soils, *professional* growers may use resistant rootstocks, e.g. in tomatoes, which are grafted onto scions of commercial cultivars. Rotation may be employed against a *Fusarium oxysporum* attack, as different forms attack different crops. Careful removal of infected and surrounding plants, e.g. in carnations, may slow down the progress of the diseases, especially if the soil area is drenched with a systemic chemical such as **carbendazin**, which reduces the infection in adjacent plants.

Bacteria

These minute organisms (see Figure 15.2) measure about 0.001 mm and occur as single cells that divide rapidly. They are important in the conversion of soil organic matter (see Chapter 18), but may, in a few parasitic species, cause serious damage or losses to horticultural plants. Some details of their classification are given in Chapter 4.

Fireblight (*Erwinia amylovora*)

Damage. This disease, which first appeared in the British Isles in 1957, can cause serious damage on members of the Rosaceae family. Individual branches wilt and the leaves rapidly turn a 'burnt' chestnut brown. When the disease reaches the main trunk, it spreads to other branches and may cause death of the tree within six weeks of first infection, the general appearance resembles a burnt tree, hence the name of the disease. Badly infected plants produce a bacterial slime on the outside of the branches in humid weather. On slicing through an infected stem, a brown stain will often be seen. Pears, hawthorn and *Cotoneaster* are commonly attacked, while apples and *Pyracantha* suffer less commonly.

Life cycle and spread. The bacterium is spread by bees as they pollinate, by harmful insects such as aphids and by small droplets of rain. Humid conditions and temperatures in excess of 18°C, which occur from June to September, favour the spread. Natural plant openings such as stomata and lenticels are common sites for infection. Flowers are the main entry point of entry in pears. The bacterial slime mentioned above is an important source of further infections. Fireblight, once notifiable nationally (see p293), must now be reported only in fruit-growing areas.

Control. The compulsory removal of the susceptible 'Laxton's Superb' pear cultivar in the 1960s eliminated a serious source of infection. Preventative measures such as removal of badly infected plants to prevent further infection, and removal of hawthorn hedges close to pear orchards, help in control. Careful pruning, 60 cm below the stained wood of early infection, may save a tree from the disease. Wounds should be sealed with protective paint, and pruning implements should be sterilized with 3 per cent **lysol**.

Bacterial canker (*Pseudomonas mors-prunorum*)

Damage. This disease affects the plant genus *Prunus* that includes ornamental species, plum, cherry, peach and apricot. Symptoms typically appear on the stem as a swollen area exuding a light brown gum (see Figure 15.18). The angle between branches is the most common site for the disease. Severe infections girdling the stems cause death of tissues above the infection, and the resulting brown foliage can resemble the damage caused by fireblight. In May and June, leaves may become

Figure 15.18 **Bacterial canker on Prunus**. Note the swollen trunk and gum oozing from the infected area

infected; dark brown leaf spots 2 mm across develop and the infected area may be blown out by heavy winds to give a 'shot-hole' effect.

Life cycle and spread. The bacteria present in the cankers are mainly carried by wind-blown rain droplets, infecting leaf scars and pruning wounds in autumn and young developing leaves in summer.

Control. Preventative control involves the use of resistant rootstocks and scions, e.g. in plums. The careful cutting out of infected tissue followed by an application of paint and the use of autumn sprays of a copper compound (Bordeaux mixture), help reduce this disease.

Soft rot (*Erwinia carotovora*)

This bacterium affects stored potatoes, carrots, bulbs and iris, where the bacterium's ability to dissolve the cell walls of the plant results in a mushy soft rot. High temperatures and humidity caused by poor ventilation promote infection through lenticels, and major losses may occur. A related strain of this bacterium causes **black leg** on potatoes in the field. Preventative control measures are important. Crops should be damaged as little as possible when harvesting, and diseased or damaged specimens should be removed before storage. Hot, humid conditions should be avoided in store. No curative measures are available.

Crown gall (*Agrobacterium tumifasciens*)

This bacterium affects apples, grapes, peaches, roses, *Euonymus* and many herbaceous plants. The disease is first seen just **above** ground level as a swollen, cancer-like structure (often about 5 cm in size) growing out of the stem. It may occasionally cause serious damage, but usually is not a very important problem. The bacterium is able to survive well in soils, and infects the plant through small wounds in the roots.

It is of special scientific interest in the area of **plant breeding**, having the ability to add its genetic information to that of the plant cell. It does this by means of a small unit of DNA called a 'plasmid'. This plasmid ability of *A. tumifasciens* has been harnessed by plant breeders to transfer genetic information between unrelated plant species. It is the properties of this bacterium that have led to the new term 'genetically modified crop' or more simply 'GM' (see Chapter 10).

Control of crown gall depends on cultural control methods, such as disease-free propagating material, avoiding wounds at planting time and budding scions to rootstocks (rather than grafting) to avoid injuries near the soil level.

Viruses

Structure and biology

Viruses are extremely small, much smaller even than bacteria (see Figure 15.2). The light microscope is unable to focus in on them, but they appear as rods or spheres when seen under an electron microscope. The virus particle is composed of a DNA or RNA core surrounded by a protective protein coat. On entering a plant cell, the virus takes over the organization of the cell nucleus in order to produce many more virus particles. Since the virus itself lacks any cytoplasm cell contents, it is often considered to be a non-living unit. Some details of its classification are given in Chapter 4.

The virus's close **association with the plant cell nucleus** presents difficulties in the production of a curative virus control chemical that does not also kill the plant. No established commercial 'viricide' has yet been produced against plant viruses.

In recent years the broad area called 'virus diseases' has been closely investigated. Virus particles have, in most cases, been isolated as the cause of disease, e.g. cucumber mosaic. Other agents of disease to be discovered are **viroids** (e.g. in chrysanthemum stunt disease) and these are smaller than viruses. *Mycoplasmas* (the cause of diseases such as aster yellows) are a group of bacteria that induce symptoms similar to those produced by viruses.

Spread. A number of organisms (**vectors**) spread viruses from plant to plant and then transmit the viruses into the plant. **Peach-potato aphid** is capable of transmitting over 200 types of virus (e.g. cucumber mosaic) to different plant species. The aphid stylet injects salivary juices containing virus into the parenchyma and phloem tissues, enabling the virus to then travel to other parts of the plant. '**Persistent virus transmission**' is seen in some vector/virus combinations such as peach-potato aphid/potato virus X, and *Xiphinema* dagger nematode/arabis mosaic where the virus is able to survive and increase within the vector's body for several weeks. In many vector/virus combinations such as plum pox, the virus survives only briefly as a contaminant on the insect's stylet. Other vector/virus combinations include bean weevils/broad bean stain virus; and *Olpidium* soil fungus/big vein agent on lettuce.

Other important methods of spread involve vegetative material (e.g. chrysanthemum stunt viroid and plum pox), infected seed (e.g. bean common mosaic virus), seed testa (e.g. tomato mosaic virus) and mechanical transmission by hand (e.g. tomato mosaic virus).

Symptoms. The presence of a damaging virus in a plant is recognizable to horticulturists only by means of its symptoms. For confirmation, they may need to consult a virologist, whose identification techniques include electron microscopy, transmission tests on sensitive plants such as *Chenopodium* species, and serological reactions using specific antiserum samples.

Leaf **mosaic**, a yellow mottling, is the most common symptom (e.g. cucumber mosaic virus). Other symptoms include leaf **distortion** into feathery shapes (cucumber mosaic virus), flower **colour streaks** (e.g. tulip break virus), fruit **blemishing** (tomato mosaic and plum pox), **internal discolouration** of tubers (tobacco rattle virus causing 'spraing' in potatoes) and **stunting** of plants (chrysanthemum stunt viroid).

Symptoms similar to those described above may be caused by misused herbicide sprays, genetic 'sports', poor soil fertility and structure (see **deficiency symptoms**) and mite damage.

In the following descriptions of major viruses, Latin names of genus and species are not included, since no consistent classification is yet accepted.

Tomato mosaic

Damage. This disease may cause serious losses in tomatoes. Infected seedlings have a stunted, spiky appearance. On more mature plants leaves have a pale green mottled appearance, or sometimes a bright yellow (aucuba) symptom. The stem may show brown streaks in summer when growing conditions are poor, a condition often resulting in death of the plant. Fruit yield and quality may be lowered, the green fruit appearing bronze, and the ripe fruit hard, making the crop unsaleable (see Figure 15.19).

Figure 15.19 **Tomato mosaic virus**. Note the orange-yellow patches on the fruit

Life cycle and spread. The virus is a rod shaped virus. The period from plant infection to symptom expression is about 15 days. The virus may survive within the seed coat (testa) or endosperm of the tomato seed. It is very easily spread by human contact as it is present in large numbers in the leaf hairs of infected plants.

Control. Heat treatment of dry seed at 70°C for 4 days by seed merchants helps remove initial infection. Infected debris, particularly roots, in the soil enables the virus to survive from crop to crop, and soil temperatures of 90°C for 10 min are normally required to kill the organism. Peat-growing bag and nutrient-film methods enable the grower to avoid this source of infection. Hands and tools should be washed in soapy water after working with infected plants. Clothing may harbour the virus.

Cultivars and rootstocks containing several factors for resistance are commonly grown, but newly arriving virus strains may overcome this resistance. A mild strain spray inoculation method has been used at the seedling stage to protect non-resistant cultivars from infection with severe strains. Great care is required to avoid mosaic-contaminated equipment when using this method.

Cucumber mosaic

Damage. Several strains of virus cause this disease. In addition to cucumber, the following may also be affected: spinach, celery, tomato, *Pelargonium* and *Petunia*. On cucumbers, a mottling of young leaves occurs (see Figure 15.20) followed by a twisting and curling of the whole foliage, and fruit may show yellow sunken areas. On the shrub *Daphne oderata*, a yellowing and slight mottle is commonly seen on infected foliage, while *Euonymus* leaves produce bright yellow leaf spots. Infected tomato leaves are reduced in size (fern-leaf symptom).

Figure 15.20 **Cucumber mosaic virus**. Note the leaf mottling

Life cycle. The virus may be spread by infected hands, but more commonly an aphid (e.g. peach-potato aphid) is involved. Many crops (e.g. lettuce, maize, *Pelargonium* and privet) and weeds (e.g. fat hen and teasel) may act as a reservoir for the virus.

Control. Since there are no curative methods for control, care must be taken to carry out preventative methods. Choice of uninfected stock is vital in vegetatively propagated plants, e.g. *Pelargonium*. Careful control of aphid vectors may be important where susceptible crops (e.g. lettuce and cucumbers) are grown in succession or next to other susceptible species. Removal of infected weeds, particularly from greenhouses, may prevent widespread infection.

Tulip break

The petals of infected tulips produce irregular coloured streaks and may appear distorted. Leaves may become light green, and plants become stunted after several years' infection. The virus is spread mechanically by knives, while three aphid vectors are known: the bulb aphid in stores, the melon aphid in greenhouses, and the peach-potato aphid outdoors and in greenhouses.

Preventative control must be used against this disease. Removal of infected plants in the field prevents a source of virus for aphid transmission. Aphid control in field, store and greenhouse further reduces the virus's spread.

Plum pox

Damage. This disease, also called 'Sharka', has increased in importance in the British Isles since 1970 after its introduction from mainland

Europe. Plums, damsons, peaches, blackthorn and ornamental plum are affected, while cherries and flowering cherries are immune. Leaf symptoms of faint interveinal yellow blotches can best be seen on leaves from the centre of the infected tree.

The most reliable symptoms, however, are found on fruit, where sunken dark blotches are seen. Ripening of infected fruit may be several weeks premature, yield losses may reach 25 per cent, and the fruit is often sour.

Life cycle. The virus is spread by several species of aphids. The speed of spread is quite slow because the virus is not able to live and multiply in the aphid. Movement of infected young plants is an important method of spread.

Control. Preventative control is the only option open to growers. Clean Ministry-certified stock should be used. Routine aphid-controlling insecticides should be applied in late spring, summer and autumn. Suspected infected trees should be reported and infected trees removed and burnt.

Chrysanthemum stunt viroid

Damage. This disease, found only on plants of the Asteraceae family and mainly on the chrysanthemum, produces a stunted plant, often only half the normal size but without any distortion. Flowers often open one week earlier than normal, and may be small and lacking in colour.

Life cycle. The virus enters gardens and nurseries through infected cuttings, and is readily transmitted by leaf contact and by handling.

Control. Symptoms may take several months to appear, thus seriously reducing the chance of early removal of the disease source. The grower must use preventative control. **Certified planting material** derived from heat-treated meristem stock (see tissue culture) reduces the risk of this disease.

Arabis mosaic

Damage. This virus infects a wide range of horticultural crops. On strawberries, yellow spots or mottling are produced on the leaves, and certain cultivars become severely stunted. On ornamental plants, e.g. *Daphne odorata*, yellow rings and lines are seen on infected leaves, and the plants may slowly die back, particularly when this virus is associated with cucumber mosaic inside the plant.

Life cycle. Several weeds, e.g. chickweed and grass spp., may harbour this disease, and in strawberries severe attacks of the disease may occur when planted into ploughed-up grassland. The virus is spread by a common soil-inhabiting nematode, *Xiphinema diversicaudatum*, which may retain the virus in its body for several months.

Control. Control of this disease can be achieved by preventative methods. Certified virus-free soft fruit planting material is available. Fumigation of soil with chemicals such as **dichloropropene**, applied well before planting time, eliminates many of the eelworm vectors. No curative chemical is available to eliminate the virus inside the plant.

Reversion disease on blackcurrants

Damage. This virus disease, caused by blackcurrant reversion virus can seriously reduce blackcurrant yields. Flower buds on infected bushes are almost hairless and appear brighter in colour than healthy buds. Infected leaves often have fewer main veins than healthy ones (see Figure 15.21). After several years of infection, the bush may cease to produce fruit.

Life cycle and spread. The virus is spread by the blackcurrant gall mite, and reversion infected plants are particularly susceptible to attack by this pest.

Control. Removal and burning of infected plants is an important form of control. Use of certified plant material, raised in areas away from infection and vectors, is strongly recommended. Control of the mite vector in spring and early summer has already been described on p225.

Figure 15.21 **Reversion disease of blackcurrant**. Note that the infected leaf (bottom) has fewer main veins and leaf lobes than the healthy leaf (top)

Physiological disorders

A **physiological disorder** is a condition in the plant resulting from a non-living (abiotic) factor such as nutrient or water being present at the incorrect level.

There are several symptoms that show on plant leaves, stems and flowers that are **not** caused by pests or diseases. The main causes are: nutrient deficiencies, excess fertilizer, frost, high temperature, lack of light, overwatering and underwatering.

Nutrient deficiencies

Each nutrient (the commonest being nitrogen, phosphorus, potassium, calcium and magnesium) is required in the correct amounts to enable the plant to carry out its chemical processes. When amounts present are too low, deficiencies begin to show, usually by means of leaf symptoms (see Chapter 21, p369).

Care should be taken to provide regular applications of a suitable fertilizer, especially during the summer months and in situations where the roots are restricted (as in pots).

Two common horticultural problems should be noted. In tomatoes and peppers, **blossom end rot** (see Figure 15.22) produces a symptom of a black, concave lesion which looks at first sight like a fungal disease. It is caused by an imbalance between potassium and calcium in the soil or compost. It occurs most often when the soil or compost is allowed to dry out while the fruits are swelling. It is seen more often in greenhouse container-grown plants than with plants growing in the open garden

Figure 15.22 **Blossom end rot in tomato**. The fruit at the opposite end from the stalk has a typical black sunken appearance

or greenhouse borders. It is most common when plants are raised in grow bags, where they have a small, shallow root run that dries out easily. Although there is no cure for blossom end rot once the symptoms begin to appear, the obvious recommendation is that fruiting crops should never be allowed to have dry roots.

A second problem is **bitter pit** in apples. Here the fruit develop many small, dark-brown, sunken pits. The tissues below are stained to depth of about 2 mm. Cultivars such as 'Bramley's Seedling' and 'Egremont Russet' are most susceptible. Young over-bearing trees show the worst effects. The disorder is caused by low calcium levels in the fruit, influenced by irregular water supply in the tree. Four recommendations are given for this problem.

- Ensure a steady water supply to the tree during dry spells.
- Mulch around the tree to help moisture retention.
- Summer prune young, vigorous trees especially when they are holding too many fruit.
- Occasionally use foliar sprays of calcium nitrate plus detergent in the evening during summer to help prevent this problem.

Excess fertilizer

When fertilizers are present at too high levels, roots are scorched and are unable to provide nutrients for the other parts of the plant, often resulting in the plant's death. This condition is described on p380 in Chapter 21. Careful consideration of the appropriate frequency and amounts of fertilizer will prevent this embarrassing situation.

Low temperatures

Plants differ in their tolerance to low temperatures. Low temperatures slow down the plant's growth. Frost often causes the above-ground parts of sensitive plants to collapse into a mess of green tissue after ice has formed inside the plant and fractured all the cells.

High temperatures

Plants may become exposed to very high temperatures in greenhouses, where growth may be weak and 'leggy'. Their leaves also may become dry and brittle, especially if they are touching the glass sides or roof of the greenhouse. Regular attention to ventilators or the use of the automatic ventilators available to amateur growers avoids this problem.

Lack of light

House plant species are sometimes placed in parts of the house unsuitable for their ideal growth. For example, a poinsettia needs high light levels. Plants outdoors may be subjected to the same oversight. *Pelargoniums* used as bedding plants should be given full sunlight and will develop a pale foliage colour if placed in a shady place. *Impatiens*, on the other hand, is able to withstand considerable shade and maintain its rich dark-green foliage.

Overwatering

Overwatering replaces the air spaces in soil and growing composts with water, thus preventing root respiration which is needed to supply energy for root growth and nutrient uptake. Overwatering symptoms may include the following.

- The whole plant may wilt, the lower leaves turn yellow and drop.
- New foliage may have brown spots.
- The whole plant may become stunted, and stems and roots become brown and decayed.

Underwatering

The plant needs sufficient water to carry nutrients around, to be present as an ingredient for making sugar, to transpire from the leaf in order to keep a desirable leaf temperature and to maintain turgidity in some plant tissues. In some plant species, leaves change from shiny to dull as a first signal of water stress and also may change from bright green to a grey green. New leaves wilt, but in species such as holly and conifers only the very youngest leaves wilt. Flowers may fade quickly and fall prematurely. Older leaves often turn brown, dry and fall off. Digging a few centimetres into the soil may indicate the need for watering with shallow rooted perennials and annual border plants. Shrubs with deep roots rarely need watering, although transplanted older shrubs may show summer water-stress for a number of years (see also Chapters 9 and 19).

Figure 15.23 **Raised oedema spots** on lower leaf surface of pelargonium. This symptom superficially resembles rust pustules

Oedema

Oedema is seen as raised corky spots on the undersurface of leaves. Species such as pelargonium (see Figure 15.23), rhododendrons, begonias, pansies, violets and some fleshy-leaved plants such as *Peperomia* are affected. Orchids can show oedema on their petals. Oedema occurs when the roots' ability to supply water exceeds the leaves' ability to release the water by transpiration. Conditions favouring oedema occur most commonly in late winter and early spring especially during extended periods of cool, cloudy weather. Warm, moist soil occurring alongside cool, moist air brings on the condition most severely. The symptoms are commonly seen in unheated greenhouses. The problem can be greatly reduced by glasshouse heating and automatic venting.

Symptoms of disease and physiological disorders

Below in Table 15.2 is a summary of the most important symptoms to help the reader 'home-in' on disease problems and physiological disorders.

Table 15.2 Some symptoms of diseases and physiological disorders

Symptom	Cause	Other cause
Leaf spot	fungus e.g. apple scab	bacterial canker (*Prunus*)
Raised leaf spots	rusts	oedema (corky spots)
White covering on leaf	powdery mildew – upper	spraying hard water
	downy mildew – lower	
Leaf yellowing	low nitrogen levels	root disease
Brown edge to the leaf	low potassium	
Leaves curl and go brown	underwatering	
Dry, crumbling leaves	plants overheated	too much fertilizer
Yellow leaf veins	low magnesium/iron	
Dark coloured leaves	low phosphorus	
Lower leaves yellow	wilt fungus	overwatering
Yellow/ green leaf mottle	virus mosaic	mutation/chimaera
Fruit spots	fungus	bitter pit (apple)
Sunken fruit lesions	blossom end rot (tomatoes)	
Bud or leaf drop	sudden change of temperature	
Stems elongated	too little light	
Whole plant wilts	severe underwatering	wilt disease, vine weevil grubs
Fluffy mould	*Botrytis* (grey)	*Penicillium* (blue)
Brown stem lesions	tomato mosaic virus	
Swollen woody stems	fungal/bacterial canker	
Oozing from woody stems	bacterial canker/fireblight	
Brown roots	root rot	overfertilizing

Check your learning

1. Define the term 'disease'.

2. State the damage caused by three named fungal diseases.

3. Describe the life cycle of one named fungal disease.

4. Describe how the life cycle in the above fungal example is related to its control.

5. Describe the available control measures for the disease in question 3.

6. State the damage caused by one named bacterial disease.

7. Describe one method of reducing the damage caused by the above-named bacterium.

8. Explain how fungal resistance to fungicides can be reduced.

9. Explain the difference in the symptoms caused by downy mildews and powdery mildews.

10. Explain what disease danger the roots of a dead tree represent to plants in its vicinity.

Further reading

Agrios, E.N. (2005). *Plant Pathology*. Academic Press.

Brown, L.V. (2008). *Applied Principles of Horticultural Science*. 3rd edn. Butterworth-Heinemann.

Buczacki, S. and Harris, K. (1998). *Guide to Pests, Diseases and Disorders of Garden Plants*. Collins.

Cooper, J.I. (1979). *Virus Diseases of Trees and Shrubs*. Institute of Terrestrial Ecology.

Fletcher, J.T. (1984). *Diseases of Greenhouse Plants*. Longman.

Greenwood, P. and Halstead, A. (2007). *Pests and Diseases*. Dorling Kindersley.

Hope, F. (1990). *Turf Culture: A Manual for Groundsmen*. Cassell.

Ingram, D.S. *et al.* (eds) (2002). *Science and the Garden*. Blackwell Science Ltd.

Lindow, E.S. *et al.* (2002). *Phyllosphere Microbiology*. American Phytopathological Society.

Scopes, N. and Stables, L. (1989). *Pest and Disease Control Handbook*. British Crop Protection Council.

Snowdon, A.L. (1991). *Post-harvest Diseases and Disorders of Fruits and Vegetables*. Wolfe Scientific.

UK Pesticide Guide (2007). CAB Publishing.

Chapter 16 Plant protection

Summary

In the three preceding chapters important weeds, pests and diseases were described with an emphasis on symptoms and damage, life cycles and with brief comments on control relevant to the particular causative organism.

This chapter includes the following topics:

- **Physical control**
- **Cultural control**
- **Biological control**
- **Chemical control**
- **Plant selection for resistance**
- **Integrated control**
- **Supervised control**
- **Legislative control**

with additional information on environmentally sustainable practices:

- **Benefits or problems and hazards associated with controls**
- **Minimization of risks to humans and the environment**
- **Natural balances and their disturbance**
- **Maintenance and restoration of balances**
- **Selection of plants to avoid plant health problems**

Figure 16.1 Biological control in a **greenhouse**

A good general principle for the gardener or grower is that they use as many different kinds of control as possible within a plant or crop cycle to bring about precise and efficient control for a pest or disease. And so a parasitic wasp may be encouraged or applied against an aphid attack, whilst also considering a cultural control such as the removal of alternate host weeds or a carefully considered pesticide if this is needed.

A second principle highlights a distinction between pests and diseases. For pests, **biological control** is a major form of control in the natural environmental habitat (and to a similar extent in the garden/commercial holding when predators and parasites are encouraged). For diseases, however, in the natural environment, **plant resistance** (rather than biological control) is the important control method. In gardens and horticultural units, plants which have not been highly bred often exhibit a high level of resistance similar to their wild relatives. But intensively bred cultivars of annual flowers, annual vegetables and fruits may largely lack this important form of control.

A third principle suggests that a programme of control against a pest or disease should consider non-pesticide controls first before relying on the alternative route. It has to be admitted that few people would disagree with the principle, but many growers will be able to cite instances when this idealistic attitude has left them with serious pest or disease problems on their hands (such as slugs in wet summers). In the 'supervised control' section of this chapter, there is a brief discussion of 'economic damage'. At relevant points in this chapter, distinction is drawn between those measures available to the amateur gardener and those used by the professional grower.

Physical control

Physical control is a material, mechanical or hand control where the **pest** is directly blocked, or destroyed.

Benefits. Physical controls are long-lasting and need little maintenance.

Limitations. Some physical methods are expensive to set up.

Warm water treatment

This method is used for pests such as stem and bulb nematodes in narcissus bulbs. Immersion of bulbs for 2 hours at 44°C controls the pest without seriously affecting bulb tissues. Chrysanthemum stools and strawberry runners may be similarly treated, using temperature and time combinations favourable to each crop. Viruses (such as aspermy virus on chrysanthemum) are more difficult to control, since viruses are more intimately associated with the plant nuclei. Virus concentrations may be greatly reduced in meristems of stock plants grown at temperatures of 40°C for about a month. This has enabled the production of tissue-cultured disease-free stock material of both edible and non-edible crops (see **tissue culture**, p177).

Flame throwers are used for the control of weeds when other methods, such as cultivation, hand-weeding or herbicidal control are not considered suitable.

Partial soil sterilization

Commercial greenhouse soils are commonly sterilized by high-pressure steam released to penetrate downwards into the soil, which is covered by heat resistant plastic sheeting (sheet steaming). The steam condenses on contact with soil particles, and moves deeper only when that layer of soil has reached steam temperature. Some active soil pests, such as symphilids, may move downwards ahead of the steam 'front'.

The temperatures required to kill most nematodes, insects, weed seeds and fungi are 45°C, 55°C, 55°C and 60°C respectively. Beneficial bacterial spores are not killed below 82°C, and therefore growers attempt to reach, but not exceed, this soil temperature. Most mycorrhizal fungi are unfortunately killed by this process.

In this way, organisms difficult to sterilize, such as fungal sclerotia, Meloidogyne and Verticillium in root debris, may be killed. Sheet steaming is effective only to depths of about 15 cm, and its effect is reduced when soil aggregates (see p311) are large and hard to penetrate, or when soils are wet and hard to heat up. When soil pests and diseases occur deep in the soil, heating **pipes** may be placed below the soil surface, as grids or spikes, to achieve a more thorough effect. The 'steam-plough' achieves a similar result, as it is winched along the greenhouse. If soil is to be used in growing composts it should be sterilized (see sterilizing equipment). The clear advantage of soil sterilization may occasionally be lost if a serious soil fungus (such as *Pythium*) is accidentally introduced into a crop where it may quickly spread in the absence of fungal competition.

Barriers

A physical **barrier** such as a fence sunk into the ground deters rabbits and deer. Fine screens placed over ventilation fans help prevent the entry of pests, such as fungus gnats, from outside a greenhouse or mushroom house. Pots placed on small stands in water-filled trays are freed from the visitations of red spider mite and adult vine weevils. Peach leaf curl is a difficult disease to control. A plastic sheet placed over the peach or almond over winter will greatly reduce both arrival of spores and the moisture needed for infection of the buds.

Traps

Pheremone traps containing a specific synthetic chemical similar to the attractant odour of a female moth are used in apple orchards to lure male codling and tortrix moths onto a sticky surface, thus enabling an accurate assessment of their numbers and therefore more effective control. Comparable traps are available against plum moth and pea moth.

A rodent trap is available that entices the rat or mouse into a container with suitable bait and then the pest is killed humanely with a high voltage charge.

Allotment owners sometimes use containers such as plastic milk bottles or jam jars sunk into the ground that when part-filled with beer, attract slugs. Between two rows of tomatoes, a 'sacrificial' row of lettuce can be grown to attract slugs (see Figure 16.2), which can then be controlled in a trap (see Figure 16.2).

Figure 16.2 (a) **Horticultural fleece** used to protect from pests (b) **Plastic guard** used against rabbits and deer (c) **Lettuce** planted to attract slugs away from tomatoes (d) **Slug trap** that avoids bird-poisoning

Deterrents

Measures here warn off the pest in a chemical or physical way.
Ultrasonic devices create high frequency sounds, unheard by humans, but offensive to animals such as rats and mice. The odour from onions inter-planted with carrots may deter the carrot fly from attacking its host crop. Marigolds planted in amongst crops deter whitefly and aphids. A spray repellent using an extract from a *Yucca* species is used to deter slug attack.

Cultural control

Cultural control is a procedure, or manipulation of the growing environment that results in pest, disease or weed control.

Horticulturists, in their everyday activities, may remove or reduce damaging organisms in many different ways and thus protect the crop. Below are described some of the more important methods used.

Benefits. Fit in with daily routines. Have a long-lasting effect.

Limitations. May be time-consuming.

Cultivation

Ploughing and **rotavating** of soils enables a physical improvement in soil structure as a preparation for the growing of crops. The improved drainage and tilth may reduce damping-off diseases, disturb annual and perennial weeds, such as chickweed and couch-grass, and expose soil pests, such as leatherjackets and cutworms, to the eager beaks of birds. Repeated **rotavation** may be necessary to deplete the underground rhizomes or tap roots of perennial weeds, such as couch-grass. **Hoeing** annual weeds is an effective method, provided the roots are fully exposed, and the soil dry enough to prevent root re-establishment.

Soil fertility

While the correct content and balance of major and minor nutrients (see Chapter 21) in the soil are recognized as vitally important for optimum crop yield and quality, it should be remembered that plant resistance to pests and diseases is also affected by nutrient levels in the plant. Excessive nitrogen levels, causing soft tissue growth, encourage the increase of insects such as peach-potato aphid, fungi such as grey mould, and bacteria such as fireblight. Adequate levels of potassium, on the other hand, help control fungal diseases, such as Fusarium wilt on plants such as peas, tomatoes and carnation, and tomato mosaic virus. Fertility provided by composted material usually provides nutrients to the plant at the correct concentrations.

Club root disease of brassicas is less damaging in soil with a pH greater than 6, and lime may be incorporated before planting these crops to achieve this aim. Amenity horticulturists apply mulches, such as composted bark, grass cuttings or straw, to bare soil in order to control annual weeds by excluding the source of light. Black polythene sheeting is used in soft fruit production to achieve a similar objective.

Crop rotation

Some important soil-borne pests and diseases attack specific crops, such as potato cyst nematode on potato and club root on cruciferous plants. As they are soil-borne, they are slow in their dispersal, but are difficult to control. By the simple method of planting a given crop in a different plot each season, such pests or diseases are excluded from their preferred host crop for a number of years, during which their numbers will slowly decline. A gardener often creates five or six plots (sometimes bounded by wooden planking) to achieve successful rotation. Plants belonging to the same plant families fit into the rotation system. They have the same sensitivity to particular pests and diseases. Potatoes, tomatoes, peppers and aubergines are all members of the Solanaceae. Melons, marrows, courgettes and cucumbers all belong to the Cucurbitaceae. When considering a rotation plan, it is advisable to confine members of the same family to the same plot in the same growing season. Rotation does not work well against unspecific problems, such as grey mould, which may attack a wide range of plant families. Rotation is also not likely to be effective against rapidly spreading organisms such as aphids.

Figure 16.3 **Rotation plots** used to reduce pest and disease problems

The **sclerotium** stage of white rot disease (*Sclerotum cepivorum*) on onion and related crops, such as leeks, garlic, chives and shallots, is able to survive in the soil for 15 years or more. It can be seen that a very long rotation period would be necessary to remove this serious disease. A six year rotation is not normally sufficient.

Planting and harvesting times

Some pests emerge from their overwintering stage at about the same time each year, such as cabbage root fly in late April. By planting early to establish tolerant brassica plants before the pest emerges a useful supplement to chemical control is achieved. The deliberate planting of early potato cultivars enables harvesting before the maturation of potato cyst nematode cysts, so that damage to the crop and the release of the nematode eggs is avoided. Annual weeds may be induced to germinate in a prepared seedbed by irrigation. After they have been controlled with a contact herbicide, such as paraquat, a crop may then be sown into the undisturbed bed or stale seedbed, with less chance of further weed germination.

Clean seed and planting material

Seed producers take stringent precautions to exclude weed seed contaminants and pests and diseases from their seed stocks. While weed seeds are, in the main, removed by mechanical separators, and insects can be killed by seed dressings, systemic fungal seed infections, such as celery leaf spot disease in celery seed, are best controlled by immersion of dry seed in a 0.2 per cent thiram solution at 30°C for 24 hours (**thiram soak treatment**). The seed is then re-dried. A similar treatment is often given for carrot and parsley seed.

Equal care is taken to monitor seed crops likely to carry virus disease (such as tomato mosaic). Curative control by dry heat at 70°C for four days is usually effective, although it may reduce subsequent seed germination rates.

Vegetative propagation material

Vegetative propagation material is used in all areas of horticulture, such as bulbs (tulips and onions), tubers (dahlias and potatoes), runners (strawberries), cuttings (chrysanthemums and many trees and shrubs) and graft scions in trees. The increase of nematodes, viruses, fungi and bacteria by vegetative propagation is a particular problem, since the organisms are inside the plant tissues, and since the plant tissues are sensitive to any drastic control measures.

Inspection of introduced material may greatly reduce the risk of this problem. Soft, puffy narcissus bulbs, chrysanthemum cuttings with an internal rot, whitefly or red spider mite on stock plants, virus on nursery stock, are all symptoms that would suggest either careful sorting, or rejection of the stocks.

Accurate and rapid methods of virus testing (using test plants, electron microscopy and staining by ELISA techniques) now enable growers to learn quickly the quality of their planting stocks. Fungal levels in cuttings (such as *Fusarium* wilt of carnations) can be routinely checked by placing plant segments in sterile nutrient culture.

The quality of vegetative material is monitored in the UK by the **Plant Health Propagation Scheme**. In particular, it supervises the provision of six quality levels of plant material (Foundation, Super Elite, Elite, A, Approved and Healthy grades) for the soft fruit, top fruit and bulb industries.

Hygienic growing

During the crop, the grower should aim to provide optimum conditions for growth. Water content of soil should be adequate for growth (see **field capacity**), but not so excessive that root diseases (such as damping off in pot plants, club root of cabbage and brown root rot of conifers) are actively encouraged.

Water sources can be analyzed for *Pythium* and *Phytophthora* species if damping off diseases are a constant problem. Covering and regular cleaning of water tanks to prevent the breeding of these fungi in rotting organic matter may be important in their control. Seed trays and pots should be washed to remove all traces of compost that might harbour damping off disease.

Conifer nursery stock grown on raised gravel beds is less likely to suffer the water-borne spread of conifer brown rot. Many protected crops are grown in isolated beds or peat modules to reduce spread of wilt-inducing organisms (such as *Fusarium* spp.). Gooseberry sawfly caterpillars can be removed in spring from leaves found on the lower centre stems of the gooseberry. This action helps prevent subsequent invasions by the pest in the summer months.

High humidity encourages many diseases. In greenhouses, the careful timing of daily overhead irrigation and of ventilation (to reduce overnight condensation on leaves or flowers) may greatly reduce levels of diseases, such as grey mould on pot plants or downy mildew on lettuce. The slow drawing-back of motorized **thermal screens** high above commercial glasshouse crops (so as to prevent condensation problems) has contributed greatly to the reduction of disease.

Reducing spread of pests and diseases

The spread of pests and diseases from plant to plant or field to field can be slowed down. Tomato mosaic virus spread may be reduced by

Figure 16.4 A **sowthistle** acting as alternate host to glasshouse whitefly

leaving suspect plants till the end of de-leafing or harvesting. Washing knives and hands regularly in warm, soapy water will reduce subsequent viral spread. Soil-borne problems, such as club root, eelworms and damping off diseases, are easily carried by boots and tractor wheels. Foot and wheel dips, containing a general chemical sterilant, such as formaldehyde, have been successfully used, especially in preventing damping off problems in greenhouses.

Alternate hosts

Alternate hosts harbouring pests and diseases should be removed where possible. A few examples of many alternate hosts are given here. Soil-borne problems, such as club root of cabbage and free-living eelworms on strawberries are harboured by shepherd's purse (*Capsella bursa-pastoris*) and chickweed (*Stellaria media*) respectively. Groundsel (*Senecio vulgaris*) is an alternate host of rust on cinerarias, while docks (*Rumex spp.*) act as a reservoir of dock sawfly, which damages young apple trees.

Removal of infected plant material

With rapid-increase problems, such as peach-potato aphid and white rust of chrysanthemum fungus in greenhouses, removal of affected leaves is practicable in the early stages of the problem, but becomes progressively unmanageable after the pest or disease has increased and dispersed throughout the plants. Slow-increase problems, such as Fusarium wilt disease on tomatoes or carnations and vine weevil larvae found damaging roots of plants such as primulas and begonias, may be removed throughout the crop cycle, but the infected roots and soil must be carefully placed in a bag to prevent dispersal of the problem. In commercial outdoor production, labour costs usually prevent such removal during the growing season. However, removal is achieved chemically in some situations. The destruction of blight-infected potato foliage with herbicide such as diquat prior to harvest reduces infection of the tubers. Burning of post-harvest leaf material and lifting of root debris after harvest (against grey mould on strawberries and club root on brassicas respectively) may help prevent problems in the next crop.

In fruit tree species such as apple, routine pruning operations may remove serious pests such as fruit tree red spider mite, and diseases such as canker and powdery mildew. Pruning should also aim to reduce the density of shoots in the centre of the tree. The reduction in humidity provides a microenvironment less favourable to disease increase.

Tree stumps harbouring serious underground diseases such as honey fungus should be removed manually or using a mechanical stump grinder. Making a feature of an infected stump by placing a bird table on it is one of the least recommended activities in gardening.

Safe practice

In physical/cultural control, some **hazards** are:

- Unsafe use of cultivation equipment, such as ploughs, rotavators, flame throwers and steam sterilization equipment, used to control weeds, pests and diseases.
- Unsafe removal of infected trees.
- Unsafe burning of infected plant material.

When using cultural control, risks can be **minimized** by:

- following guidelines for the safe use of ploughs and rotavators to avoid damage to humans and adjoining crops or plantings;
- following guidelines for the safe use of flame throwers and steam sterilization equipment;
- taking safety precautions when removing infected shrubs and trees;
- carefully transferring infected plants to dumps or compost sites, thus avoiding the spread of infections.

Biological control

Biological control is the use of natural enemies to reduce the damage caused by a pest (or disease).

Benefits. Non-toxic, no build-up of resistant pests and diseases.

Limitations. Needs careful introduction and knowledge of life cycles. Can easily be affected by pesticides.

There are two sources of 'natural enemies' to pests, the local species and the exotic ones. Many pests of outdoor horticultural crops such as peach-potato aphid are **indigenous** (i.e. they are present in wild plant communities in the UK). Such pests are often reduced in nature by other organisms which, as predators, eat the pest, or, as parasites, lay eggs within the pest. These beneficial organisms, found also on horticultural crops, are to be encouraged and in some cases are deliberately introduced. A range of important organisms useful in horticulture is now described in some detail.

Indigenous predators and parasites

Wild birds

It has been shown that a pair of blue tits can consume 10 000 caterpillars and one million aphids in a 12 month period. The installation of tit boxes is a worthwhile activity. Wrens, thrushes and blackbirds similarly contribute to the control of garden insects.

Hedgehogs

Hedgehogs belong to the insectivore group of mammals, but are omnivorous. Although their preferred diet is insects (up to 200 g per day) they will eat slugs. Care must be taken that they are not exposed to dead slugs which have consumed slug bait containing methiocarb or methaldehyde, as these would be toxic to the hedgehog. Hedgehogs are encouraged to enter gardens by means of small holes cut into the base of a fence panel. Within gardens, heaps of logs and piles of leaf litter in a quiet location are suitable for their daytime and overwintering retreat. Wooden hedgehog shelters are commercially available.

The black-kneed capsid

The **black-kneed capsid** (*Blapharidopterus angulatus*) is an insect found on fruit trees alongside its pestilent relative, the common green capsid. It eats more than 1000 fruit tree red spider mites per year. Its eggs are laid in August and survive the winter. Winter washes used by professional horticulturalists against apple pests and diseases often kill off this useful insect. The closely related **anthocorid bugs**, such as *Anthocoris nemorum*, are predators on a wide range of pests, such as aphids, thrips, caterpillars and mites, and have recently been used for biological control in greenhouses.

Lacewings

Lacewings, such as *Chrysopa carnea*, lay several hundred eggs per year on the end of fine stalks, located on leaves. Several are useful horticultural predators, their hairy larvae eating aphids and mite pests, often reaching the prey in leaf folds where ladybirds cannot reach.

Ladybird beetle

The 40 species of **ladybird** beetle are a welcome sight to the professional horticulturist and lay person alike. Almost all are predatory. The red two-spot ladybird (*Adalia bipunctata*) emerges from the soil in spring, mates and lays about 1000 elongated yellow eggs on the leaves of a range of weeds, such as nettles, and crops such as beans, throughout the growing season. Both the emerging slate-grey and yellow larvae and the adults feed on a range of aphid species. Wooden ladybird shelters and towers are now available to encourage the overwintering of these useful predators.

A worrying development in the last few years has been the rapid spread and increase of the **harlequin ladybird** from South-East Asia. This species is larger (6–8 mm long) and rounder than the two-spot species (4–5 mm). It has a wider food range than other ladybird species, consuming other ladybird's eggs and larvae, and eggs and caterpillars of moths. Furthermore, it is able to bite humans and be a nuisance in houses when it comes out of hibernation.

Carabid beetles

The **ground beetle** (such as *Bembidion lampros*), a 2 cm long black species (see Figure 16.5), is one of many active carabid beetles that actively predates on soil pests such as root fly eggs, greatly reducing their numbers.

Hoverflies

Superficially resembling wasps, these are commonly seen darting or hovering above flowers in summer. Several of the 250 British species, such as *Syrphus ribesii*, lay eggs in the midst of aphid colonies, and the legless light-green coloured grubs consume large numbers of aphids.

Figure 16.5 (a) Predatory **ground beetle** (b) **Ichneumon wasp** parasitic on caterpillars

The flowers of some garden plants are especially useful in providing pollen for the adults and therefore encouraging aphid control in the garden. Summer flowering examples are poached-egg plant (*Limnanthes douglasii*), baby-blue-eyes (*Nemophila menziesii*) and Californian poppy (*Romneya coulteri*). Later summer and autumn examples are *Phacelia tanacetifolium* and ice plant (*Sedum spectabile*).

Mites and spiders

Predatory **mites** such as *Typhlodromus pyri* eat fruit tree red spider mite and contribute importantly to its control. The numerous species of web-forming and hunting **spiders** help in a very important but unspecific way in the reduction of all forms of insects.

Wasps

The much maligned **common wasp** (*Vespula vulgaris*) is a voracious spring and summer predator on caterpillars, which are fed in a paralyzed state to the developing wasp grubs. **Digger wasps** also help control caterpillar numbers and benefit from dead hollow stems of garden plant which they use as nests all year round.

There are about 3000 **parasitic wasp** species of the families Ichneumonidae, Braconidae and Chalcidae found on other insects in Britain. Ichneumons (*Opion spp.* see Figure 16.5) lay eggs in many moth caterpillars. The braconid wasp (*Apanteles glomeratus*) lays about 150 eggs inside a cabbage white caterpillar and the parasites pupate outside the pest's dead body as yellow cocoon masses. The chalcid (*Aphelinus mali*) parasitizes woolly aphid on apples.

Figure 16.6 Swollen **aphid** parasitized by tiny wasp

The spiracles of insects provide access to specialized **parasitic fungi**, particularly under damp conditions. In some years, aphid numbers are quickly reduced by the infection of the fungus *Entomophthora aphidis*, while codling moth caterpillars on apple may be enveloped by *Beauveria bassiana*. Cabbage white caterpillar populations are occasionally much reduced by a **virus**, which causes them to burst.

Increased attention is being given by horticulturalists to the careful selection of pesticides (if they are needed) to avoid unnecessary destruction of indigenous predator and parasite numbers (see also p271).

In recent years, commercial firms have begun to make available **ready-to-use products** containing indigenous predators or parasites to outdoor growers. Examples are two-spot ladybird, lacewing larvae and three nematode parasites (against slugs, vine weevil larvae and flea beetles (see also Table 16.1).

Table 16.1 Biological control organisms reared commercially for use in horticulture

Pest or disease	Crop	Control	Type
Aphids	G	*Aphidoletes aphidimyza*	Midge
	G	*Aphidius spp.*	Wasp
	Ch	*Verticillium lecani*	Fungus
Caterpillars	B	*Trichogramma brassicae*	Wasp
	G	*Bacillus thuringiensis*	Bacterium extract
Flea beetle	B	*Howardula phyllotreta*	Nematode
Glasshouse whitefly	T	*Macrolophus caliginosus*	Anthocorid bug
	G	*Encarsia formosa*	Wasp
Leaf miner	T, C, F	*Dacnusa sibirica*	Wasp
	T, C, F	*Diglyphus isaea*	Wasp
Leaf hopper	T	*Anagrus atomus*	Wasp
Mealy bug	G	*Cryptolaemus montrouzieri*	Ladybird
	G	*Leptomastix dactylopii*	Wasp
Red spider mite	G	*Phytoseiulus persimilis*	Mite
	G	*Amblyseius calfornicus*	Mite
	G	*Feltiella acarisuga*	Midge
	T, C, F	*Therodiplosis persicae*	Midge
Sciarid fly	G	*Steinernema feltiae*	Nematode
	G	*Hypoaspis miles*	Mite
	G	*Bacillus thuringiensis strain*	Bacterial extract
Silver leaf fungus	Tf	*Trichoderma viride*	Fungus
Slugs	G	*Phasmarhabditis hermaphrodita*	Nematode
Thrips	C, F, P	*Neoseiulus cucumeris*	Mite
	C, F	*Orius laevigatus*	Anthocorid bug
Tomato mosaic virus	T	mild strain of virus	
Vine weevil	G	*Steinernema capsicarpae*	Nematode

Key; B: brassicas; T: tomato; C: cucumber; P: sweet peppers; F: flowers; Tf: top fruit; Ch: chrysanthemum; G: general use.

Exotic predators and parasites

In greenhouses and polythene tunnels, high temperatures often all year round and sub-tropical species of plants bring with them exotic pests and diseases. Further, the increase of both pests and diseases is much quicker than comparable pests or diseases growing outdoors. Also, these greenhouse inhabiting organisms have, over the last half-century, developed resistance to almost all available pesticides.

Biological control of exotic pests requires exotic predators and parasites. And so the health of the major greenhouse crops is in large measure due to two organisms: a South American mite which eats all stages of the glasshouse red spider mite, and a tiny South-East Asian wasp that parasitizes the glasshouse whitefly.

The conditions in a greenhouse have two **advantages** for biological control. Firstly, the environment is relatively isolated so that the controlling organisms are not likely to disappear. Secondly, the glasshouse environment is relatively stable and allows biological control to be more measured (than in the outdoor situation) with interactions between pests and their predator or parasite being more predictable.

Almost all commercial production of glasshouse crops in the UK now uses biological control. The two commonest biological control organisms are described below in some detail. Much more information is available from commercial companies or from the Internet. A more extensive listing of commercially available biological control species is given in Table 16.1.

Phytoseiulus persimilis (see Figure 16.7)

This is a 1 mm globular, deep orange, predatory tropical mite used in greenhouse production to control glasshouse red spider mite (see p224). It is raised on spider mite-infected beans and then evenly distributed throughout the crop, such as cucumbers, at the rate of about one predator per plant. Some growers who have suffered repeatedly from the pest first introduce the red spider mite throughout the crop at the rate of about five mites per plant a week before predator application, thus maintaining even levels of pest–predator interaction. The predator's short egg–adult development period (7 days), laying potential (50 eggs per life cycle) and appetite (five pest adults eaten per day), explain its extremely efficient action.

Figure 16.7 Glasshouse **red spider mite predator.** (a) Phytoseiulus predator eating glasshouse red spider mite (b) Eggs and young of Phytoseiulus (c) Application of Phytoseiulus to crop

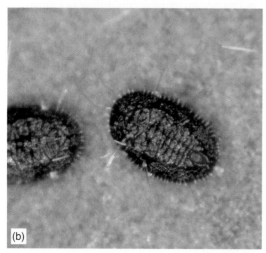

Figure 16.8 Glasshouse **whitefly parasite**. (a) Encarsia wasp laying an egg into a whitefly scale (b) Parasitized whitefly scale has turned black in colour (c) Application of Encarsia to a crop (as blackened whitefly scales)

Encarsia formosa (see Figure 16.8)

This is a small (2 mm) wasp which lays an egg into the glasshouse whitefly **scale** (see p210), causing it to turn black and eventually to release another wasp. This parasite is raised commercially on whitefly-infested tobacco plants. It is introduced to the crop, such as tomato, at a rate of about 100 blackened scales per 100 plants. The parasite's introduction to the crop is most successful when the whitefly levels are low (recommended less than one whitefly per 10 plants). Its mobility (about 5 m) and successful parasitism are most effective at temperatures greater than 22°C when its egg-laying ability exceeds that of the whitefly.

The wasp lays most of its 60 or more eggs within a few days of emergence from the black scale. Thus, a series of weekly applications from late February onwards ensures that viable eggs are laid whenever the susceptible whitefly scale stage is present. The appearance of newly infected **black scales** on leaves is often taken as an indication that parasite introductions can be stopped.

An understanding of each pest's and each biological control organism's life cycle is vital to ensure success in control. A combination of biological methods may be used on some crops, such as chrysanthemums, tomatoes, peppers, aubergines and cucumbers in order to simultaneously control a range of organisms occurring on the crop at the same time (see Table 16.1, and **integrated control**, p292). Several specialist firms now have contracts to apply biological control organisms to greenhouse units. There are several practical points that confront growers in both the outdoor and the glasshouse situation.

Safe practice

The main problems with biological control are:

- Unsuccessful application of biological control organisms that lead to a severe pest problem.
- Introduction of a biological control organism that subsequently kills desirable or beneficial organisms in the environment.

These can be **minimized** by:

- understanding both the pest's and predator/parasites' life cycles in order to achieve reliable control;
- carefully choosing the best predator or parasite for the problem pest or disease concerned;
- taking care that environmentally useful species are not subject to the attacks of the predators and parasites.

In most horticultural situations, there are important examples of **natural balance** between species.

- With pests, their naturally occurring predators and parasites are an important form of crop protection (see page 271).
- With diseases, naturally occurring predators and parasites are less common, but the nutritional condition of the plant and the resulting naturally occurring bacterial and fungal populations on leaf, stem and root surfaces (see phyllosphere p235 and rhizosphere p322) often help slow a disease's progress.
- The garden represents a complex situation. There may be plant species present from every continent (see p73), and any of these plant species may be accompanied by a specific pest from its country of origin. Plant species that have been present in the UK for many years (such as apple) often have beneficial predators and parasites introduced accidentally or deliberately, from their country of origin, that limit pest numbers. It is quite likely, however, that for more recently imported plant species, there may not be appropriate predators or parasites to control an introduced pest occurring on the plant species in the British Isles.

Some horticultural practices can **disturb natural balances**.

- In a natural habitat such as woodland, a **climax population** of plants and animals develops (see p52). Here, a complex balance exists between indigenous pests and their predators/parasites. The food webs (see p53) include several types of predator/parasite found on each plant species that limit (but do not eliminate) the pests. This development of food webs is **not** achieved to such an extent in most gardens since the natural succession of wild plant species mentioned above is not desirable to gardeners as they aim for optimum production of edible crops or for an aesthetic layout of decorative plants free from weeds (see also Chapter 3).
- Regular movement or removal of cultivated plants and weeds without particular thought to the natural balance between predator/parasites and pests will make pest attacks more likely in the garden/nursery situation.
- The removal of the rotting hollow stems of herbaceous perennials and branches of decaying wood which are common sheltering sites for predatory beetles and centipedes reduces their control potential.
- In a similar way, removal of old plants such as brassicas or bedding plants in autumn may take away the parasitized aphids or caterpillars that would normally serve as the next year's control measures.
- The absence in gardens of plant species acting as a pollen food source to adults such as hoverflies may delay the emergence of their predatory larvae amongst aphid populations.
- The lack of good soil structure (see p310) resulting from poor cultivation or inadequate incorporation of organic matter in a garden may hinder the movement of predatory animals in their search for soil pests.
- A poor physical preparation of soil, and lack of attention to pH and nutrient levels in soil may result in poor soil microbial action (see p321).

- The repeated planting of crops or annual bedding plants into the same area of soil often leads to serious attacks of persistent soil-borne pests or diseases. Notable examples are club root disease on brassicas (see p228) and potato cyst nematode pest on potatoes (see p244). A comparable situation is found when young trees and shrubs (such as roses) are planted into a soil previously occupied by an old specimen of the same plant species, with the resulting problem called '**replant disease**' caused by high level of *Pythium* fungus (see p246).
- The unconsidered use of **pesticides** may result in a rapid decrease in predators and parasites and may considerably delay their appearance and build-up the following growing season.

The natural balances of organisms can be **maintained** and **restored** in order to reduce pesticide use. At the private garden level, there are an increasing number of practices being used that encourage natural balances in order to reduce pesticide use. These physical and cultural methods have been described earlier.

Chemical control

> **Chemical control** is the use of a chemical substance intended to prevent or kill a destructive pest, disease or weed.

Benefits. Produce a rapid control. Are easily accessible.

Limitations. Can be dangerous to humans, animals and plants. Can cause resistant strains of pests and diseases to develop.

In past centuries pests, such as apple woolly aphid, were sprayed with natural products, such as turpentine and soap, while weeds were removed by hand. In the nineteenth century, the chance development of Bordeaux mixture from inorganic copper sulphate and slaked lime, and in the early twentieth century the expansion of the organic chemical industry, enabled a change of emphasis in crop protection from cultural to chemical control.

The word '**pesticide**' is used in this book to cover all crop protection chemicals, which include **herbicides** (for weeds), **insecticides** (for insects), **acaricides** (for mites), **nematicides** (for nematodes) and **fungicides** (for fungi). About 2.5 million tonnes of crop protection chemicals are used worldwide each year, about 40 per cent being herbicides, about 40 per cent insecticides and about 20 per cent fungicides. Health and Safety aspects of chemical control are described at the end of the 'chemical control' section.

Chemical sterilization

This involves the use of substances toxic to most living organisms and must be used only by specialist operators and professional horticulturists. The chemical's toxicity to plants also means that they can only be applied to soil or compost that has no crops.

With the recent discontinuation of methyl bromide, two remaining soil applied ingredients are **dazomet** (applied outdoors and in protected crops as a granule against soil-borne insects, fungi, and weed seeds) and **dichloropropene** (applied outdoors as a vapour-releasing liquid by an injection apparatus against nematodes). The fumigant action of these substances is prolonged by rolling the soil after application. Precautions such as rotavating the soil need to be taken several weeks after application, to release any chemical residues before succeeding crops can be planted.

Professional horticulturists sterilize greenhouse structures using toxic compounds, such as **formaldehyde** and burning **sulphur**. Common pests and diseases, such as whitefly, red spider mite and grey mould, may be greatly reduced by this intercrop method of control.

European legislation saw the withdrawal of **methyl bromide** in 2005 on the grounds of human and environmental safety. Whilst being very effective as a sterilant of growing media, methyl bromide had three serious drawbacks. It was very poisonous to man and animals. As a gas, it found its way into the atmosphere. Lastly, its chemical similarity to the chlorinated hydrocarbons used in refrigerator coolants meant that it was held partly responsible for the 'global warming' phenomenon. These three factors combined to rule out its continued use in horticulture.

Active ingredients

Each container of commercial pesticide contains several ingredients. The active ingredient's role is to kill the weed, pest or disease. More detailed lists of the range of active ingredients can be found in government literature. The other constituents of pesticides are described under **formulation** (see page 284).

Herbicides

Herbicides that are applied to the seedbed or growing crop must have a **selective** action, i.e. kill the weed but leave the crop undamaged. This selective action may succeed for one of several reasons. Chemicals often affect different plant families in different ways. The broad-leaved turf weed, daisy (*Bellis perennis*, a member of the Asteraceae) is controlled by **2,4-D**, leaving the turf grasses (Graminae) unaffected.

Sometimes plant species are affected by different concentrations of the chemical to a degree that can be exploited. The correct concentration of selective chemicals may be vital if the crop is to remain unharmed.

It can thus be seen that a concentration of 25 ppm of propyzamide applied to lettuce would leave the crop unaffected, but control all the weeds except groundsel.

The following relative values (parts per million, ppm) for the amount of **propyzamide** herbicide required to kill different plant species illustrate this point:

Crops		**Weeds**	
• carrot	0.8	• knotgrass	0.08
• cabbage	1.0	• black nightshade	0.2
• lettuce	78.0	• fat hen	0.2
		• pennycress	0.6
		• groundsel	78.0

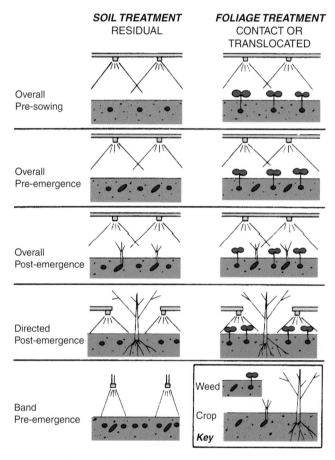

Figure 16.9 Types of herbicide action. (Reproduced by permission of Blackwell Scientific Publications)

A third form of selectivity operates by correct timing of herbicide application. A seedbed with crop seeds deep enough below and weed seeds germinating at the surface may receive a contact chemical, such as **paraquat**, which permits germination of the crop without weed competition. A similar effect is achieved when a residual herbicide, such as **propachlor**, is sprayed onto the soil surface to await weed seed germination. The situations for weed control are summarized in Figure 16.9.

Herbicides may conveniently be divided into two main groups: the **foliage-acting**, and the **soil-acting** (residual) chemicals.

Foliage-acting herbicides

These enter the leaf through fine pores in the cuticle or the stomata. The herbicide may move through the vascular system (translocated chemicals) to all parts of the plant before killing plant cells, or it may kill on contact with the leaf. Four active ingredients are described, each belonging to a different chemical group, and each having a different effect on weeds:

- **Glufosinate-ammonium** is a non-selective, non-residual herbicide available to both amateur and professional growers. It is commonly used to kill top growth of a wide spectrum of weeds in stale seedbeds, after harvest, in bush and tree fruit, or in waste land. It is translocated from leaves to the roots. It is quickly absorbed in damp soils, thus allowing planting soon after its application. It should not be used on foliage of vegetables, annual bedding plants or on turf. Similarly, after spraying this chemical, it is crucial that soles of boots contaminated with the chemical do not inadvertently walk across grass.
- **2,4-D** is available to both amateur and professional growers. It is an auxin and causes uncontrolled abnormal growth on leaves, stems and roots of broad-leaved weeds, which eventually die. It is a useful

selective herbicide on turf because the protected meristems of grasses can survive unaffected. It must be kept well away from nearby border plants and from some crops, such as tomatoes, which are extremely sensitive to minute quantities of the chemical. In formulations for the private garden, 2,4-D is mixed with other ingredients, such as **dichlorprop**, to give a broader spectrum of weed control.

- **Amitrole** is available to professional growers. It is used in similar situations to glufosinate-ammonium, but is more residual, surviving in the soil for several weeks. It stops photosynthesis, scorching both grass and broad-leaved weeds (**non-selective**). It is especially useful on uncropped land and, when applied in autumn, it is translocated to underground rhizomes of couch that are then killed. It should not be sprayed onto the foliage of growing plants.
- **Glyphosate** is available to professional and amateur growers. It enters the foliage of actively growing annual and perennial weeds (**unselective**) and is translocated (see p130) to underground organs, subsequently killing them. It is commonly used several weeks before drilling or planting of crops, around perennial plants such as apples or in established nursery stock trees. Glyphosate is inactivated in soils (particularly peats), thus preventing damage to newly sown crops. It may cause damage if spray-drift to adjoining plants or fields occurs.

Soil-acting herbicides

These are either sprayed onto the soil surface or soil incorporated (see Figure 16.9). They must be persistent (**residual**) for several weeks or months to kill the seedling before or after it emerges. Root hairs are the main point of entry. Increased rates may be necessary for peat soils (since they inactivate some herbicides). The chemical may be applied as a spray or granule before the crop is sown (**pre-sowing stage**), before the crop emerges (**pre-emergence**) or, with more selective chemicals, after the crop emerges (**post-emergence**). Three active ingredients are described, each belonging to a different chemical group, and each having a different effect on weeds:

- **Sodium chlorate** is a residual inorganic ingredient available to amateur and professional growers for total weed control in non-crop situations. It enters leaves and roots and has a rapid action. No plantings can be made until six months after the chemical has been applied.
- **Chlorpropham**, available only to professional growers, is a relatively insoluble compound. It is applied as a pre-emergent spray to control many germinating weeds species, such as chickweed, in crops such as bulbs, onions, carrots and lettuce. It usually persists for less than three months in the soil. In light, porous soils with low organic matter, its rapid penetration to underlying seeds make it an unsuitable chemical. Earthworm numbers may be reduced by its presence.
- **Propachlor**, available only to professional growers, is a relatively insoluble compound; it is applied as a pre-sowing or pre-emergent spray to control a wide variety of annual weeds in brassicas, strawberries, onions and leeks. For weeds in established herbaceous borders (such as rose), the granular formulation gives a residual protection against most germinating broad-leaved and grass weeds.

Mixtures

The horticulturist must deal with a wide range of annual and perennial weeds. The somewhat specialized action of some of the herbicide active ingredients previously described may be inadequate for the control of a broad weed spectrum. For example, in the case of chlorpropham, the addition of diuron enables an improved control of charlock and groundsel, while a different formulation containing chlorpropham plus linuron is designed to have greater contact action and thus control both established and germinating weeds in bulb crops. Careful selection of the most suitable mixture of active ingredients is therefore necessary for a particular crop/weed situation.

Insecticides and acaricides

The insects and mites have three main points of weakness for attack by pesticides which are as follows:

- their waxy exoskeletons (see Chapter 14) may be penetrated by wax-dissolving contact chemicals;
- their abdominal spiracles allow fumigant chemicals to enter tracheae;
- their digestive systems, in coping with the large food quantities required for growth, may take in stomach poisons.

Four groups of insecticides are described (details of associated hazards to spraying operators and the general public are discussed later in this chapter):

- An insecticide manufactured from **natural plant extracts** is approved for amateur and professional growers. It contains **alginates/polysaccharides** and acts by blocking the spiracles of pests such as aphids, thrips and mites. It has been given clearance for use by organic growers.
- **Pirimicarb** is available to professional growers. It belongs to the carbamate group. It enters the pest as a stomach poison, acting on the insect's nervous system. It is slightly systemic in plants, and controls many aphid species without affecting beneficial ladybirds. **Aldicarb**, a related chemical, combines a soil action against nematodes with a systemic, broad-spectrum activity against foliar pests, such as aphids, whitefly, leaf miners, mites and nematodes of ornamental plants.
- **Bifenthrin** is available to amateur and professional growers. It belongs to the pyrethroid group. It has both cuticle and stomach action. It is effective against caterpillars, aphids and mites outdoors. It is residual in soils for a period of up to eight months.
- A fourth group of insecticides contains potassium salts of **fatty acids**. It is available to amateur and professional growers. It works by contact action dissolving the cuticle of pests such as aphids, whitefly spider mites, mealy bugs and scale insects.

Nematicides

No active ingredients are, at present, available exclusively for nematode control. Soil-inhabiting stages of cyst nematodes, stem and bulb eelworm, and some ectoparasitic root eelworms are effectively reduced by soil incorporation of granular pesticides, such as **aldicarb**, at planting time of crops such as potatoes and onions. This group of chemicals acts systemically on leaf nematodes of plants such as chrysanthemum and dahlia.

Fungicides

Fungicides must act against the disease, but not seriously interfere with plant activity. **Protectant** chemicals prevent the entry of hyphae into roots and the germination of spores into leaves and other aerial organs (see Figure 15.4). **Systemic** chemicals enter roots, stems and leaves, and are translocated to sites where they may affect hyphal growth and prevent spore production. Although there are many fungicidal chemical groups, four are chosen here as examples:

- **Inorganic chemicals** contain no carbon. Two chemicals are available to amateur and professional growers. Commercially formulated compounds of copper salts mixed with slaked lime (**Bordeaux mixture**) form a protective barrier to fungi such as potato blight when sprayed onto the leaf. Fine-grained (colloidal) **sulphur** controls powdery mildews and apple scab.
- Organic chemicals contain carbon. **Mancozeb** (dithiocarbamate group) and related synthetic compounds act protectively on a wide range of quite different foliar diseases, such as downy mildews, celery leaf spot and rusts, by preventing spore germination. Mancozeb is available to both amateur and professional growers.
- **Carbendazim** (benzimidazole group) is available only to professional growers. It is an example of a systemic ingredient, which moves upwards through the plant's xylem tissues, slowing hyphal growth and spore production of fungal wilts, powdery mildews and many leaf spot organisms. Damping off, potato blight and downy mildews are **not** controlled by chemicals within this chemical group. Many different systemic groups are now used in horticulture.
- **Myclobutanil** belongs to the conazole group. It is available to both amateur and professional growers. It is protectant and systemic, on powdery mildews, black spot of rose and apple scab.

Resistance to pesticides

The development of resistant individuals from the millions of susceptible weeds, pests and diseases occurs most rapidly when exposure to a particular chemical is continuous or when a pesticide acts against only one body process of the organism. Resistance, e.g. in powdery mildews, to one member, e.g. carbendazin, of a chemical group confers resistance to other chemicals in the same benzimidazole group. Growers should therefore follow the strategy of alternating between different groups and not simply changing active ingredients. Particular care should be taken with systemic chemicals that present to the organism inside the plant a relatively weak concentration against which the organism can develop resistance. Increase in dosage of the chemical will not, in general, provide a better control against resistant strains. Biological control, unlike chemical control, does not create resistant pests.

Formulations

Active ingredients are mixed with other ingredients to increase the efficiency and ease of application, prolong the period of effectiveness or reduce the damaging effects on plants and man. The whole product (formulation) in its bottle or packet is given a trade name, which often differs from the name of the active ingredient. The main formulations are as follows:

- **Liquids** (emulsifiable concentrates) contain a light oil or paraffin base in which the active ingredient is dissolved. Detergent-like substances (emulsifiers) present in the concentrate enable a stable emulsion to be produced when the formulation is diluted with water. In this way, the correct concentration is achieved throughout the spraying operation. Long chain molecular compounds (wetter/spreaders) in the formulation help to stick the active ingredients onto the leaf after spraying, particularly on smooth, waxy leaves such as cabbage.
- **Wettable powders** containing extremely small particles of active ingredient and wetting agents form a stable suspension for only a short period of time when diluted in the spray tank. Continuous stirring or shaking of the diluted formulation is thus required. An inert filler of clay-like material is usually present in the formulation to ease the original grinding of particles, and also to help increase the shelf life of the product. It is suggested that this formulation is mixed to a thin paste before pouring through the filter of the sprayer. This prevents the formation of lumps that may block the nozzles.
- **Dusts** are applied dry to leaves or soil, and thus require less precision in grinding of the constituent particles and less wetting agent.
- **Seed dressings** protect the seed and seedling against pests and diseases. A low percentage of active ingredient, such as **iprodione** applied in an inert clay-like filler or liquid reduces the risk of chemical damage to the delicate germinating seed.
- **Baits** contain attractant ingredients, e.g. bran and sugar, mixed with the active ingredient, e.g. **methiocarb**, both of which are eaten by the pest, such as slugs.
- **Granules** formulated to a size of about 1.0 mm contain inert filler, such as pumice or charcoal, onto which the active ingredient is coated. Granules may act as soil sterilants (e.g. **dazomet**), residual soil herbicide (e.g. **dichlobenil**), residual insecticide (e.g. **chlorpyrifos**), or broad-spectrum soil nematicide and insecticide (e.g. **aldicarb**). Granular formulations normally present fewer hazards to the operator and fewer spray-drift problems.

Labels on commercial formulations give details of the active ingredient contained in the product. Application rates for different crops are included. DEFRA approves pesticide products for effectiveness.

Phytotoxicity (or plant damage) may occur when pesticides are unthinkingly applied to plants. Soil applied insecticides, such as **aldicarb**, can cause pot plants, such as begonias, to go yellow if used at more than the recommended rate. Plants growing in greenhouses are more susceptible because their leaf cuticle is thinner than plants growing at cooler temperatures. Careful examination of the pesticide (particularly herbicide) packet labels often prevents this form of damage.

Application of herbicides and pesticides

This subject is described in detail in machinery texts. However, certain basic principles related to the covering of the leaf and soil by sprays will be mentioned. The application of liquids and wettable powders by means of sprayers may be adjusted in terms of pressure and nozzle type to provide the required spray rate. Cone nozzles produce a turbulent spray pattern suitable for fungicide and insecticide use, while fan nozzles produce a flat spray pattern for herbicide application. In periods of active plant growth fortnightly sprays may be necessary to control pests and diseases on newly expanding foliage.

- **High volume** sprayers apply the diluted chemical at rates of 600–1000 l/ha in order to cover the whole leaf surface with droplets of 0.04–0.10 mm diameter. Cover of the under-leaf surface with pesticides may be poor if nozzles are not directed horizontally or upwards. Soil applied chemicals, such as herbicides or drenches, may be sprayed at a larger droplet size, 0.25–0.5 mm in diameter, through a selected fan nozzle. The correct height of the sprayer boom above the plant is essential for downward-directed nozzles if the spray pattern is to be evenly distributed. Savings can be achieved by band spraying herbicides in narrow strips over the crop to leave the inter-row for mechanical cultivation.

Figure 16.10 **Knapsack sprayer** being used in nursery stock

- **Medium volume** (200–600 l/ha) and low volume (50–200 l/ha) equipment, such as knapsack sprayers, apply herbicides and pesticides onto the leaf at a lower droplet density, and in tree crops, mist blower equipment creates turbulence, and therefore increased spray travel, by means of a power driven fan. **Ultra-low volume** sprays (up to 50 l/ha) are dispersed on leaving the sprayer by a rapidly rotating disc which then throws regular-sized droplets into the air. Larger droplets (about 0.2 mm) are created by herbicide sprayers to prevent spray-drift problems, while smaller droplets (about 0.1 mm) allow good penetration and leaf cover for insecticide and fungicide use.

Fogging machines used in greenhouses and stores produce very fine droplets (about 0.015 mm diameter) by thermal and mechanical methods, and use small volumes of concentrated formulation (less than 1 l in 400 m^3) which act as fumigants in the air, and as contact pesticides when deposited on the leaf surface. **Dust and granule applicators** spread the formulations evenly over the foliage or ground surface. When mounted on seed drills and/or fertilizer applicators, granules may be incorporated into the soil. Care must be taken to ensure good distribution to prevent pesticide damage to germinating seeds or planting material.

Safe practice

In chemical control, the **hazards** are:

- possible acute poisoning of humans, pets, farm animals, bees, and wild animals;
- possible accumulation of pesticides that lead to toxic levels in humans, pets, farm animals, bees and wild animals;
- possible cancer inducing effects in humans;
- possible damage to cultivated and wild plants especially by herbicides;
- possible contamination of streams and dams;
- possible development of strains of rodents, insects, mites, and fungi, resistant to pesticides.

When using chemical control, risks can be **minimized** by:

- restricting chemical applications to only those situations that justify such a control measure. In many instances, other controls measures may be preferable and less hazardous;
- carefully choosing the least hazardous chemical to effectively control the problem organism;
- carefully reading the instructions on the product label;
- carefully choosing the correct clothing, where necessary;
- carefully measuring the correct amount of concentrate water (where relevant);
- calculating (where appropriate) the amount of pesticide and water necessary for application to the crop area in question;
- carefully mixing the two, avoiding spillage on to skin, clothing and the surrounding area;

- carefully applying the product so that the same area is not covered more than once, at any one time;
- carefully applying the product under suitable dry, wind-free weather conditions;
- carefully applying the product so that other humans, beneficial animals, waterways and adjacent plantings are avoided;
- carefully avoiding spray drift, especially with herbicides;
- carefully storing pesticides in a secure, safe, dry place away from children and pets.

Toxicity aspects of pesticides

The basic biochemical similarities between all groups of plants and animals means that any potential chemical chosen for its action against a weed, pest or disease may also be toxic to humans, pets, horticultural species and wildlife animals and plants. Prospective pesticides therefore have to go through a thorough examination over a period of several years to determine whether there are any dangers. This is carried out by the chemical companies and by contracted independent organizations. The evidence is scrutinized by government committees before there can be any possibility of the product's commercial release. This ensures that no damage will occur to non-target species (particularly humans) if safety precautions are followed. During 2005, 360 pesticide commercial products were withdrawn from use in agriculture and horticulture. During the same period, 160 products were approved.

Acute toxicity to humans

An important indicator of the safety of an ingredient is the lethal dose figure (LD50) by ingestion of a chemical. It expresses the amount of active ingredient required to kill 50 per cent of a population of animals and is expressed as mg/kg of animal tissue. This oral LD50 is used as an indicator for establishing the precautions needed for a grower to safely mix and apply a product. The lower the LD50 figure for a chemical, the more toxic it is. To put toxicity levels into some perspective, five everyday substances are presented in Table 16.2 alongside a range of five pesticides/growth regulators available to amateur and professional growers.

Other aspects of toxicity

Acute toxicity is not the only property of a potential pesticide to be assessed. Its chronic (long lasting) aspect must also be tested. For example, its **survival time** on the surface of the leaf may influence its suitability, particularly on leaf crops, such as lettuce, which have a large surface area of pesticide deposit and which are eaten fresh. Pesticides must also be checked against their ability to cause **irritation** and **allergies** in humans and their ability to cause cancer. An active ingredient may be particularly toxic to other mammals, fish, earthworms, bees and predatory animals. When testing active ingredients, research workers remember very well the havoc that chemicals such as DDT caused in killing animals at the top of food chain (see also p52).

Table 16.2 A comparison of LD50s in households, private gardens, and commercial units

Material	Use	Oral LD50 mg/kg
Aldicarb (most toxic)	Insecticide/nematicide (CU)	0.5
Nicotine	Cigarettes (H)	50
	Insecticide (CU)	50
Methiocarb	Slug pellets (G, CU)	100
Caffeine	Beverage (H)	190
Deltamethrin	Insecticide (CU)	50
2,4-D	Herbicide (G, CU)	450
Bifenthrin	Insecticide (G, CU)	630
Metalaxyl	Fungicide (CU)	670
NAA	Plant growth regulator (G, CU)	1000
Common salt	Food additive (H)	1000
Mancozeb	Fungicide (G, CU)	5000
Alcohol	Beverage (H)	5000
Glyphosate	Herbicide (CU)	5000
Fatty acids	Insecticide (G, CU)	5000
Vitamin C (least toxic)	Food ingredient (H)	12000

Key: Household – H; Garden – G; Commercial Unit – CU

Product label

The 'statutory area' on the label present on each packet or bottle of pesticide must provide the following details:

- The 'field' of use, whether agriculture, horticulture, home garden or animal.
- The plant species, crop or situation where treatment is permitted.
- The maximum dose or concentration.
- The maximum number of treatments.
- The latest time of application or harvest interval (days between application and harvest).
- Any specific restrictions, such as clothing required, temperature at which application should be made. (The nature of the protective clothing stated on the label commonly reflects the LD50 status of the ingredient.)
- A reminder to read all other safety precautions on the label and directions for use.

The **amateur gardener** does not need to pass a proficiency test for pesticide usage. Active ingredients have been selected with care to ensure that no danger from toxicity is present. In 2008 there is a choice of 24 active ingredients (4 insecticides/acaricides, 4 fungicides, 1 animal repellent, 5 slug control chemicals, 9 herbicides, and 1 growth regulator for plant propagation). The **professional horticulturist** may need to use pesticides which have special requirements in terms of their storage, mixing and application. Three main items of legislation come into play for them.

The first item of legislation focuses on the skill and understanding of the operator as they approach a chosen pesticide application. This was seen in the **Health and Safety Regulations 1975** which was summarized in the government '**Poisonous Chemicals on the Farm**' leaflet. This document specifies the correct procedures for pesticide use. A detailed register must be kept of spraying operations and any dizziness or illness reported. Correct washing facilities must be provided. A lockable dry store is necessary to keep chemicals safe. Warnings of spraying operations should be prominently displayed. A suitable fabric coverall suit with a hood must be used to protect most of the body from diluted pesticide. Rubberized suits should be used in conditions of greater danger, such as in an enclosed greenhouse environment, when dealing with ultra-low volume spray or when applying upward-directed sprays into orchards. Rubber boots should be worn inside the legs of the suit. Thick gauge gauntlets are worn outside the suit when dealing with concentrates, but inside when spraying. Face shields should be worn when mixing toxic concentrates. A face mask covering the mouth and/or nose, and capable of filtering out less toxic active ingredients may need to be used for spraying, but a respirator with its large filter is required for toxic products, particularly when used in greenhouses where toxic fume levels build up.

With regard to wildlife, pesticides should not be sprayed near ponds and streams unless designed for aquatic weed control. Crops frequented by bees, such as apples and beans, should be sprayed with insecticides only in the evening when most of the insects' foraging has ceased. Beekeepers should be informed of spraying operations.

The second item of legislation was the important, wide ranging, **Food and Environmental Protection Act 1985 (FEPA)** which highlighted public and government concern about pesticide dangers, and the need for a UK-wide improvement in responsible pesticide usage. The Act further required that chemical manufacturers, distributors and professional horticulturists should be able to demonstrate skills in the choice and careful management of pesticides. The Act also sought to make information about pesticides more available to the public.

A very practical aspect of FEPA is the specific requirement that all professional personnel involved with pesticides should demonstrate a high level of competence. To this end, the Act requires anyone intending to apply a pesticide to have passed two tests. The first (PA 1) test assesses knowledge in the following subject areas; legislation, places of special environmental value, safe use of pesticides, keeping records of products used and applications performed, storage of pesticides, cleaning of equipment, protective clothing and appliances, disposal of unwanted pesticide, and dealing with contamination and poisoning incidents. The second test (PA 6) assesses practical ability in pesticide application (normally by means of a knapsack sprayer or tractor mounted field sprayer). This involves proficiency in choosing a product for a specified job, calculating the amount of product and volume of water needed for a given area of land, using the correct clothing and equipment for mixing a concentrate and for the application of diluted product, performing the spraying operation, disposing of excess spray liquid, and the cleaning of spray equipment.

A third item of legislation was the **Control of Substance Hazardous to Health (COSHH) Regulations 2002** which formalized further the responsibility of the pesticide operator to assess whether each pesticide application was necessary. Once a decision to use a pesticide has been made, further investigation should lead to the choice of the most appropriate and safe active ingredients available, and the most appropriate clothing to wear.

A fourth development relating to pesticides has been the **Voluntary Initiative**, set up in 2001 by the farming and crop protection industries in association with the UK government to minimize the impact of pesticides on the environment.

In the **commercial horticultural** sector, the number of active ingredients available for use is much greater than in the private garden area. In 2008 there are about 145 active ingredients (33 insecticides/acaricides, 2 nematicides, 39 fungicides, 6 sterilants/fumigants, 11 animal repellants, 2 slug control chemicals, 46 herbicides and 16 growth regulators). Many active ingredients with low LD50 values which were used in the recent past have been banned as too dangerous. The chemical sodium cyanide, which is used to control rabbits, has an oral LD50 of 5 mg/kg and must now be applied to rabbit burrows only by licensed operators. Once the approval by the UK government Pesticide Safety Directorate has been given, further details for the product are then formalized following the guidelines given in Regulations under the Food and Environmental Protection Act 1985 (**FEPA**), to ensure the pesticide's transport, storage and application do not endanger humans and wildlife.

Selection for plant resistance

Genetic answers to plant health problems

Wild plants show high levels of resistance to most pests and diseases. In the search for high yields and extremes of flower shape and colour, plant breeders have often failed to include this **wild plant resistance**. However, in crops such as antirrhinum, lettuce and tomatoes, one or more resistance genes have been deliberately incorporated to give protection against rust, downy mildew and tomato mosaic virus respectively. However, the disease organisms competing against the resistance may overcome the genetic barriers and the crop thus again becomes infected. Growers may sow a sequence of cultivars (such as in lettuce), each with different resistance genes, in order that the disease organism (such as downy mildew) is constantly exposed to a new resistance barrier, and thus limit the disease.

Examination of the genetics of wild plant resistance usually shows that there are several (often many) genes contributing to the overall resistance effect. The complex nature of the resistance prevents the frequent development of new strains of diseases that could seriously affect the plant. Gardeners and professional horticulturist alike are increasingly looking to choose cultivars with proven long-term resistance as a feature that is as important to them as yield and plant quality, etc.

Recently introduced cultivars of cabbage such as 'Kalaxy' are claimed to have stable resistance to club root, while cultivars of potatoes such as 'Sarpo Axona' have until now shown very good leaf resistance and very good tuber resistance to potato blight. Looking at plant resistance with a slightly wider view, it should be noted that some strains of fungus are specific to particular families of plant. For example, a *Fusarium oxysporum* wilt strain attacking tomatoes (member of the Solanaceae family) will not carry over to a subsequently planted crop of pinks and carnations (members of the Caryophyllaceae).

Vegetatively propagated species, such as potatoes, and tree crops, such as apple, which remain genetically unaltered for many years are now being bred with high levels of 'wild plant' resistance (to blight and powdery mildew respectively on potato and apple) so that the crops may resist these serious problems more permanently.

Crop resistance to insects is now being more seriously considered by plant breeders. Some lettuce cultivars are resistant to lettuce root aphid (*Pemphigus bursarius*). A few carrot cultivars have some resistance to carrot fly (*Psila rosae*). A new apple cultivar has shown resistance to apple aphids. Table 16.3 illustrates some vegetable cultivars showing their resistance to various diseases and pests.

Table 16.3 Some examples of resistance to pests and diseases present in vegetable crops

Crop	Cultivar	Pest or disease
Aubergine	'Bonica'	Cucumber mosaic
Bean, dwarf	'Forum'	Anthracnose
Brussels sprout	'Saxon'	Powdery mildew
		Turnip mosaic virus
		White blister rust
Carrot	'Fly away'	Carrot fly
Chinese cabbage	'Harmony'	Club root
Cucumber	'Burpee Hybrid'	Cucumber mosaic
		Downy mildew
Leek	'Poristo'	Viruses
Lettuce	'Beatrice'	Root aphid
		Downy mildew
Pea	'Onward'	Fusarium wilt
Pepper	'Bell Boy'	Tomato mosaic virus
Potato	'Sarpo Axona'	Blight
Tomato	'Primato'	Tomato mosaic virus
		Fusarium wilt
		Verticillium wilt

Integrated control

Integrated control involves the use of natural pest enemies together with cultural, physical and chemical controls.

In integrated control the main **hazard** is the unintentional killing of biological control organisms by pesticides (which are being used for pests that cannot be controlled biologically). When using integrated control, **risks** can be minimized by choosing the appropriate pesticides that do not harm biological control organisms.

Integrated control, increasingly termed '**Integrated Pest Management**' or IPM, requires the grower to understand all types of control measures, particularly biological and chemical, in order that they complement each other. In greenhouse production of cucumbers, the *Encarsia formosa* parasite and *Phytoseiulus persimilis* predator are used for whitefly and red spider mite control respectively. However, the other harmful pest and disease species must be controlled, often by chemical means, without killing the parasites and predators. Drenches of systemic insecticide, such as **pirimicarb** against aphids, soil insecticides, such as **deltamethrin** against thrips pupae, and systemic fungicide drenches, such as **carbendazim** against wilt diseases and powdery mildew, are all applied away from the sensitive biological control organisms. Similarly, high volume sprays of selective chemicals, such as **iprodione** against grey mould, *Bacillus thuringiensis* extracts against caterpillars, have little or no effect on the parasite and predator. Similar considerations may be given in control of apple pests and diseases. Reduced usage of extremely toxic winter washes and increased use of selective caterpillar and powdery mildew control by chemicals, such as **diflubenzuron** and **fenarimol** respectively, allow the almost unhindered build-up of beneficial organisms, such as predatory capsid and mites.

The methods of **organic growers** emphasize the non-chemical practices in plant protection (as well as in soil fertility). Hedges are developed within 100 m of production areas and are clipped only one year in four to maintain natural predators and parasites. Rotations are closely followed to enable soil-borne pest or disease decline, while encouraging soil fertility. Resistant cultivars of plant are chosen and judicious use of mechanical cultivations and flame weeding enables pests, diseases and weeds to be exposed or buried. A restricted choice of pesticide products such as **pyrethrins**, **derris**, **metaldehyde** (with repellant), **sulphur**, **copper salts** and **soft soap** are allowed to be applied, should the need arise. *Bacillus thuringiensis* extract, and **also pheremone** attractants, are similarly used. Table 16.4 gives a list of permitted substances.

Supervised control

Supervised control involves the careful assessment of pest, disease and weed levels in order to achieve effective control.

Most plants can tolerate low levels of pest and disease damage without yield reduction, unless the damage is to parts of the plant that become unacceptable (such as fruits for the supermarket trade). The term '**economic threshold**' is used to summarize this concept. Cucumbers, for example, require more than 30 per cent leaf area affected by red spider mite before **economic damage** occurs in terms of yield loss. This enables methods of control that depend on some damage being

Table 16.4 Some permitted products for pest and disease control in organic crop production

Preparations on basis of metaldehyde containing a repellent to higher animal species and as far as possible applied within traps. For control of slugs
Preparations of pyrethrins extracted from *Chrysanthemum cinerariaefolium*, containing possibly a synergist for insect control
Preparations from *Derris elliptica* for insect control
Preparations from *Quassia amara* for aphid control
Preparations from *Ryania speciosa* for insect control
Sulphur for fungal diseases
Bordeaux mixture (copper based) for fungal and bacterial diseases
Burgundy mixture (copper based) for fungal diseases
Potassium soap (soft soap) for insect control
Pheremone traps for evaluating insect numbers
Bacillus thuringiensis preparations for caterpillar control

done to ensure continued success, such as the use of predators. Damage assessments are used in apple orchards to decide whether control measures are necessary. Thus, at green-cluster stage (before flowers emerge) chemical sprays are considered only when an average of half the observed buds has five aphids per bud. Similarly, an average of three winter moth larvae per bud-cluster merit control at late blossom time. Pheremone traps enable the precise time of maximum codling moth emergence to be determined in early June. Catches of less than 10 moths per trap per week do not warrant control. DEFRA issue **spray warning** information to growers when serious pests, such as carrot root fly, and diseases, such as potato blight, are likely to occur. Supervised control may greatly reduce pesticide costs.

Legislative control

Before 1877 no legal measures were available in the UK to prevent importation of plants infested with pests such as Colorado beetle. Measures taken at ports from that year onwards were brought together in the 1927 Destructive Insects and Pests Acts, empowering government officials to inspect and, if necessary, refuse plant imports. Within this Act, the 'Sale of Diseased Plants Order' placed on the grower the responsibility for recognition and reporting of serious pests and diseases, such as blackcurrant gall mite and silver leaf of plums. Lack of education and enforcement led to the need for specific orders relevant to particular current problems, such as in 1958 the fireblight-susceptible pear cultivar, 'Laxton's Superb', was declared **notifiable** and prohibited under the Fireblight Disease Order.

Recent orders have helped prevent outbreaks of white rust on chrysanthemums, plum pox virus and two American leaf miner species; less success has been achieved with the Western flower thrip organism. Further importation legislation under the '1967 Plant Health Act' prohibits the landing of any non-indigenous pest or disease by aircraft or post, and the 'Importation of Plants, Produce and Potatoes Order

1971' specifically names prohibited crops, such as plum rootstocks from eastern Europe, crops without a **phytosanitary** certificate, such as potato tubers from Europe, and crops that first need to be examined by inspectors, such as Acacia shrubs and apricot seeds.

The carrying out of these orders is supervised by the Plant Health Branch of the Department for Environment, Food and Rural Affairs (DEFRA). Complete success in preventing the introduction of damaging organisms may be limited by dishonest importations and by the difficulty of detection of some diseases, especially viruses. European Council directives, such as 77/93/EEC and 2000/29/EC, are moving member nations towards a unified approach in reducing the transfer of infected plant material across national boundaries, and the UK Plant Health Order (2005) implements the EC legislation. The Weeds Act 1959 places a legal obligation on each grower to prevent the spread of weeds such as creeping thistle, spear thistle, curled dock, broad-leaved dock and ragwort. In addition, under the UK Wildlife and Countryside Act (1981), it can be an offence to plant Japanese knotweed (*Polygonum japonicum*) and giant hog-weed (*Heracleum mantegazzianum*).

Check your learning

1. Define the terms 'biological control', 'chemical control', 'physical control' and 'cultural control'.

2. State the benefits and limitations of using biological control.

3. Describe one physical method of plant protection.

4. Describe one cultural method of plant protection.

5. Describe one biological method of plant protection.

6. Describe one chemical method of plant protection.

7. Name five types of hazard which may be encountered with pesticide use.

8. Describe an example of the term 'economic threshold' used in plant protection.

9. Describe an example of an 'integrated control' used in plant protection.

10. Describe an example of warm water treatment used in plant protection.

Further reading

Alford, D. (2000). *Pest and Disease Management Handbook*. Blackwell Science Ltd.

Brown, L.V. (2008). *Applied Principles of Horticultural Science*. 3rd edn. Butterworth-Heinemann.

Cremlyn, R.J. (1991). *Agrochemicals, Preparation and Mode of Action*. Wiley.

Debach, P. (1991). *Biological Control by Natural Enemies*. Cambridge University Press.

Dent, D. (1995). *Integrated Pest Management*. Chapman & Hall.

Hance, R.J. and Holly, K. (1990). *Weed Control Handbook*. Vol. 1. Blackwell Scientific Publications.

Helyer, N. *et al.* (2003). *A Colour Handbook of Biological Control in Plant Protection*. Manson Publishing.

Hussey, N.W. and Scopes, N. (1985). *Biological Pest Control*. Blandford Press.

Ingram, D.S. *et al.* (2002). *Science and the Garden*. Blackwell Science Ltd.

Ivens, G.W. (1989). *Plant Protection in the Garden*. British Crop Protection Council.

Lloyd, J. (1997). *Plant Health Care for Woody Ornamentals*.

Whitehead, R. (ed.) (2007). *UK Pesticide Guide*. CAB Publishing.

Chapter 17   Physical properties of soil

Summary

This chapter includes the following topics:

- Soil profiles
- Topsoil and subsoil
- Characteristics of sand, silt and clay
- Soil texture
- Soil structure
- Cultivations
- Bed systems

with additional information on the following:

- Plant requirements
- Composition of soils
- Formation of soils
- Natural soil profiles
- Soils of the British Isles
- Management of main soil types

Figure 17.1 **Formation of soil** begins with the weathering rocks which includes the powerful force of the sea.

Plant requirements

The growing tip of the root wriggles through the growing medium following the line of least resistance. Roots are able to enter cracks that are or can be readily opened up to about 0.2 mm in diameter, which is about the thickness of a pencil line. Compacted soils severely restrict **root exploration**, but once into these narrow channels the root is able to overcome great resistance to increase its diameter. Anything which reduces root exploration and activity can limit plant growth. When this happens action must be taken to remove the obstruction to root growth or to supply adequate **air**, **water** and **nutrients** through the restricted root volume.

The root normally provides the **anchorage** needed to secure the plant in the soil. Plants, notably trees with a full leaf canopy, become vulnerable if their roots are in loose material, in soil made fluid by high water content or are restricted, e.g. shallow roots over rock strata close to the surface. Until their roots have penetrated extensively into the surrounding soil, transplants are very susceptible to **wind rocking**: water uptake remains limited as roots become detached from the soil and delicate root growth is broken off. The plant may be left less upright.

In order to grow and take up water and nutrients the root must have an **energy** supply. A constant supply of energy is only possible so long as oxygen is brought to the site of uptake (see respiration). Consequently the soil spaces around the root must contain air as well as water. There must be good **gaseous exchange** between the atmosphere around the root and the soil surface. This may sometimes be achieved by the selection of plants that have modifications of their structure that enables this to occur throughout the plant tissues (see adaptations), but it is normally a result of maintaining a suitable soil structure. A lack of oxygen or a build-up of carbon dioxide will reduce the root's activity. Furthermore, in these conditions anaerobic bacteria will proliferate, many produce toxins such as ethylene. In warm summer conditions roots can be killed back after one or two days in waterlogged soils.

Composition of soils

Mineral soils form in layers of rock fragments over the Earth's surface. They are made up of mineral matter comprising **sand**, **silt** and **clay** particles. There is also a small quantity of organic matter which is the part derived from living organisms. This framework of solid material retains water and gases in the gaps or **pore space**. The water contains dissolved materials including plant nutrients and oxygen and is known as the **soil solution**. The **soil atmosphere** normally comprises nitrogen, rather less oxygen and rather more carbon dioxide than in normal air, and traces of other gases. Finally, a soil capable of sustaining plants is alive with **micro-organisms** (organisms, such as

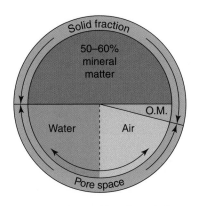

Figure 17.2 Composition of a typical cultivated soil. The solid fraction of the soil is made up of mineral (50–60 per cent) and organic (1–5 per cent) matter. This leaves a total pore space of 35–50 per cent that is filled by air and water, the proportions of which vary constantly.

bacteria, fungi and nematodes, too small to be seen with the naked eye). Larger organisms such as earthworms and insects are also normally present (see p321).

The composition of a typical mineral soil is given in Figure 17.2, which also illustrates the variation that can occur. The content of the pore space varies continually as the soil dries out and is rewetted. The spaces can be altered by the compaction or 'opening up' of the soil which in turn has a significant effect on the proportions of air and water being held.

Over a longer period the organic matter level can vary. The composition of the soil can be influenced by many factors and under cultivation these have to be managed to provide a suitable root environment. Organic soils have considerably higher organic matter content and are dealt with in Chapter 18.

Formation of soils

The Earth formed from a ball of molten rock minerals. The least dense rocks floated on the top and as they cooled the surface layer of granite, with basalt just below, solidified to form the Earth's crust. The Earth's surface has had a long and turbulent history during which it has frequently fractured, crumpled, lifted and fallen, with more molten material being pushed up from below through the breaks in the crust and in volcanoes.

Weathering

Weathering is the breakdown of rocks.

From the moment that rocks are exposed to the atmosphere, they are subject to weathering. This breakdown of the rocks is brought about by the effect of chemical, physical and biological factors.

Chemical weathering is mainly brought about by the action of **carbonic acid** that is produced wherever carbon dioxide and water mix, as in rainfall. Some rock minerals dissolve and are washed away. Others are altered by various chemical reactions, most of which occur when the rock surface is exposed to the atmosphere. All but the inert parts of rock are eventually decomposed and the rock crumbles as new minerals are formed and soluble material is released. **Oxidation** is particularly important in the formation of iron oxides, which give soils their red and yellow (when aerobic), or blue and grey colours (in anaerobic conditions).

Physical or mechanical weathering processes break the rock into smaller and smaller particles without any change in the chemical character of the minerals. This occurs on exposed rock surfaces along with chemical weathering but, in contrast, has little effect on rocks protected by layers of soil.

The main agents of physical weathering are frost, heat, water, wind and ice. In temperate zones, frost is a major weathering agent. Water percolates into cracks in the rock and expands on freezing. The pressures created shatter the rock and, as the water melts, a new surface is exposed to weathering. In hot climates the rock surface can become very much hotter than the underlying layers. The strains created by the different amounts of expansion and the alternate expansion and contraction cause fragments of rock to flake off the surface; this is sometimes known as the 'onion skin' effect. Moving water or wind carries fragments of rock that rub against other rocks and rock fragments, wearing them down. Where there are glaciers the rock is worn away by the 'scrubbing brush' effect of a huge mass of ice loaded with stones and boulders bearing down on the underlying rock.

Biological weathering is attributable to organisms such as mosses, ferns and flowering plants which fragment rock by both chemical and physical means, e.g. they produce carbon dioxide which, in conjunction with water, forms carbonic acid; roots penetrate cracks in the rock and, as they grow thicker, they exert pressure which further opens up the cracks.

Rocks

Erosion is the movement of rock fragments and soil.

Figure 17.3 **Rocks**. Granite: pink (left) silver (top) sandstone (right) slate (bottom)

Igneous rocks are those formed from the molten material of the Earth's crust. All other rock types, as well as soil, are ultimately derived from them. When examined closely, most igneous rocks can be seen to be a mixture of crystals. **Granite** is one of the commonest and contains crystals of quartz, white and shiny, felspars that are grey or pink, and micas, which are shiny black (see Figure 17.3). Many of these crystalline materials have a limited use in landscaping as formal structures rather than in the construction of rock gardens; more commonly they are used in monuments and building facades.

As granite is weathered ('rotted') the felspars are converted to kaolinite (one of the many forms of clay) and soluble potassium, a plant nutrient. Similarly, the mica present is chemically changed to form clay and yield soluble minerals. Whilst the many types of clay retain much of the potassium, sodium, calcium, etc., the soluble material is carried by water to the sea making the sea 'salty'. The inert quartz grains are released and form sand grains.

Sedimentary rock is derived from accumulated fragments of rock. Most have been formed in the sea or lakes to which agents of **erosion** carry weathered rock. Organisms in the seas with shells die and accumulate on the bottom of the sea. Layers of sediment build up and, under pressure and slow chemical change, eventually become rock strata such as shale, chalk or limestone. In subsequent earth movements much of it has been raised up above sea level and weathered again. Similarly, the sand grains that accumulate to great depths in desert areas eventually become sandstones (Figure 17.3).

Moving water and winds are able to carry rock particles and are thus important agents of erosion. As their velocity increases the '**load**' they are able to carry increases substantially. The fast-moving water in streams is able to carry large particles, but in the slower-moving rivers some of the load is dropped. The particles settle out in order of size (see settling velocities). This leads to the **sorting** of rock fragments, i.e. material is moved and deposited according to particle size. By the time the rivers have reached the sea or lakes only the finest sands, silts and clays are in the water. As the river slows on meeting the sea or lake all but clay is dropped. The clay eventually settles slowly in the quieter waters of the sea or lake. Moving ice is also an agent of erosion, but the load dropped on melting consists of unsorted particles known as **boulder clay** or till.

The type of sedimentary rock formed depends on the nature of its ingredients. Sandstones, siltstones and mudstones are examples of sedimentary rocks derived from sorted particles in which characteristic layers are readily seen. Limestones are formed from the accumulation of shells (see Figure 17.4) or the precipitation of materials from solution mixed with varying amounts of deposited mud. Chalk is a particularly pure form derived from the calcium carbonate remains of minute organisms that lived in seas in former times. Many of these are attractive materials for use in hard landscaping, where care should be taken to align the strata (layers) for a natural effect.

Metamorphic rock is formed from igneous or sedimentary rocks. The extreme pressures and temperatures associated with movements and fracturing in the Earth's crust or the effect of huge depths of rock on underlying strata over very long periods of time has altered them. Slate is formed from shale, quartzite from sandstone, and marble from limestone. Metamorphic rock tends to be more resistant to weathering than the original rock.

Figure 17.4 **Limestone**

Natural soil profiles

Sedentary soils

Sedentary soils develop in the material gradually weathered from the underlying rock. True sedentary soils are uncommon because most loose rock is eroded, but the same process can be seen where great depths of transported material have formed the parent material, as in the boulder clays left behind after the Ice Ages. A hole dug in such a soil shows the gradual transition from unweathered rock to organicmatter rich topsoil (Figure 17.5). Under cultivation a distinctive topsoil develops in the plough zone.

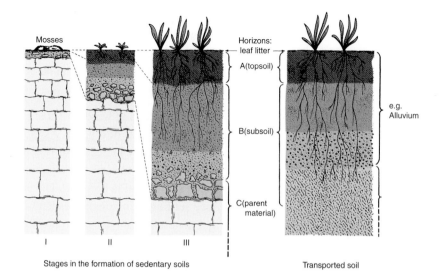

Stages in the formation of sedentary soils

Transported soil

Figure 17.5 The development from a young soil consisting of a few fragments of rock particles to a deep sedentary soil is shown alongside a transported soil. A subsoil, topsoil and leaf litter layer can be identified in each soil. Simple plants such as lichens and mosses establish on rocks or fragments to be succeeded by higher plants as soil depth and organic matter levels increase.

Transported soils

Once rock fragments and soil particles are created they become subject to **erosion**. **Transported soils** are those that form in eroded material that has been carried from sites of weathering, sometimes many hundreds of miles away from where deposition has occurred. They can be recognized by the definite boundary between the eroded material and the underlying rock and its associated rock fragments. Where more than one soil material has been transported to the site, as in many river valleys, several distinct layers can be seen. The right-hand part of Figure 17.5 shows an example. How they are moved depends on where the loose material lies:

- **Gravity** affects anything on a slope. On steep sides, e.g. cliffs, particles fall and accumulate at the bottom to form heaps of rock called 'scree'.
- On gentler slopes particles are helped downhill by **rainsplash**. Raindrops striking soil dislodge loose particles that tend to move downhill. As a result, surface soil is slowly removed from higher ground and accumulates at the bottom of slopes. This means that soils on slopes tend to be shallow, whereas at the bottom deep, transported soils develop, known as **colluvial** soils.
- **Glaciers** carry vast quantities of rock downhill and deposit their load at the 'snout' (terminal moraines). Of more significance is the enormous load that was left behind when the glaciers retreated after the last Ice Age (10 000 years ago). This is known as 'till' or 'boulder clay' (it comprises boulders down to clay size particles).
- Material washed away in **running water** eventually settles out according to particle size. The river valley bottoms become covered with material (alluvium) in which **alluvial** soils develop.

- **Wind** removes dry sands and silts that are not 'bound in' to the soil. The soils that develop from wind-blown deposits are known as 'loess' or 'brick-earth'.

Many of these transported soils provide ideal rooting conditions for horticultural crops because they tend to be deep, loose and open. Most are easily cultivated. However, those that have a high silt or fine sand content, notably the brick-earths, may be prone to compaction.

Soil development

The nature of a new soil (regosols) is largely determined by the rock minerals from which it is formed, but it continues to undergo changes under the influence of **climate**, **vegetation**, **topography** and **drainage**. These interact over **time** to give rise to characteristic soil profiles in different parts of the world. The soils that develop can be described in terms of the characteristics of the different horizons (layers) that make up the soil profile.

A **soil horizon** is a specific layer in the soil seen by digging a **soil pit**. The layers revealed make up the **soil profile**.

The 'O' or 'L' horizon is the organic matter found on top of the mineral soil and commonly referred to as the litter layer. The upper layer of the soil, from which components are normally washed downwards, is the 'A' horizon. This is usually recognized by its darker colouring, which is a result of the significant levels of humus present. The lighter layer below it, where finer materials tend to accumulate, is the 'B' or illuvial horizon. Under cultivation, the 'A' horizon broadly aligns with the 'topsoil' and the 'B' with the 'subsoil'. The parent material below these is the 'C' horizon and where there is an underlying unweathered rock layer it is often known as bedrock.

Soils of the British Isles

In the British Isles four main types of mineral soil are found; **brown earths**, **gleys**, **rendzinas** and **podsols** (see Figure 17.6). **Peats** develop in waterlogged conditions (see organic soils p328).

Brown earth soils develop in the well-drained medium to heavy soils in the lowlands of the British Isles. They are associated with a climax vegetation (p52) of broad-leaved woodland especially oak, ash and sycamore, the roots of which have ensured that nutrients moving down the soil profile are captured and returned to the soil via the leaf fall. Surplus water does not accumulate and the soil remains aerobic for most of the year. The plentiful earthworms incorporate the deep litter layers. The resultant dark A horizon ('topsoil') rich in organic matter merges gradually into a bright brown and deep B horizon ('subsoil'). The soil structures that develop in the surface layers are granular and rounded fine blocky in which there is an excellent balance of air and water and into which roots can readily penetrate.

Brown earths are usually mildly acid (pH 5.5 to 6.5), but **acid brown earths** (pH 5.5 to 4.5) can develop on lighter textured soils in wetter areas

Figure 17.6 **Soil types**: (a) Podsol; (b) Gley; (c) Brown earth; (d) Rendzina.

Figure 17.7 **Soil colours**. Brown (top), Mottled, Gley (bottom).

(800–1000 mm rain per year) especially under beech or birch woodland. There is less earthworm activity, with a resultant reduced incorporation of organic matter down the profile. The soil structure is usually less satisfactory and clay particles that work their way down can form clay pans (see p313) in the B horizon. These can be productive soils if ameliorated with lime and fertilizer (see Chapter 21).

Gleys occur in poorly drained soils (see p343). **Surface water gleys** are found where the percolation of water is restricted by the poor structure in the A or B horizon to produce a **perched water table** (see p340). This is typically where the subsoil is heavy and impervious, especially in wetter regions. Oxygen in the waterlogged soil is depleted and, in these anaerobic conditions below the water table, the iron oxides that colour the soil become dull grey or bluey (in aerobic conditions the iron oxides are rust coloured). The extent of the waterlogging that the soil has been subjected to as the water table fluctuates can be judged from the degree to which it has become completely grey; usually there is a rusty mottle present, indicating that aerobic conditions exist in the soil for part of the year (see Figure 17.7). Plants growing in them are often shallow rooted and suffer from drought in dry periods. These soils are only productive after they have been drained, limed and fertilized.

Ground water gleys develop where there is a permanent water table that is very near the surface of the soil, so that to lower the water table drainage has to be undertaken on a regional basis, e.g. Romney Marsh. Drainage pipes can only be used when the water can be run to a ditch with a water level below that desired in the field (see p344); for some areas this can only be achieved by maintaining an artificially low level by the use of pumps (powered in former times by windmills).

Podsols (from the Russian 'under-ash') are strongly leached, very acid soils that develop on freely draining soils, such as coarse sands and gravels, commonly under heather or pine or spruce forest in high rainfall areas. Because of the high acidity levels, earthworms are absent so there is a build-up of the litter layer. Poorly decomposed organic matter that is not incorporated (a 'mor' humus) is characteristic of this soil type. Some of the organic matter combines with the iron in the top layers to form soluble compounds which are leached ('podsolization') to leave a grey ('ash-like') A horizon (all that remains are bleached sand grains). These compounds become insoluble again in the conditions that prevail in the B horizon, where organic matter accumulates to create a dark or black horizon below which is an iron rich red layer. The iron compounds that accumulate can form a strongly cemented 'iron pan'. As a continuous pan that water (and roots) cannot penetrate is formed, a waterlogged area develops and peat can form at the surface. Podsols are prone to drought and are 'hungry' soils that require considerable ongoing inputs of lime and manure to make them productive. They are of little use in horticulture except for the growing of acid loving trees and shrubs.

Rendzinas are very thin dark-brown, sometimes black, soils with a strong granular structure sitting directly on chalk or limestone. They are typical of the soils on the steeper slopes of chalk or limestone hills under grass. Shallow soils develop because of the continued erosion on the slopes, which also keeps these soils heavily charged with lime. Where the soils become deeper on the less severe slopes it is common for the A horizons to become acid, as the lime is leached downwards. They are well drained because of the slope and because of the porous nature of the underlying rock. Rendzinas are not suitable for most horticultural purposes because the high lime content causes induced nutrient deficiencies (see p372). Roots are severely restricted by the shallow soils and vulnerable to drought.

Topsoil and subsoil

Topsoil is the uppermost layer of soil normally moved during cultivation. Typically it is 10 to 40 cm deep and darkened by the decomposed organic matter it contains.

Subsoil is the layer below that normally cultivated and lighter in colour because of its low organic matter level.

In natural conditions organic matter is more abundant at the surface and declines in concentration with depth (see p327), whereas under cultivation the organic matter is redistributed to create a distinct topsoil and subsoil with the boundary at 'plough depth'. The dark colour that the decomposed organic matter gives the soil makes the boundary obvious by the colour change.

Soil components

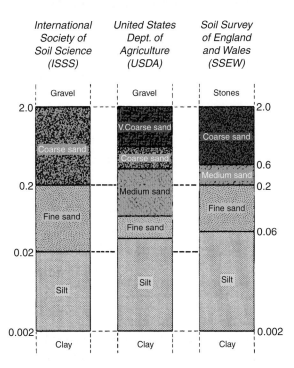

Figure 17.8 **Particle size classes** (diameters in mm)

The solid parts of soils consist of mineral matter derived from rocks and organic matter derived from living organisms. Levels and characteristics of organic matter are dealt with in Chapter 18. Most soils have predominantly mineral particles that vary enormously in size from boulders, stones and gravels down to the smaller soil particles ('fine earth'); sand, silt and clay.

Particle size classes

There is a continuous range of particle sizes, but it is convenient to divide them into classes. Three major classification systems in use today are those of the International Society of Soil Science (ISSS), United States Department of Agriculture (USDA) and the Soil Survey of England and Wales (SSEW). These are illustrated in Figures 17.8 and 17.9. In this text the SSEW scale used by the Agricultural Development and Advisory Service of England and Wales (ADAS) is adopted. In each case, soil is considered to consist of

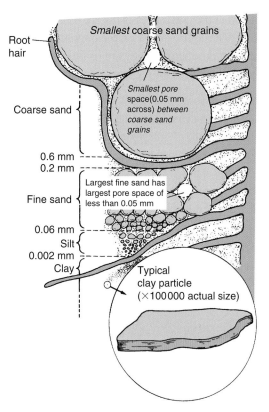

Figure 17.9 **The relative sizes of coarse sand, fine sand, silt and clay particles** (based on SSEW classification) with root hairs drawn alongside for comparison. Note that even the smallest pore spaces between unaggregated spherical coarse sand grains still allow water to be drawn out by gravity and allow some air in at field capacity, whereas most pores between unaggregated fine sand grains remain water-filled (pores less than 0.05 mm diameter)

those particles that are less than 2 mm in diameter. The silt and clay particles are sometimes referred to as 'fines'.

Sand

Sands are gritty to the touch; even fine sand has an abrasive feel. Sand is mainly composed of quartz. (Although any particle of this size is a sand grain, it is most often quartz because, unlike other minerals, it resists weathering.)

The shape of the particles varies from the rough and angular sand to more weathered, rounded grains. They are frequently coated with iron oxides, giving sand colours from very pale yellow to rich, rusty brown. Silver sand has no iron oxide covering. Chemically most of the sand grains are **inert**; they neither release nor hold on to plant nutrients and they are not cohesive.

The influence of sand on the soil is mainly physical, and as such the size of the particles is the important factor. As the particles become smaller and the volume of individual grains decreases, the **surface area** of the same quantity of sand becomes greater (see Figure 17.10). Sand grains are non-porous so their water-holding properties are directly related to their surface areas. It can be readily seen that, since water will not flow through gaps less than about 0.05 mm in diameter, there are very big differences in the drainage characteristics of coarse and fine sands (Figure 17.9). Consequently, soils dominated by coarse sand are usually free draining but have poor water retention, whereas those composed mainly of unaggregated fine

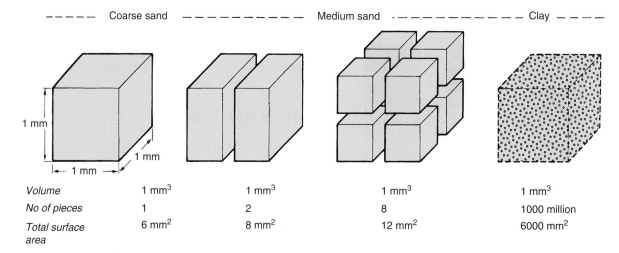

Figure 17.10 **Surface area of soil particles**. The effect of sub-dividing a cube corresponding in size to a grain of coarse sand. The same volume of medium sand is made up of over eight times more pieces that have a total surface area more than double that of coarse sand. It requires over a thousand million of the largest clay particles to make up the volume of one grain of coarse sand and their total surface area is approximately six thousand times greater.

Sand grains are particles between 0.06 mm and 2.0 mm in diameter.

Clay particles are those less than 0.002 mm in diameter.

sand hold large quantities of water against gravity. The water held on all sand particles is readily removed by roots (see available water p342).

Clay

There are many different clay minerals, e.g. kaolinite, montmorillinite, vermiculite, and mica, all of which are derived from rock minerals by chemical weathering.

Most clay minerals have a layered crystalline structure and are plate-like in appearance (Figure 17.9). The clay particles have surface charges that give clay its very important and characteristic property of **cation exchange**. The charges are predominantly negative which means that the clay platelets attract positively charged **cations** in the soil solution. These include the nutrients potassium, as K^+; ammonia, as NH_4^+; magnesium, as Mg^{++}; and calcium, as Ca^{++}, as well as hydrogen and aluminium ions. These **ions** (p371) are held in an **exchangeable** way so that they remain available to plants, but are prevented from being leached unless displaced by other cations. The greater the **cation exchange capacity**, the greater the reserves of cations held this way.

Hydrogen and aluminium cations make the soil acid. The other cations Ca^{++}, Mg^{++}, K^+ and Na^+ are called **bases** and make soils more alkaline (see soil pH (p356)). The proportion of the cation exchange capacity occupied by bases is known as its **percentage base saturation**. A soil's **buffering capacity**, i.e. its ability to resist changes in soil pH, also depends on these surface reactions. The presence of high levels of exchangeable aluminium and hydrogen means that very large quantities of calcium, in the form of lime, are required to raise the pH of acid clays. In contrast, only small quantities of lime are needed to raise the pH of a sand by the same amount (see liming p361).

The clay particles are so small that the minute electrical forces on the surface become dominant (Figure 17.10); thus clay and water mixtures behave as colloids (see Table 17.1). This gives clay soils properties of cohesion, plasticity, shrinkage and swelling. The small particles can pack and stick together very closely and in a continuous mass they restrict water movement.

The water-holding capacity of clay-dominated soils is very high because of the large surface areas and because many of the particles are porous. However, a high proportion of the water is held too tightly for roots to extract (see p342). Moist balls of clay are plastic, i.e. can be moulded. On drying, they harden and some shrink. In the soil, the cracks that form on shrinking become an important network of drainage channels. The cracks remain open until the soils are re-wetted and the clay swells. Humus and calcium appear to combine with clay in such a way that when the combination dries, extensive cracking occurs and favourable growing conditions result. Some clays are non- shrinking and are consequently more difficult to manage although they present less problems with regard to building foundations.

Characteristics of small particles

Taking a cube and then cutting it up into smaller cubes readily demonstrates the relationship between volume and surface area (see Figure 17.10). While the total volume is the same in all the small cubes compared with the original large cube, the sides of each are smaller, but the total surface area is much greater because new surfaces have been exposed. Many soil particle characteristics are directly related to particle size and in particular to surface area.

Note that as a particle is sub-divided the total surface area increases; the surface area doubles each time the sides of the individual pieces are reduced by half. In particles that are less than 0.001 mm in diameter, which includes most clay, colloidal properties are observed; most notably the properties of their surface dominate their chemical behaviour. **Colloids** are mixtures that are in permanent suspension.

Table 17.1 Colloidal systems

Mixture	Colloidal system	Examples
Solid dispersed in gas	Solid aerosol	Smoke, dusts
Solid dispersed in liquid	Sol or gel	Paste, clay, humus, protoplasm
Liquid dispersed in gas	Liquid aerosol	Mist, fog
Liquid dispersed in liquid	Emulsion	Milk
Gas dispersed in gas		The atmosphere

Water based colloids, such as clay, are 'runny' when mixed with plenty of water (a 'sol') but with less water they are stiff and jelly-like (a 'gel'), e.g. paste. As they dry, these mixtures become sticky and eventually hard. Many natural materials such as gelatin, starch, gums and protoplasm (mainly protein and water) are colloidal.

Silt

Most in this size range are inert and non-porous like sands, but many particles, including felspar fragments, have the properties of clay. Soils dominated by silt do have a small cation exchange capacity, but in the main they behave more like a very fine sand. They have very good water-holding capacity and plants can take up a high proportion of this water.

Silt particles are those between 0.002 mm and 0.06 mm in diameter.

When wetted they have a distinctly soapy or silky feel. Silt soils are made up of particles that readily pack closely together, but have little ability to form stable crumbs (see soil aggregates p311). This makes them particularly difficult to manage.

Stones and gravel

Stones are particles larger than 2 mm in diameter.

Particles bigger than sand are commonly known as grit, gravel, pebbles, cobbles and boulders, according to size and shape. The effect of stones on cultivated areas depends on the type of stone, size and the proportion in the soil. The presence of even a few stones larger than 20 cm prevents cultivation. Stones in general have detrimental effects on mechanized work; ploughshares, tines and tyres are worn more quickly, especially if the stones are hard and sharp such as broken flint. Stones interfere with drilling of seeds and the harvesting of roots. Close cutting of grasses is more hazardous where there are protruding stones. Mole draining becomes less effective in stony soils and large stones make it impossible.

Stones can accumulate in layers and become interlocked to form **stone pans**. In very gravelly soils the water-holding capacity is much reduced and the increased leaching leads to acid patches. Nutrient reserves are also reduced by the dilution of the soil with inert material. However, stones can help water infiltration, protect the surface from capping, and check erosion by wind or water.

Soil texture

Soil texture can usefully be defined as the relative proportions of the sand, silt and clay particles in the soil.

Soil texture describes the mineral composition of a soil. In most cultivated soils the mineral content forms the framework and exerts a major influence on its characteristics. Examples of different textures are given in Figure 17.11.

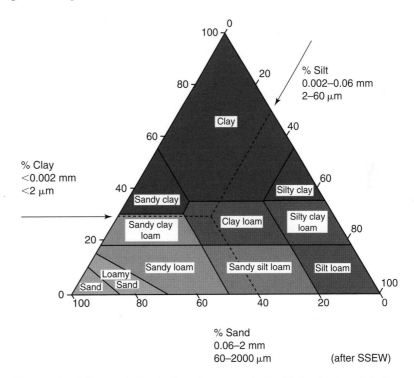

Figure 17.11 **Soil textural triangle**. The soil texture can be identified on this type of chart when at least two of the major size of fractions are known, e.g. 40 per cent sand, 30 per cent silt and 30 per cent clay is a clay loam (SSEW Soil-Particle Size Classification).

Texture can be considered to be a fixed characteristic and provides a useful guide to a soil's potential. Fine-textured soils such as clays, clay loam, silts and fine sands have good water-holding properties, whereas coarse textured soils have low water-holding capacity but good drainage. This also means that **soil temperatures** are closely related to soil texture, because water has a very much higher specific heat value than soil minerals. Consequently freely draining, coarse sand warms up more quickly in the spring but is also more vulnerable to frosts than wetter soils.

Soils with high clay content have good general nutrient retention, whereas nutrients are readily lost from sandy soils, especially those with a high coarse sand fraction. The application rate of pesticides and herbicides is often related to soil texture. The power requirement to cultivate a clay soil is very much greater than that for a sandy soil. The expression 'heavy' for clay and 'light' for sandy soils is derived from this difference in working properties rather than the actual weight of the soil. The texture of a soil also influences the soil structure and soil cultivations.

The addition of a calcareous clay to a sandy topsoil, a practice known as **marling**, can improve its water-holding capacity, as well as reducing wind erosion, but it requires the incorporation of 500 tonnes of dry clay per hectare to convert it to a sandy loam. The practice of adding clay is now largely confined to the building of cricket squares. To 'lighten' a clay loam topsoil to a sandy loam more than 2000 tonnes of dry sand is needed on each hectare (roughly the same volume of sand has to be added as the volume to be changed). The addition of smaller quantities of sand is often an expensive exercise to no effect; at worst it can make the resultant soil more difficult to manage.

Texturing by feel

A more practical method of determining soil texture, especially in the field, is by feel. This can, with experience, be a very accurate means of distinguishing between over thirty categories. A ball of soil about the size of a walnut is moistened and worked between the fingers to remove particles greater than 2 mm and to break down the soil crumbs. It is essential that this preparation is thorough or the effect of the silt and clay particles will be masked. The characteristics of the different mixtures of sand, silt and clay enable the texture to be determined:

- **Sands** are soils that have little cohesion. Sand has little tendency to bind even when wetted and it cannot be rolled out into a 'worm'.
- **Loamy sand** has sufficient cohesiveness to be rolled into a 'worm', but it readily falls apart.

What is generally known as loam moulds readily into a cohesive ball and it has no dominant feel of grittiness, silkiness or stickiness:

- **Sandy loam** – if grittiness is detected and the ball is readily deformed.
- **Silty loam** – if it is readily deformed but has a silky texture.
- **Clay loams** bind together strongly, do not readily deform, and take a polish when rubbed with the finger.

Clays bind together and are very difficult to deform. A clay soil readily takes a very marked polish but it is:

- **Silty clay** if there is also a feeling of silkiness, or
- **Sandy clay** if grittiness is evident.

Wherever grittiness is detected, the designation sand can be further qualified by stating whether it is coarse, medium or fine sand, e.g. coarse sandy loam. Table 17.2 shows the range of textural groupings commonly used.

Determining texture by feel has the limitation that the influence of organic matter and chalk cannot be eliminated. Chalk tends to give a soil a silky or gritty feel depending on fineness, but the fact that a soil is known to be chalky should not influence the texturing. Its textural class may be prefixed '**calcareous**', e.g. calcareous silty clay. **Organic matter**

Table 17.2 **Soil texture** classification based on hand-texturing

Textural class	Symbol	Textural group
Coarse sand	CS	
Sand	S	
Fine sand	FS	
Very fine sand	VFS	
Loamy coarse sand	LCS	
Loamy sand	LS	
Loamy fine sand	LFS	Very light soils
Coarse sandy loam	CSL	
Loamy very fine sand	LVFS	
Sandy loam	SL	Light soils
Fine sandy loam	FSL	
Very fine sandy loam	VFSL	
Silty loam	ZyL	Medium soils
Loam	L	
Sandy clay loam	SCL	
Clay loam	CL	
Silt loam	ZyCL	Heavy soils
Silty clay loam	ZyCL	
Sandy clay	SC	
Clay	C	Very heavy soils
Silty clay	ZyC	

tends to increase the cohesiveness of light soils, reduce the cohesiveness of heavy soils, and large quantities can impart a silky or greasy feel. The prefix '**organic**' can be used for describing mineral soils with 15–20 per cent organic matter. Soils with 20–35 per cent organic matter are **peaty loams**, 35–50 per cent organic matter **loamy peats** and soils with greater than 50 per cent organic matter are termed **peaty**. **Peats** are almost pure organic matter (see organic soils p320).

Mechanical analysis of soils

Soil texture can be determined by finding the particle size distribution. There are several methods, but all depend on the complete separation of the particles, the destruction of organic matter and the removal of particles greater than 2 mm in diameter. Sieving can separate the stones, coarse sand, medium sand and fine sand fractions.

Finer particles are usually separated by taking advantage of their different **settling velocities** when in suspension. The settling velocity of a particle depends on its density and radius, the viscosity and density of the liquid and the acceleration due to gravity; the method is simplified by assuming that soil particles are spherical and have the same density and the investigations are conducted in water at 20°C.

Particles that are less than 0.001 mm in diameter are kept permanently in suspension by the bombardment of vibrating water molecules and are referred to as **colloids**, e.g. most clay particles. All sand particles will have fallen more than 10 cm after 50 seconds, so a sample taken at that depth can be used to calculate the clay plus silt left in the suspension. Similarly, other fractions can be calculated until the sand, silt and clay are determined. The soil texture can be deduced from this information using a **textural triangle** (Figure 17.11), which is the basis of identifying soil types.

Soil structure

> **Soil structure** is the arrangement of particles in the soil.

In order to provide a suitable root environment for cultivated plants the soil must be constructed in such a way as to allow good gaseous exchange, whilst holding adequate reserves of available water. There should be a high water infiltration rate, free drainage and an interconnected network of spaces allowing roots to find water and nutrients without hindrance. There should be no large cavities that prevent thorough contact between soil and roots and allow roots to dry out in the seedbed. The soil should be managed so that erosion is minimized. Good structural stability should be maintained so that the structure does not deteriorate and limit crop growth.

Porosity

The plant roots and soil organisms live in the pores between the solid components of the growing medium. In the same way that a house is mainly judged by the living accommodation created by the bricks, wood, plaster, cement, etc., so a soil is evaluated by examining the spaces created.

> The key to managing most growing media is in maintaining a high proportion of **air-filled pores** without restricting water supply.

Pores greater than about 0.05 mm in diameter, called **macropores**, can drain easily to allow in air within hours of being saturated (i.e. fully wetted), whereas the smaller pores, **micropores** continue to contain only water. The roots remove more water from these micropores allowing more air back into the soil (see soil water p338). Ideally, there should be a mixture of pore sizes allowing good water holding, free drainage, gaseous exchange and thorough root exploration, as shown in Figure 17.12.

An important indicator of a satisfactory growing medium is its air-filled porosity or air capacity, i.e. the percentage volume filled with air when it has completed draining, having been saturated with water.

Bulk density is the mass of soil per unit volume and it can be measured by taking a core of soil of known volume and weighing it after thorough drying. In normal mineral soils results are usually between 1.0 and 1.6 g/ml. The difference is largely attributable to variation in total pore space. Finer textured soils tend to have more pore

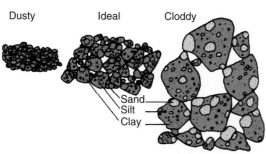

Dusty   Ideal   Cloddy

Sand
Silt
Clay

Figure 17.12 Tilth. The ideal tilth for most seedbeds is made up of soil aggregates between 0.5 and 5 mm diameter. Within these crumbs are predominantly small pores (less than 0.05 mm) that hold water and between the crumbs are large pores (greater than 0.05 mm) that allow easy water movement and contain air when soil is at field capacity (×5 actual size).

space and therefore lower bulk density than sands, but for all soils higher values indicate greater packing or **compaction**.

This information is not only useful to diagnose compaction problems, but can also be used to calculate the weight of soil in a given volume. Assuming a cultivated soil to have a bulk density of 1.0 g/ml, the weight of dry soil in one hectare to a plough depth of 15 cm is 1500 tonnes; when compacted the same volume weighs 2400 tonnes. Similarly, 1 m³ of a typical topsoil with a bulk density of 1.0 will weigh 1 tonne (1000 kg) when dry and up to half as much again when moist.

Soil structures

The pore space does not depend solely upon the size of the soil particles as shown in Figure 17.8, because they are normally grouped together. These **aggregates**, or peds, are groups of particles held together by the adhesive properties of clay and humus. The ideal arrangement of small and large pores for establishing plants is illustrated in Figure 17.12 alongside a dusty tilth with too few large pores and a cloddy tilth that has too many large pores.

A soil with a **simple structure** is one in which there is no observable aggregation. If this is because none of the soil particles are joined together, as in sands or loamy sands with low organic matter levels, it is described as **single grain** structure. Where all the particles are joined with no natural lines of weakness the structure is said to be **massive**.

A **weakly developed** structure is one in which aggregation is indistinct and the soil, when disturbed, breaks into very few whole aggregates, but a lot of unaggregated material. This tends to occur in loamy sands and sandy loams. Soils with a high clay content form **strongly-developed** structures in which there are obvious lines of weakness and, when disturbed, aggregates fall away undamaged. The prismatic, angular blocky, round blocky, crumbs and platy structures which are found in soils are illustrated in Figure 17.13.

Development of soil structures

Soil structure develops as the result of the action on the soil components of natural **structure-forming agents**, freezing and thawing, wetting and drying, root growth, soil organisms, as well as the influence of cultivation:

- **frost** leads to the shattering of clods by producing a frost mould. It is largely confined to the surface layers and is advantageous in the management of clays;
- **drying** soil can affect the whole rooting depth. Cracks usually open up in heavier soils as the clay shrinks;
- **earthworms** and other soil organisms play an important part in loosening soil, maintaining the network of drainage channels and stabilizing the soil structure;

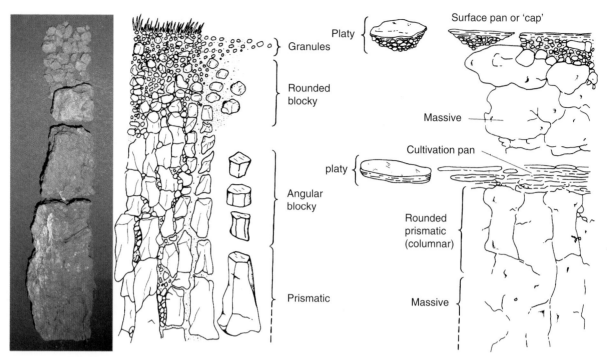

Figure 17.13 Soil structures. The soil profile on the left is composed of soil particles aggregated into structures that produce good growing conditions. Examples of structures that create a poor rooting environment are shown in the profile on the right.

- **roots** have a major effect on the drying of deeper soil layers, but they also play an important part in soil structure by growing into the cracks and keeping them open. They help establish the natural fracture lines. In strong structures, a close-fitting arrangement of **prismatic** (see Figure 17.13) or **angular blocky** aggregates is readily seen.

In soils with low clay content the roots are vital in maintaining an open structure. The exploring roots probe the soil, opening up channels where the soil is loose enough and producing sideways pressure as they grow. On death, the root leaves behind channels stabilized by its decomposed tissue for other roots to follow. Fine **granular** structures are developed under pastures by the action of the fibrous rooting over many years. The soil structure is greatly improved by the rootball. Its physical influence is most easily appreciated by shaking out the soil crumbs from around the root of a tuft of well-established grass and comparing them with the structure of soil taken from a nearby bare patch.

Freshly exposed land is often referred to as **raw**; when weathered it becomes mellow. Once **mellow**, a seedbed is more easily prepared. The weathering process and influence of cultivation tend to produce rounded blocky **structures** and **rounded granules** in the cultivated zone.

Cultivation of soil by hand or mechanized implements is undertaken to produce a suitable rooting environment for plants, to destroy pests and weeds and to mix in plant residues, manures and fertilizers (see p374). However, the use of cultivators can lead to the formation of platy layers or **pans**, which are characterized by the lack of vertical cracks and form an obstruction to root and water movement. The surface of soils is also compacted to create surface capping by associated traffic, whether by feet or tyres, if undertaken in the wrong conditions (see soil consistency p342).

Natural pans develop in some soils as a result of fine material cementing a layer of soil together. In some sandy soils rich in iron oxide, these oxides cement together a layer of sand where there has been a fluctuating water table, to produce an **iron pan**.

Structural stability

> The **structural stability** of soil refers to its ability to resist deformation when wet.

Soil aggregates with little or no stability collapse spontaneously as they soak up water, i.e. they slake. Those high in fine sand or silt are particularly vulnerable to slaking. Aggregates with better stability maintain their shape when wetted for a short time, but gradually pieces fall off if left immersed in water. Aggregates with **good structural stability** are able to resist damage when wet unless vigorously disturbed. Soils with a high level of clay content have better stability than those with low levels. Stability is also increased by the presence of calcium carbonate (chalk), iron oxides, and, most importantly, **humus** (see p326).

Tilth

> **Tilth** is the structure of the top 50 mm of the soil.

The soil surface or seedbed should be carefully managed to produce the required crumb structure.

Sandy soils are easily broken down to the right size with cultivation equipment. Heavier soils are less easy to cultivate and benefit from weathering to produce a frost 'mould'.

Figure 17.14 **Soil cap**.

The fineness of a seedbed should be related to the size of seeds, but ideally consists of granules or crumbs between 0.5 and 5 mm in diameter (see Figure 17.12). Cloddy surfaces lead to poor germination, as well as poor results from soil herbicide treatments. The rain on the soil surface breaks down tilth. As soil crumbs break up, the particles fill in the gaps; this reduces infiltration rates. As the surface dries, a **cap** or **pan** is formed (Figure 17.14). Thus fine 'dusty' tilths should be avoided and the soil crumbs should be stable so that they can withstand the effect of rain until plants are established. This is particularly important on fine sandy and silty soils, which tend to have poor structural stability. In general, fine tilths should be avoided outdoors until well into spring when conditions are becoming more favourable and growth through any developing cap is rapid.

Cultivations

In temperate areas, the conventional preparation of land for planting is a thorough disturbance of the top 20–30 cm of soil. Digging or ploughing buries residues of previous plantings and weeds, and with repeated passes of rakes or harrows a suitable **tilth** is created (see p313). This procedure is very demanding on energy, labour and time. Many of the cultivations tend to interfere with the natural structure-forming agents and when undertaken at the wrong time they create pans or leave a bare, loose soil vulnerable to erosion (see soil structure p311).

Some of the compaction problems are overcome by cultivating in **beds** which confines traffic to well-defined paths between the growing areas. The advent of effective herbicides has, in certain cases, enabled the inversion of soil to be eliminated. The use of powered implements has speeded up work and reduced the number of tillage passes. In some areas of horticulture the adoption of **minimum** or **zero tillage** has preserved natural structure while beneficially concentrating organic matter levels in the surface layers and reducing wind and water erosion.

Ploughing and digging

Ploughing and digging are used to loosen and invert the soil. The land is broken up into clods and an increased area is exposed to weathering. As the soil is inverted, weeds, plant residues and bulky manures are incorporated. The depth of ploughing or digging should be related to the depth of topsoil, because bringing up the subsoil reduces fertility in the vital top layers, seriously affecting germination of seeds and establishment of plants. If deeper layers are to be loosened a **subsoiler** should be used. In plastic soil conditions the plough can smear the soils, more so if the wheels of the tractor spin in the furrow bottom. These **plough pans** tend to develop with successive ploughing to the same depth. Ploughing at different depths or attaching a subsoil tine can reduce their incidence. Digging with a spade does not produce a cultivation pan and is still used on small areas. **Spading machines** or **rotary diggers** imitate the digging action without the disadvantages of ploughing, but tend to be very slow.

Rotary cultivators

Rotary cultivators are used to create a tilth on uncultivated or on roughly prepared ground. The type of tilth produced depends on suitable adjustment of forward speed, rotor speed, blade design and layout, shield angle and depth of working. The 'hoe' blade is normally used for seedbed production, but does have the disadvantage of smearing plastic soils at the cultivation depth, producing a **rotovation pan** . 'Pick' tines produce a rougher tilth, but less readily cause a pan. Subsoil tines can be fitted to prevent these pans developing.

Harrowing and raking

Harrowing and raking are methods of levelling soils, incorporating fertilizers and producing a suitable tilth on the roughly prepared ground. The soil must be in a **friable** condition (see p342) for this operation and it is made easier if the top layers have been suitably weathered. The impact of the tines breaks the clods. The number of passes to create a

Figure 17.15 **Seedbed preparation**. Several stages of the process are completed in a single pass

seedbed has been reduced by the use of rotavators and other equipment that completes several of the stages in one pass (see Figure 17.15).

'Progressive' type cultivators were introduced essentially to loosen coarse structured clays by drawing through the soil banks of tines, increasing in depth from the front, to cultivate the soil from the top down in one pass. Although this requires powerful tractors to pull, especially if subsoiling tines are attached, it is a time-saving operation and the recompaction inherent in multiple pass methods is reduced. These cultivators should not be used on well-structured soils where full depth loosening is unnecessary.

'Under-loosening' cultivators have been designed to loosen compact topsoils without disturbing the surface, which ensures a level, clod-free and organic-matter-rich tilth. Used under the right conditions these implements improve water movement and plant growth. However, loosened soils are more susceptible to compaction and consequently the equipment should only be used when compaction is known to be present.

Subsoiling

Subsoiling is used to improve soil structure below plough depth by drawing a heavy cultivation tine through the soil to establish a system of deep cracks in compacted zones. This helps the downward movement of water, circulation of air and penetration of roots (see Figure 17.16). The operation is most effective when the subsoil is friable and the surface

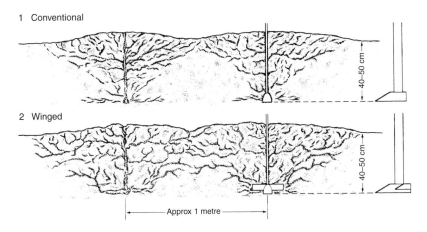

Figure 17.16 **Subsoiling**. The subsoiler is drawn through the soil to burst open compacted zones. It leaves cracks which remain open to improve aeration, drainage, and root penetration. The cracks created should link up with artificial drainage systems unless the lower layers are naturally free draining

dry enough to be able to withstand the heavy tractor that is needed (see loadbearing).

Effective subsoiling is made easier if the top surface is loosened by prior cultivation. Although the draught is higher, subsoil disturbance is increased substantially by attaching inclined blades or wings. Successful subsoiling is accompanied by a lift in the soil surface (soil heave) which usually makes it unsuitable for improving conditions in playing fields.

Subsoiling should only be used when the cause of any waterlogging is related to a soil structure fault (see also drainage). Slow subsoil permeability caused by high clay content is usually rectified with **mole drainage** (see p345). If the soil is too sandy or stony, a subsoiler can be used so long as the cracks created lead the water into a natural or artificial drainage system. Subsoilers used in the right conditions readily burst massive structures and soil pans created by machinery, but some natural pans are too strong for normal equipment. The problem of cultivation pans can be dealt with by using conventional subsoilers or by attaching small subsoil tines to the cultivation equipment. This tends to increase the power requirement but eliminates the pan as it is created.

Management of main soil types

Sandy soils

Sandy soils are usually considered to be easily cultivated, but serious problems can occur because the particles readily pack together, especially when organic matter levels are low. Consequently many sandy soils are difficult to firm adequately without causing over-compaction. Pans near the surface caused by traffic and deeper cultivation pans frequently occur on sandy soils, resulting in reduced rooting and water movement. **Subsoiling** is frequently undertaken on a routine basis every 4–6 years, although the need can be reduced by keeping machinery off land while it has low **load-bearing** strength (see p342) and by encouraging natural structure-forming agents.

Coarse sands have low water-holding capacity, which makes them vulnerable to drought, particularly in drier areas. This is not such a disadvantage if irrigation equipment is installed and water is readily obtainable. In many categories of horticulture there is a demand for soils with **good workability**. Coarse sands, loamy sands and sandy loams have the advantage of good porosity and can be cultivated at field capacity. Sands tend to go acid rapidly and are vulnerable to overliming because of their **low buffering capacity** (see p359).

Silts and fine sands

These can be very productive soils because of their good water-holding capacity and, while organic matter levels are kept above 4 per cent, their ease of working. However silts and fine sands present soil management problems, especially when used for intensive plantings, because they have **weak structure**, are vulnerable to **surface capping**, and are easily compacted to form **massive structures**. To achieve their high potential, efficient drainage is vital to maximize the rooting depth. Fine tilths in the open should be avoided, especially in autumn and early spring, because frosts and heavy rainfall reduce the size of surface crumbs. For the same reason, care should be taken with irrigation droplet size that, if too large, can damage the surface structure. Improving soil structure is not easy after winter root crop harvesting or orchard spraying on wet soils, because low clay content results in very little cracking during subsequent wetting and drying cycles. Improvement therefore depends on other natural structure forming agents or on subsoiling.

Clay soils

Clay soils tend to be **slow draining**, **slow to warm** up in spring, and have **poor working properties** (see soil consistency p342). A serious limitation is that the soil is still plastic at field capacity, which delays soil preparation until it has dried by evaporation. Permanent plantings are established to avoid the need to rework the soil. Playing surfaces created over clays have severe limitations, particularly when required for use in all-weather conditions. Where high standards have to be maintained, as in golf greens, fine turf is established in a suitable growing medium overlying the original soil. However, a high clay content is an advantage for the preparation of cricket squares where a hard, even surface is required but is played on only in drier weather. Increasingly, heavily used areas are replaced by artificial surfaces.

Horticultural cropping of clays is limited to summer cabbage, Brussels sprouts and to some top fruit in areas where the water table does not restrict rooting depth. Under-drainage is normally necessary. In wetter areas most clay soils are put down to grass. Timeliness, encouraging the annual drying cycle of the soil profile and maximizing the effect of weathering to help cultivations are essential for successful management of clay soils.

Peat soils

Peat soils (see p328) have very many advantages over mineral soils for intensive vegetable and outdoor flower production. Fenland soils and Lancashire Moss of England; peatlands of the midland counties of Ireland; the 'muck' soils of North America; and similar soils in the Netherlands, Germany, Poland and Russia have proved valuable when their limitations to commercial cropping have been overcome.

Well-drained peat at the correct pH is an **excellent root environment**. It has a very much higher water-holding capacity than the same volume of soil and yet gaseous exchange is good. Root development is uninhibited because friable peat offers hardly any mechanical resistance to root penetration. This leads to high quality root crops that are easily cleaned. These cultivated peat lands **warm up quickly** at the surface because the sun's energy is efficiently absorbed by their dark colour, with consequent rapid crop growth. These soils have a very **low power requirement** for cultivation, are free of stones, and can be worked over a wide moisture range.

Plant nutrition is complicated by natural **trace element deficiencies** and the effect of pH on plant **nutrient availability**. Peat has **poor load-bearing** characteristics and specialized equipment is often needed to harvest in wet conditions. Whilst peat warms up quickly on sunny days, its dark surface makes it vulnerable to air frost because it acts as an efficient radiator. Firming the surface and keeping it moist combat this. Weeds grow well and their control is made more difficult by the ability of peat to absorb and **neutralize soil-acting herbicides**. The high organic matter levels also make the peats and sandy peats **vulnerable to wind erosion** in spring when the surface dries out and there is no crop canopy to protect it.

Check your learning

1. Describe the profile of a typical mineral soil.

2. Describe the differences between topsoil and subsoil.

3. Summarize the differences between the main soil types.

4. State the differences between soil texture and structure.

5. State how structure in cultivated soils is: a) improved; b) damaged.

6. Explain how soil development occurs on poorly drained soils in a temperate area.

Further reading

Bell, B. and Cousins, S. (1997). *Machinery for Horticulture*. 2nd edn. Old Pond Publishing.

Brown, L.V. (2008). *Applied Principles of Horticultural Science*. 3rd edn. Butterworth-Heinemann.

Castle, D.A. *et al.* (1984). *Field Drainage Principles and Practices*. Batsford.

Coker, E.G. (1970). *Horticultural Science and Soils*. Macmillan.

Culpin, C. (1992). *Farm Machinery*. 2nd edn. Blackwell Science.

Davies, D.G., Eagle, D.T., and Finley, J.B. (1993). *Soil Management*. 5th edn. Farming Press Books and Videos.

Handreck, K. and Black, N. (1989). *Growing Media for Ornamental Plants and Turf*. New South Wales University Press.

Ingram, D.S. *et al.* (eds). (2002). *Science and the Garden*. Blackwell Science Ltd.

Munns, D.N. and Singer, M.J. (2005). *Soils: An Introduction*. 6th edn pb. Prentice Hall College.

Simpson, K. (1983). *Soil*. Longmans Handbooks in Agriculture.

Chapter 18   Soil organic matter

Summary

This chapter includes the following topics:

- Organic matter in soil
- Decomposition of organic matter
- Nutrient cycles
- Humus
- Benefits of organic matter
- Bulky organic matter
- Methods of composting
- Green manures

with additional information on the following:

- The rhizosphere
- Carbon to nitrogen ratio
- Organic matter levels in the soil
- Organic soils
- Mulching

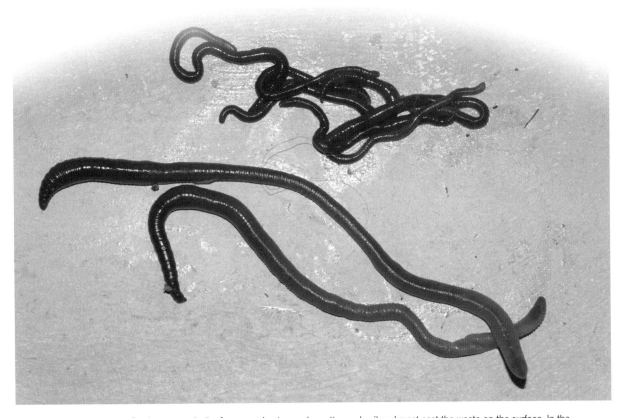

Figure 18.1 **Earthworms**. Casting worms in the foreground eat organic matter and soil and most cast the waste on the surface. In the background are worms that only eat organic matter which we find in compost heaps and wormeries

Organic matter in soil

The main **types of organic matter** in the soil are

- living organisms;
- dead, but recognizable;
- dead but decomposed;
- humus.

A typical mineral soil contains between 2 and 5 per cent organic matter. This is made up of **living organisms** such as plant roots, earthworms, insects, fungi and bacteria. On death these then decompose along with any other organic matter that is incorporated, either naturally such as leaves or by the addition of organic matter from elsewhere such as compost, farmyard manure, spent mushroom compost, coir and bark. Many of the living organisms are responsible for the decomposition of the **dead organic matter**. This is eventually broken down into its component parts becoming carbon dioxide, water, and minerals; all of which is recycled. There also persists for a very long time a group of organic compounds collectively known as **humus**.

Living organisms in the soil

As in any other plant and animal community the organisms that live in the soil form part of the **food webs** (see p53). The main types present in any soil are the **primary producers** which are those capable of utilizing the sun's energy directly, synthesizing their own food by photosynthesis, such as green plants (see photosynthesis), the **primary consumers** which are those organisms which feed directly on plant material, and **secondary consumers** which feed only on animal material. In practice, there are some organisms that feed on both plants and animals and also parasites living on organisms in all categories many of which are pests or diseases of horticultural plants.

Decomposers are an important group, which have the special function within a community of breaking down dead or decaying matter into simpler substances with the release of inorganic salts, making them available once more to the primary producers. **Primary decomposers** are those organisms that attack the freshly dead organic matter. These include earthworms and some species of arthropods and fungi. Fungi are particularly important in the initial decomposition of fibrous and woody material. **Secondary decomposers** are those organisms that live on the waste products of other decomposers and include bacteria and many species of fungi.

Plant roots

These are important as contributors to the organic matter levels in the soil. They move soil particles as they penetrate the soil and grow in size. This rearrangement changes the sizes and shapes of soil aggregates and when these roots die and decompose, a channel is left which provides drainage and aeration. Root channels are formed over and over again unless the soil becomes too dense for roots to penetrate. Roots absorb water from soils and dry it, causing those with a high clay content to shrink and crack. This helps develop and improve structures on heavier soils (see p 311).

Earthworms

There are ten common species of earthworm in Britain that vary in size from *Lumbricus terrestris*, which can be in excess of 25 cm, to the many small species less than 3 cm long (see Figure 18.1). The main food of earthworms is dead plant remains. Casting species of earthworms are those that eat soil, as well as organic matter, and their excreta consist of intimately mixed, partially digested, finely divided organic matter and soil. Many species never produce casts and only two species regularly cast on the surface giving the **worm casts** that are a problem on fine grass areas, particularly in the autumn (see Figure 18.2). It has been estimated that in English pastures the production of casts each year is 20–40 t/ha, the equivalent of 5 mm of soil deposited annually. This surface casting also leads to the incorporation of the leaf litter and the burying of stones. However, *L. terrestris* is the organism mainly responsible for the **burying** of large quantities of litter by dragging plant material down its burrows.

The **network of burrows** which develops as a result of worm activity is an important factor in maintaining a good structure, particularly in uncultivated areas and in soils of low clay content. Some species live entirely in the surface layers of the soil others move vertically establishing almost permanent burrows down to two metres.

Earthworm activity and distribution is largely governed by moisture levels, soil pH, temperature, organic matter and soil type. Most species tend to be more abundant in soils where there are good reserves of calcium. Earthworm populations are usually lower on the more acid soils, but most thrive in those near neutral. Worm numbers decrease in dry conditions, but they can take avoiding action by burrowing to more moist soil or by hibernating. Each species has its optimum temperature range; for *L. terrestris* this is about 10°C, which is typical of soil temperatures in the spring and autumn in the UK. Soils with low organic matter levels support only small populations of worms. In contrast, compost heaps and stacks of farmyard manure have high populations. In oak and beech woods where the fallen leaves are palatable to worms, their populations are large and they can remove a high proportion of the annual leaf-fall. This also happens in orchards unless harmful chemicals such as copper have reduced earthworm populations. Light and medium loams support a higher total population than clays, peat and gravelly soils.

Slugs, **snails** and **arthropods** (such as millipedes, springtails and mites), and **nematodes** are also found in high numbers and play an important part in the decomposition of organic matter. Several species are also horticultural pests (see Chapter 14).

Bacteria

Bacteria are present in soils in vast numbers. About 1000 million or more occur in each gram of fertile soil. Consequently, despite their microscopic size, the top 150 mm of fertile topsoil carries

Figure 18.2 **Earthworm cast**: comprise a mix of organic matter and finely divided soil which is good for the garden except where it is cast on the surface of turf

about one tonne of bacteria per hectare. There are many different species of bacteria to be found in the soil and most play a part in the **decomposition of organic matter**. Many bacteria attack minerals; this leads to the weathering of rock debris and the release of plant nutrients. **Detoxification** of pesticides and herbicides is an important activity of the bacterial population of cultivated soils.

Soil bacteria are inactive at temperatures below 6°C, but their activities increase with rising temperature up to a maximum of 35°C. Bacteria which are actively growing are killed at temperatures above 82°C, but several species can form thick-walled resting spores under adverse conditions. These spores are very resistant to heat and they survive temperatures up to 120°C. **Partial sterilization** of soil can kill the actively growing bacteria, but not the bacterial spores. The growth rate and multiplication also depends upon the **food supply**. High organic matter levels support high bacteria populations so long as a balanced range of nutrients is present. Bacteria thrive in a range of **pH 5.5–7.5**; fungi tend to dominate the more acid soils. Aerobic conditions should be maintained because the beneficial organisms, as well as plant roots, require oxygen, whereas many of the bacteria that thrive under anaerobic conditions are detrimental.

Fungi

The majority of fungi live saprophytically on soil organic matter. Some species are capable only of utilizing simple and easily decomposable organic matter whereas others attack cellulose as well. There are some important fungi that can decompose **lignin**, making them one of the few primary decomposers of wood and fibrous plant material. Several fungi in the soil are parasites and examples of these are discussed in Chapter 15. Fungi appear to be able to tolerate acid conditions and low calcium better than other micro-organisms and are abundant in both neutral and acid soils. Most are well adapted to survive in dry soils, but few thrive in very wet conditions. Their numbers are high in soils rich in plant residues but decline rapidly as the readily decomposable material disappears. The bacteria persist longer, where present, and eventually consume the fungal remains.

The rhizosphere

The rhizosphere is a zone in the soil that is influenced by roots. Living roots change the atmosphere around them by using up oxygen and producing carbon dioxide (see respiration p118). Roots exude a variety of organic compounds that hold water and form a coating that bridges the gap between root and nearby soil particles. Micro-organisms occur in greatly increased numbers and are more active in proximity to roots. Some actually invade the root cells where they live as **symbionts**. The *Rhizobium spp.* of bacteria lives symbiotically with many legumes (see nitrogen cycle p366).

Symbiotic associations involving plant roots and fungi are known as **mycorrhizae** (see Figure 18.3). There is considerable interest in

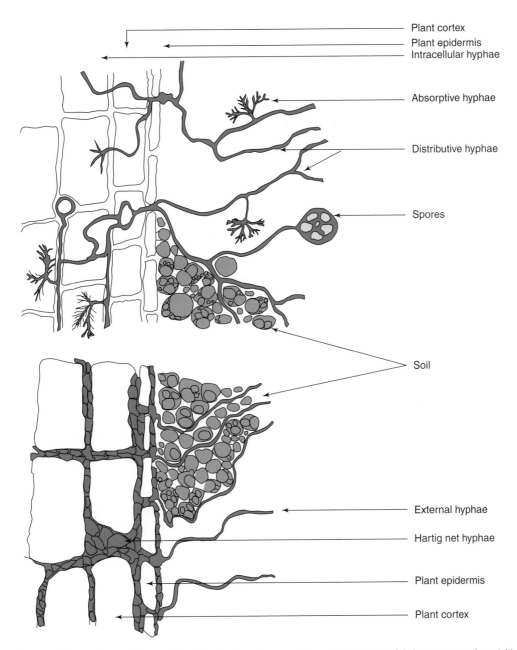

Plant cortex
Plant epidermis
Intracellular hyphae

Absorptive hyphae

Distributive hyphae

Spores

Soil

External hyphae

Hartig net hyphae

Plant epidermis

Plant cortex

Figure 18.3 Mycorrhizal structures. Ectomycorrhizae, top found mainly around tree roots have most of their structure on the outside whereas the Endomycorrhiza have most of their hyphae on the inside of the very wide range of plants with which they are symbiotic

exploiting the potential of **mycorrhizae**, which appear to be associated with a high proportion of plants especially in less fertile soils. In this symbiotic relationship the fungus obtains its carbohydrate requirements from the plant. In turn, the plant gains greater access to nutrients in the soil, especially phosphates, through the increased surface area for absorption and because the fungus appears to utilize sources not available to higher plants. Most woodland trees have fungi covering their roots and penetrating the epidermis. Orchids and heathers have an even closer association in which the fungi invade the root and coil up within the cells. The association appears to be necessary for the successful development of the seedlings. Mycorrhizal plants generally appear to be more tolerant of transplantation and this is thought to be an important factor for orchard and container grown ornamentals.

Nutrient cycles

All the plant nutrients are in continuous circulation between plants, animals, the soil and the air. The processes contributing to the production of simpler inorganic substances, such as ammonia, nitrites, nitrates, sulphates and phosphates, are sometimes referred to as **mineralization**. Mineralization yields chemicals that are readily taken up by plants from the soil solution. The formation of humus, organic residues of a resistant nature, is known as **humification**. Both mineralization and humification are intimately tied up in the same decomposition process, but the terms help identify the end product being studied. Likewise it is possible to follow the circulation of carbon in the carbon cycle and nitrogen in the nitrogen cycle, although these nutrient cycles along with all the others are interrelated.

The carbon cycle

Green plants obtain their carbon from the carbon dioxide in the atmosphere and, during the process of photosynthesis, are able to fix the carbon, converting it into sugar. Some carbon is returned to the atmosphere by the green plants themselves during respiration, but most is incorporated into plant tissue as carbohydrates, proteins, fats, etc. The carbon incorporated into the plant structure is recycled and eventually released as carbon dioxide, as illustrated in Figure 18.4.

All living organisms in this food web release carbon dioxide as they respire. The sugars, cellulose, starch and proteins of **succulent** plant tissue, as found in young plants, are rapidly decomposed to yield plant nutrients and have only a short-term effect. In contrast, the **lignified tissue** of older plants rots more slowly. Besides the release of nutrients, **humus** is formed from this fibrous and woody material, which has a long-term effect on the soil. Plants grown in the vicinity of vigorously decomposing vegetation, e.g. cucumbers in straw bales, live in a carbon dioxide enriched atmosphere. Carbon dioxide is also released on combustion of all organic matter, including the fossil fuels such as coal and oil. Organic materials such as paraffin or propane, which do not produce harmful gases when burned cleanly, are used in protected culture for **carbon dioxide enrichment** (p113).

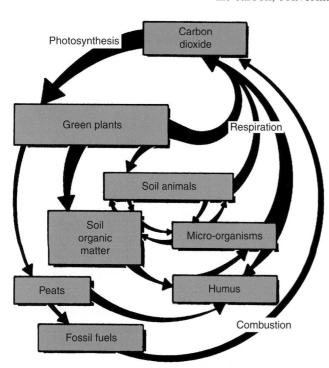

Figure 18.4 Carbon cycle. The recycling of the element carbon by organisms is illustrated. Note how all the carbon in organic matter is eventually released as carbon dioxide by respiration or combustion. Green plants convert the carbon dioxide by photosynthesis into sugars which forms the basis of all the organic substances required by plants but also animals and micro-organisms

The nitrogen cycle

The nitrogen cycle similarly follows the fate of nitrogen in its many forms in the plant, the soil and the atmosphere (see p366).

The sulphur cycle

Sulphur is an essential constituent of plants that accumulates in the soil in organic forms (see page 370). This sulphur does not become available to plants until aerobic micro-organisms mineralize the organic form to produce soluble **sulphates**. Under anaerobic conditions there are micro-organisms which utilize organic sulphur and produce hydrogen sulphide, which has a characteristic smell of bad eggs often evident in waterlogged soils in warm conditions.

Carbon to nitrogen ratio (C:N)

All nutrients play a part in all nutrient cycles simply because all organisms need the same range of nutrients to be active. Normally there are adequate quantities of nutrients, with the exception of carbon or nitrogen, which are needed in relatively large quantities. A shortage of nitrogenous material would lead to a hold-up in the nitrogen cycle, but would also slow down the carbon cycle, i.e. the decomposition of organic matter is slowed because the micro-organisms concerned suffer a shortage of *one* of their essential nutrients. A useful way of expressing the relative amounts of the two important plant foods is in the carbon to nitrogen (C:N) ratio.

Plant material has relatively wide C:N ratios, but those of micro-organisms are much narrower. This is because micro-organisms utilize about three quarters of the carbon in plants during decomposition as an energy source. The carbon utilized this way is released as carbon dioxide, whereas, usually, all the nitrogen is incorporated in the microbial body protein. This concentrates the nitrogen in the new organism that is living on the plant material.

Sometimes the C:N ratio is so wide that some nitrogen is drawn from the soil and 'locked up' in the microbial tissue. This is what happens when **straw** (and similar fibrous or woody material such as wood chips and bark) with a ratio of 60:1 is dug into the soil (see Table 18.1). For example, if one thousand 12 kg bales of straw are dug into one hectare of land then the addition to the soil will be 12 000 kg of straw containing 4800 kg of carbon and 80 kg of nitrogen. Three-quarters of the carbon (3600 kg) is utilized for energy and lost as carbon dioxide and a quarter (1200 kg) is incorporated over several months into microbial tissue. Microbial tissue has a C:N ratio of about 8:1, which means that by the time the straw is used up some 150 kg of nitrogen is locked up with the 1200 kg of carbon in the micro-organisms. Since there was only 80 kg of nitrogen in the straw put on the land, the other 70 kg has been 'robbed' from the soil. This nitrogen is rendered unavailable to plants ('locked up') until the micro-organisms die and decompose. To ensure rapid decomposition or to prevent a detrimental effect on crops the addition of straw must be accompanied by the addition of nitrogen.

Nitrogen is released during decomposition if the organic material has a C:N ratio narrower than 30:1, such as young plant material, or with nitrogen-supplemented plant material such as farmyard manure (FYM).

Dead organic matter in the soil

The dead organic matter has an important effect on the soil. The fresh, still recognizable material physically 'opens up' the soil, improving aeration. Active micro-organisms gradually decompose this material until it consists of unrecognizable plant and micro-organism remains. This finer material has less physical effect, but usually improves the water holding capacity of the soil.

In general, succulent ('green', leafy) organic matter decomposes very rapidly, so long as conditions are right, so has only a short-term physical effect, but yields nutrients, especially nitrogen compounds. The fibrous or woody ('brown') plant material tends to decompose very slowly so its physical effect persists, but nutrient contributions are low. The distinction between the 'green' and 'brown' organic matter is a crude but useful one when composting (p333).

Humus

This process of decomposition continues until all the organic matter is reduced to carbon dioxide, water, minerals and humus. The **humus** arises from a small proportion of the fibrous ('brown') organic matter which is highly resistant to decomposition; the lignin and other resistant chemicals form a collection of humic acids which forms a black colloidal (jelly-like) material. The humus coats soil particles and gives topsoil its characteristic dark colour.

This colloidal material has a high **cation exchange capacity** and therefore can make a major contribution to the retention of exchangeable cations, especially on soils low in clay (see sands p304). It also adheres strongly to mineral particles, which makes it a valuable agent in soil **aggregation**. In sandy soils it provides a means of sticking particles together, whereas in clays it forms a clay-humus complex that makes the heavier soils more likely to crumble. Its presence in the soil crumbs makes them more stable, i.e. more able to resist collapse when wetted, and it increases the range of soil **consistency** (see p342). Bacteria eventually decompose humus so the amount in the soil is very dependent on the continued addition of appropriate bulky organic matter.

Organic matter levels

The routine laboratory method for estimating **organic matter** levels depends upon finding the total carbon content of the soil. A simpler method is to dry a sample of soil and burn off the organic matter. After

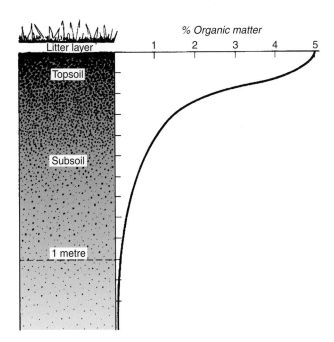

Figure 18.5 Distribution of organic matter in an uncultivated soil. Organic matter content of soil decreases from the soil surface downwards. Note that the topsoil is significantly richer in humus, which gives it a characteristically darker colour

cooling, the soil can be re-weighed and the loss in weight represents destroyed organic matter. These methods give an overall total of soil organic matter excluding the larger soil animals.

Most topsoil contains between 2 per cent and 5 per cent of organic matter, whereas subsoil usually contains less than 2 per cent. The distribution of organic matter under grass in normal temperate areas is shown in Figure 18.5.

The organic matter is concentrated in the topsoil because most of the roots occur in this zone and the plant residues tend to be added to the surface, forming the leaf litter layer. The organic matter level in any part of the soil depends upon how much fresh material is added compared with the rate of decomposition. It is stable when these two processes are balanced and the equilibrium reached is determined mainly by climate, soil type and treatment under cultivation.

Climate

Climate affects both the amount of organic matter added and the rate of decomposition. Below 6°C there is no microbial activity, but it increases with increasing temperature so long as conditions are otherwise favourable. In dry areas there is not only less plant growth, resulting in less organic matter being added to the soil, but also less microbial action. In **warm climates**, where there is adequate moisture, low organic matter results from very much increased decomposition. In **cooler areas** there tends to be an accumulation of organic matter because the decreased plant growth is more than offset by the reduced micro-organism activity that occurs over the long winter periods. Organic matter also tends to accumulate in **wetter conditions**. Where waterlogging is prevalent, 10–20 per cent organic matter levels develop; where waterlogging is permanent organic matter accumulates to give rise to **peat**.

Soil type

Generally, soils with higher clay contents have higher organic matter levels. Coarse sands and sandy loams tend to be warmer than finer textured soils and have better aeration, which results in higher microbial activity. Such soils often support less plant growth because of poor fertility and poor water holding capacity. These factors combine to give soils with low organic matter levels. In cultivation these same soils become a problem unless large quantities of organic matter are applied at frequent intervals to maintain adequate humus levels. Such soils are often referred to as 'hungry soils' because of their high demand for manure. On finer-textured soils the higher fine sand, silt or clay content increases water holding capacity. This reduces soil temperature, resulting in less microbial activity. The presence of clay directly reduces the rate of decomposition because it combines with humus and protects it from microbial attack.

Cultivation

Once soil has been cultivated a distinct boundary between topsoil and subsoil is developed as the concentration of organic matter in the surface layers is evened out (see Figures 18.5 and 17.5). On first cultivation the increased aeration and nutrients stimulate micro-organisms and a new equilibrium with lower soil organic matter levels prevails. Once under cultivation grasses and high-producing legumes tend to increase organic matter levels, but most crops, particularly those in which complete plant removal occurs, lead to decreased levels. Only large, regular dressings of bulky organic matter such as compost, straw, farmyard manure or leaf mould can improve or maintain the level of soil organic matter on cultivated soils.

Organic matter can accumulate under grass and form a mat on the surface where the carbon cycle is slowed because of nutrient deficiency usually induced by surface soil acidity or excess phosphate levels. This is part of the reason for the development of 'thatch' in turf (see Figure 18.6).

Figure 18.6 **Thatch**. This shows the build-up of organic matter in the surface of the turf

Organic soils

While all soils contain some organic matter, most are classified as mineral soils. However, at levels above about 15 per cent (when the organic matter present dominates the soil properties) they become classified as 'organic soils', e.g. organic clay loam. They develop where decomposition is slow because the activity of micro-organisms is reduced by cold, acidity or waterlogged conditions. **Peaty soils** are those where organic matter content is greater than 50 per cent; if content is more than 95 per cent the soil is considered to be a **peat**.

Peat is formed from partially decomposed plant material. This usually develops in waterlogged conditions where decomposition rates are low. There are great differences between peats because of the range of species of plants involved, which in turn depends on the conditions where they occur. Some peat is formed in shallow water, as found in poorly drained depressions or infilling lakes. In such circumstances the water drains from surrounding mineral soils and consequently has sufficient nutrients to support vegetation, often dominated by sedges, giving rise to **sedge peat**. As the waterlogged area, pond or lake becomes full of humified organic matter it forms a bog, moor or fen. In wetter areas, **sphagnum moss**, which is able to live on the very low nutrient levels that prevail, grows on top of the infilled wet land.

The dead vegetation becomes very acid and decomposes slowly. It builds up to form a high moor; sphagnum moss growing on top of very slowly decomposing sphagnum moss.

Some of the peatlands that are enriched with minerals prove very valuable when drained, e.g. the fenlands of eastern England. They are easily worked to produce vegetables and other high value crops. Unfortunately the increased aeration allows the organic matter to be decomposed at a rate faster than it can be replenished. Furthermore, when the surface dries out, the light particles are vulnerable to wind erosion. Consequently the soil level of these areas is falling at a rate of about three metres every hundred years. This can be checked by keeping the water table as high as possible and providing protection against wind.

Benefits of organic matter

Organic matter plays an important part in the management of soils. The main benefits are:

- **living organisms** in the soil play their part in the conversion of plant and animal debris to minerals and humus;
- *Rhizobia* and *Azotobacter spp.* fix gaseous nitrogen;
- plant roots, earthworms and other burrowing organisms improve the soil structure;
- many types of bacteria play an important role in the detoxification of harmful organic materials such as pesticides and herbicides;
- **dead organic matter** is food for soil organisms and increases microbial activity;
- dead but recognizable organic matter physically opens up the soil and improves aeration;
- fine, unrecognizable organic matter helps improve the water holding capacity of the soil;
- decomposing organic matter provides a source of dilute slow release fertilizer;
- **humus** coats soil particles with a black colloid and modifies their characteristics:
  - darker soils warm up faster in the spring;
  - organic matter improves water-holding capacity;
  - cation exchange capacity is increased, which can reduce the leaching of cations from the profile;
  - on sandy and silty soils the humus enables stable crumbs to be formed;
  - the surface charges on humus are capable of combining with the clay particles, thereby making heavy soils less sticky and more friable.

Addition of organic matter

It is normal in horticulture to return plant **residues** to cultivated areas where possible. Whether or not the plant remains are worked into the soil in which they have been grown depends upon their nature. The residue of some crops, such as tomatoes in the greenhouse, is removed to reduce disease carry over and because it cannot easily be incorporated into the soil. Other crops, such as hops, are removed for harvesting and some of the processed remains, spent hops, can be returned or used elsewhere. Wherever organic matter is removed, whether it is just the marketed part, such as top fruit from the orchard, or virtually the whole crop, such as cucumbers from a greenhouse, the nutrients removed must be replaced to maintain fertility (see also fertilizers).

Open, easily worked soils are created by the addition of large quantities of bulky organic matter. Clay soils are made easier to manage and their working range increased by the addition of organic matter. Stable, well-structured sands and silts are only possible under intensive cultivation if high humus levels are maintained by the addition of large quantities of bulky organic matter.

Bulky organic matter, such as compost, straw, farmyard manure and peat, is an important means of maintaining organic matter and humus levels. It also 'opens up' the soil, i.e. improves porosity. The main problem is finding cheap enough sources because their bulk makes transport and handling a major part of the cost. They can be evaluated on the basis of their effect on the physical properties of soil and their small, nutrient content.

Straw

Straw is an agricultural crop residue readily available in many parts of the country, but care should be taken to avoid straw with harmful **herbicide residues**. It is ploughed in or composted and then worked in. There appears to be no advantage in composting if allowance is made for the demand on nitrogen by soil bacteria. About 6 kg of nitrogen fertilizer needs to be added for each tonne of straw for composting to preventing soil robbing (see p325). Chopping the straw facilitates its incorporation and while not decomposed it can open up soils. On decomposition it yields very little nutrient for plant use, but makes an important contribution to maintaining soil humus levels. Straw bales suitably composted on site are the basis of producing an open growing medium for cucumbers.

Farmyard manure (FYM)

This is the traditional material used to maintain and improve soil fertility. It consists of straw or other bedding, mixed with animal faeces and urine. The exact value of this material depends upon the proportions of the ingredients, the degree of decomposition and the method of storage. Samples vary considerably. Much of the FYM is rotted down in the first growing season, but almost half survives for another year and half of that goes on to a third season and so on. A full range of nutrients is released into the soil and the addition of major nutrients should be allowed for when calculating **fertilizer requirements**. The continued release of large quantities of nitrogen can be a problem, especially on unplanted ground in the autumn, when the nitrates formed are leached deep into the soil over the winter and can pollute waterways.

FYM is most valued for its ability to provide organic matter and humus for maintaining or improving soil structure. As with any bulky organic matter, FYM must be worked into soils where conditions are favourable for continued decomposition to occur. Where fresh organic matter is worked into wet and compacted soils, the need for oxygen outstrips supply and anaerobic conditions develop to the detriment of any plants present. Where this occurs a foul smell (see sulphur p325) and grey colourings occur. FYM should not be worked in deep, especially on heavy soil.

Horticultural peats

Sphagnum moss peats have a fibrous texture, high porosity, high water retention and a low pH. They are used extensively in horticulture as a source of bulky organic matter and are particularly valued as an ingredient of potting composts because, with their stability, excellent porosity and high water retention, they can be used to create an almost ideal root environment.

Sedge peats tend to contain more plant nutrients than sphagnum moss. They are darker, more decomposed and have a higher pH level, but also have a slightly lower water-holding capacity. They tend to be used for making peat blocks. Considerable efforts are being made to find alternative materials to replace peat in order to avoid destroying valued wetland habitats from which they are harvested.

Leaves

Leaf mould is made from rotted leaves of deciduous trees. It is low in nutrients because nitrogen and phosphate are withdrawn from the leaves before they fall and potassium is readily leached from the ageing leaf. They are often composted separately from other organic matter and much valued in ornamental horticulture for a variety of uses, such as an attractive mulch, or when well rotted down, as a compost ingredient. They are commonly composted in mesh cages, but many achieve success by putting them in polythene bags well punched with holes. The leaves alone have a high C:N ratio so decomposition is slow and it is not usually until the second year that the dark-brown crumbly material is produced, although the process can be speeded up by shredding the leaves first.

Unless they are from trees growing in very acid conditions, the leaves are rich in calcium and the leaf mould made from them should not be used with calcifuge plants. **Pine needles** are covered with a protective layer that slows down decomposition. They are low in calcium and the resins present are converted to acids. This extremely acid litter is almost resistant to decomposition. It is valued in the propagation and growing of calcifuge plants, such as rhododendrons and heathers, and as a material for constructing decorative pathways.

Air-dried digested sludge

This consists of sewage sludge that has been fermented in sealed tanks, drained and stacked to dry. The harmful organisms and the objectionable smells of raw sewage are eliminated in this process. It provides a useful source of organic matter, but is low in potash. Advice should be taken before using sewage sludges because in some regions they contain high quantities of heavy metals, such as zinc, nickel and cadmium that can accumulate in the soil to levels toxic to plants.

Leys

The practice of **ley** farming involves grassing down areas and is common where arable crop production can be closely integrated with livestock. At the end of the ley period the grass or grass and clover sward is ploughed in. The root action of the grasses and the increased organic matter levels can improve the structure and workability of problem soils. There are some pest problems peculiar to cropping after grass that should be borne in mind (see wireworms), and generally the ley enterprise has to be profitable in its own right to justify its place in a horticultural rotation. It is practiced in some vegetable production and nursery stock areas.

Green manures

Unlike leys, green manuring is the practice of growing a cover crop primarily to incorporate in the soil. It is undertaken to:

Green manuring is the practice of growing plants primarily to develop and maintain soil structure and fertility.

- provide organic matter which can improve soil structure, aeration, water-holding capacity and, on decomposition, increase micro-organism activity in the soil;
- add some nutrients, especially nitrogen (depending on the plants involved), for the following crop;
- take up and store nitrogen that would otherwise be leached from bare soil over the winter period;
- deep rooted plants can bring up nutrients which have become unavailable to shallower plants;
- suppress weeds;
- provide cover to protect the soil from wind or water erosion;
- provide flowers for pollinating insects.

The seeds for green manuring are typically broadcast sown in the autumn when there are no other overwintering plants, but it can be undertaken at other times when the ground is to be left bare for several weeks instead of planting bedding or taking a catch crop. The plants are then dug or ploughed in when the land is needed again.

Plants used are typically agricultural crops that cover the ground quickly and yield a large amount of leaf to incorporate. The choice of plants needs to take into account the time of sowing, growth rate, soil type, winter hardiness, as well as particular characteristics of the species involved, e.g. legumes which fix nitrogen. Most commonly used are:

- **legumes** including bitter lupins, clovers, fenugreek, tares and trefoils;
- **non-legumes** including buckwheat, mustard, phacelia and rye.

Green manuring has many benefits, but there are some points to note in their management. If the plants are left to the stage when they become fibrous or woody, e.g. when allowing flowering to help pollinators, they will not provide extra nitrogen but are likely to 'rob' the soil of it (see C:N ratio p325). There can be difficulties when the following planting requires a fine seedbed, especially if this is to be early in the season; alternative approaches might be to cut and compost the foliage, cut or hoe off and use as a mulch or grow a plant killed by cold and remove the residue. Whilst it is highly valued in organic gardening, the value of the result when the cost of seeds, time and energy is taken into account is less clear cut in other systems.

Composting

Compost is a dark, soil-like material made of decomposed organic matter. Many gardeners depend on composting as a means of using garden refuse to maintain organic matter levels in their soils. On a larger scale there is interest in the use of composted **town refuse** for horticultural purposes. Many councils are now collecting 'green waste' and supplying composting equipment to encourage householders to recycle organic matter, as well as paper, glass and metals. Horticulturists are increasingly concerned with the recycling of wastes and attention is being given to modern composting methods. It is fundamental to successful organic growing.

Composting refers to the rotting down of plant residues before they are applied to soils.

For successful composting, conditions must be favourable for the decomposers. The material must be moist and well aerated throughout. As the heap is built, separate layers of lime and nitrogen are added as necessary to ensure the correct pH and C:N ratio. Organic waste brought together in large enough quantities under ideal conditions and turned regularly can be composted in two to three months. It is an exothermic process (heat is given off in the reactions) so enough heat can be generated to take the temperature to over 70°C within seven days, with the advantage of killing harmful organisms and weed seeds. The high temperatures can lead to a loss of ammonia (nitrogen).

In order to achieve a mix that allows adequate aeration, it is convenient to distinguish between 'green' (leafy or 'tender') and 'brown' (fibrous or 'tough') materials and combine in approximately equal measure (see Table 18.1). Note that shredded cardboard and paper can be added as 'brown' which is useful for recycling waste, but inks should be kept to a minimum. Decomposition is quicker if the ingredients are shredded to increase availability to organisms.

Table 18.1 Compost ingredients

Proposed ingredient	Category
Cardboard	brown
Farmyard manures	intermediate
Fibrous prunings	brown
Haulm (old plants)	brown
Hedge clippings	brown
Herbaceous plants (old)	brown
Grass mowings	green
Grass – long	brown
Kitchen (plant) waste	green
Nettles – young	green
Nettles – old	brown
Paper	brown
Seaweed	green
Straw	brown
Woody prunings	brown

When very large quantities are available the ingredients can be heaped up on a concreted base. This makes it easy to use power equipment to turn the ingredients to maintain good aeration and to mix in the cooler outer layers to ensure all parts heat up and decompose rapidly.

Garden methods

Most gardeners will not be able to obtain enough components at any one time to create the ideal composting process. An alternative approach is to build the heap over time. This is normally done in a slatted bin with a front that opens for access (see Figure 18.7). There should be an open base over soil to allow organisms and air in. A suitable cover is needed to keep some warmth in and rain off once the process has started. This method can produce good compost, but tends to take many months or even years to complete. Because it does not heat up adequately, care should be taken with regard to weeds, pests and diseases which are not killed in the process.

Figure 18.7 Compost bins. A typical set of bins large enough for efficient composting, slatted to allow in air and to allow easy access to add new material and to turn the contents

As much material as possible should be collected and prepared for composting. It should be chopped or shredded, 'green' and 'brown' mixed and water added to the heap. It is difficult to be successful with batches of less than one cubic metre at a time (when less than this the cooling at the surface is greater than the heating at the centre where decomposition is proceeding). It is advantageous to have a second bin alongside so the compost heap can be turned and loosened more easily on a regular basis to maintain good aeration.

Compost tumblers

These are containers that can be rotated on an axis to provide an easy method of turning small batches to create compost in a relatively short time. Batches can heat up sufficiently to kill off weeds and diseases and the enclosed container deters vermin. The compost ingredients should be gathered together and the tumbler filled in a short space of time. Nothing is added until the batch is completed.

Worm composting and wormeries

Worm composting lends itself to handling small quantities which can be added as they arise, such as kitchen waste, especially over the winter period when there is little plant material to accumulate. Compost worms (*Eisenia foetida*), also known as brandling or tiger worms, feed on organic matter. Whilst these can be purchased they are readily found in rotting vegetation such as compost heaps.

The container can be a plastic dustbin, usually equipped with a tap to drain off liquids which can be diluted and used as liquid feed for plants. Smaller containers, wide rather than narrow, can be made out wood, ideally with some insulation to maintain temperatures. Put in a 10 cm layer of sand and cover with a polythene sheet. Bedding material, such as well rotted compost or farmyard manure, may be needed for the worms to live in until the system gets going. Spread chopped waste to a depth of 5 cm. Add 100 or so worms and cover with wet newspaper to keep out the light and maintain moisture levels. A lid is needed to keep out the rain. Ideally, temperatures should be maintained between 20°C and 25°C and the pH kept between 6 and 8; lime can be added if the compost becomes too acid. The worms eat the vegetation as it starts to rot which means that once in balance there is no smell.

The compost is removed when ready; the decomposing top layer is separated off and used to start the next run. The compost is spread out to dry in the sun and the worms are recovered by placing a wet newspaper on the compost where they will congregate under it.

On a larger scale, wormeries are used to compost farmyard manure with continuous systems available that separate the composted material from the worms which can be recycled, with surpluses being available as animal feed.

Mulching

Many organic materials are used as mulches including farmyard manure, leaf mould, bark, compost, lawn clippings and spent mushroom compost.

Mulches are materials applied to the surface of the soil to suppress weeds, modify soil temperatures, reduce water loss, protect the soil surface and reduce erosion.

Organic mulches increase earthworm activity at the surface, which promotes better and more stable soil structure in the top layers. Soil compaction by water droplets is reduced and, as the organic mulches are incorporated, the soil structure can be improved. If thick enough mulches can suppress weed growth, but it is counter-productive to introduce a material that contains weeds. Likewise, care should be taken not to introduce pests and diseases or use a material such as compost where slugs can be a problem.

When organic matter is added as a mulch it is acting, in effect, as an extra layer of loose soil. Thus, water loss from the soil surface is reduced because it is covered with a dry layer (see evaporation). Soil temperatures lag behind the surface temperatures because of its insulating properties, with the greater lag at greater depth. They tend to reduce soil temperatures in the summer, but retain warmth later in the autumn.

Manufactured materials, such as paper, metal foil or, most commonly, polythene, are also used. In response to the demand for this type of material, woven polypropothene mulches are available. Whilst these have very little insulating effect they are particularly effective in reducing water loss by evaporation at the surface (see water conservation). The colour of the mulch is important because light-coloured material will reflect radiation whereas dark material will absorb it and can thus lead to earlier cropping by warming up the soil earlier.

Check your learning

1. Describe how organic matter is decomposed.

2. State the conditions required for the rapid breakdown of organic matter.

3. List the benefits of humus in soils of different types.

4. Describe one method of composting that can be adopted in a small garden.

5. Compare organic mulches with alternatives that can be used in the garden.

Further reading

Bragg, N. (1991). *Peatland: Its Alternatives*. Horticultural Development Council.

Brinton, W.F. (1990). *Green Manuring: Principles and Practice of Natural Soil Improvement*. Woods End Agricultural Inst.

Brown, L.V. (2008). *Applied Principles of Horticultural Science*. 3rd edn. Butterworth-Heinemann.

Caplan, B. (ed.) (1992). *The Complete Manual of Organic Gardening*. Hodder.

Edwards, C.A. and Lofty, J.R. (1977). *Biology of Earthworms*. 2nd edn. Halsted Press.

Fedor, J. (2006). *Organic Gardening*. Frances Lincoln.

Hills, L.D. (1977). *Organic Gardening*. Penguin Books Ltd.

Jackson, R.M. and Raw, F. (1973). *Life in the Soil*. Edward Arnold.

Killham, K. (1994). *Soil Ecology*. CUP.

Lowenfels, J. and Lewis, W. (2006). *Teaming with Microbes: A Gardener's Guide to the Soil Food Web*. Timber Press.

Paul, E.A. (2007). *Soil Microbiology, Ecology and Biochemistry*. 3rd edn.

Pears, P. and Strickland, S. (1999). *Organic Gardening*. 2nd edn. RHS. Octopus Publishing.

Postgate, J. (1998). *Nitrogen Fixation – Studies in Biology No. 92*. 3rd edn. Edward Arnold.

Robinson, D.W. and Lamb, L.J.D. (eds) (1982). *Peat in Horticulture*. HEA/Academic Press.

Russell, E.J. (1957). *The World of the Soil*. Collins New Naturalist.

Shepherd, A. (2007). *The Organic Garden: Green Gardening for a Healthy Planet*. Collins.

Chapter 19   Soil water

Summary

This chapter includes the following topics:

- **The wetting of a dry soil**
- **Saturation point**
- **Field capacity**
- **Symptoms of poor drainage**
- **Drainage of soil**
- **The drying of a wet soil**
- **Permanent wilting point**
- **Irrigation**

with additional information on the following:

- **Rainfall**
- **Water tables**
- **Soil consistency**
- **Soil moisture deficit**
- **Water quality**
- **Water conservation**

Figure 19.1 **Drainage pipes**. Modern clay pipes shown in the centre are butted up close together. An older method is shown below it and modern plastic piping is shown above. In the top left is a view of the modern clay pipe alongside the old 'horse shoe' tile (it would have sat on a 'mug plate' to prevent moving water washing away soil) and at the bottom are smaller examples of this type of old pipe

Wetting of a dry soil

Rainfall is recorded with a rain gauge (see Figure 2.13) and is measured in millimetres of water. Thus '1 mm of rain' is the amount of water covering any area to a depth of 1 mm. Therefore '1 mm of rain' on one hectare of land is equivalent to $10\,m^3$ or 10 000 litres of water per hectare (area $10\,000\,m^2 \times$ depth 0.001 m). As rain falls on a dry surface the water either soaks in (**infiltration**) or runs off over the surface as **surface run-off**. Accumulation of water on the surface (ponding) is a result of infiltration rates slower than rainfall. **Ponding** leads to soil capping, which further reduces infiltration rates. Soil surfaces can be protected with mulches (see p335) and care should be taken with water application rates during irrigation.

Saturated soils

The **saturation point** of a soil is when water has filled all the soil pores.

As water soaks into the dry soil, air is forced out of the surface layers which become **saturated** (waterlogged).

As water continues to enter the soil it moves steadily downwards, with a sharp boundary between the saturated zone and the dry, air-filled layers, as shown in Figure 19.2. So long as water continues to soak into the soil, this wetting front moves to greater depths and air is forced out of this zone.

When rainfall ceases the water in the larger soil pores continues to move downwards under the influence of gravity. Water is held in the soil in the form of water films around all the soil particles and aggregates. Forces in the surface of the water films, surface tension, hold water to the soil particles against the forces of gravity and the suction force of roots.

As the volume of water decreases, its surface area and hence its **surface tension** becomes proportionally greater until, in very thin films of water, it prevents the reduced volume of water from being removed by gravity. A useful comparison can be seen when your hands are lifted from a bowl of water. They drip until the forces in the surface of the thin film become equal to the forces of gravity acting on the remaining small volume of water over the hands.

Figure 19.2 **Wetting front**. As water is added to a dry soil it soaks into the soil with a clear line that can be seen between the unchanged (dry) soil and the saturated soil above

Field capacity (FC)

As **gravitational water** (sometimes referred to as 'excess water') is removed, air returns in its place. On sandy soils this may take a matter of hours after the rain has stopped, but may take far longer on clay where the process may continue for many days. The soil is then said to be at **field capacity** (FC). More precisely, it is a soil that has been saturated, then allowed to drain freely without evaporation until drainage effectively ceases. In practice it is assessed after two days.

Gravitational water is the water that can be removed by the force of gravity.

Field capacity is the amount of water the soil can hold against the force of gravity.

At field capacity the **micropores** (those less than about 0.05 mm diameter) remain full of water; whereas in the **macropores** (greater than about 0.05 mm) air replaces the gravitational water, as illustrated in

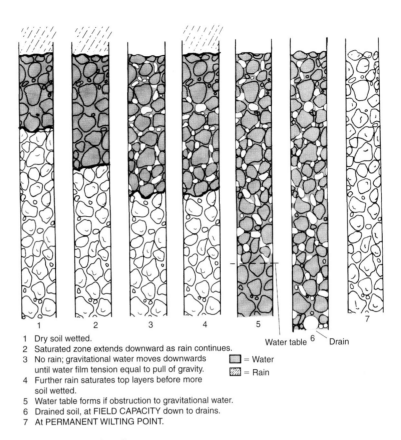

1 Dry soil wetted.
2 Saturated zone extends downward as rain continues.
3 No rain; gravitational water moves downwards until water film tension equal to pull of gravity.
4 Further rain saturates top layers before more soil wetted.
5 Water table forms if obstruction to gravitational water.
6 Drained soil, at FIELD CAPACITY down to drains.
7 At PERMANENT WILTING POINT.

Water table 6 Drain

□ = Water
▨ = Rain

Figure 19.3 Water in the soil

Table 19.1 **Soil water holding capacity**: the amount of water in a given depth of soil at field capacity can be calculated by simple proportion

Soil texture	Water held in 300 mm soil depth (mm)		
	at field capacity (FC) i.e. water holding capacity (WHC)	at permanent wilting point (PWP)	Available water (AW)
Coarse sand	26	1	25
Fine sand	65	5	60
Coarse sandy loam	42	2	40
Fine sandy loam	65	5	60
Silty loam	65	5	60
Clay loam	65	10	55
Clay	65	15	50
Peat	120	30	90

Figure 19.3. The air expelled has been replaced by 'fresh' air which is higher in oxygen and lower in carbon dioxide content.

The amount of water held at field capacity is known as the **water-holding capacity (WHC)** or moisture-holding capacity (MHC). Examples are given in Table 19.1 The WHC is expressed in millimetres of water for a given depth of soil. Thus a silty loam soil 300 mm deep holds 65 mm of water when at field capacity. Conversely, if a silty loam had become completely dry to 300 mm depth, it would require 65 mm of rain or irrigation water to return it to field capacity; about an average month's rainfall in many parts of the British Isles. Since 1 mm of water is equivalent to 10 m^3/ha, a hectare of silty loam would hold 650 m^3 water in the top 300 mm when at field capacity. The principle described enables water-holding capacity or irrigation requirement to be determined for any soil depth. The amount of water required to return a soil to field capacity is called the **soil moisture deficit** (SMD).

Water tables

Groundwater occurs where the soil and underlying parent material are saturated (see Figure 19.3) and the **water table** marks the top of this saturated zone, which fluctuates over the seasons, normally being much higher in winter. In wetlands the water table is very near the soil surface and the land is not suitable for horticulture until the water table of the whole area is lowered (see drainage). Where water flows down the soil profile and is impeded by an impermeable layer, such as saturated clay or silty clay, a **perched** (or crown) **water table** is formed. Water from above cannot drain through the impermeable barrier and so a saturated zone builds up above it. **Springs** appear at a point on the landscape where an overlying porous material meets an impermeable layer at the soil surface, e.g. where chalk hills or gravel mounds overlie clay.

Capillary rise

Water is drawn upwards from the water table through a continuous network of pores. The height to which water will rise and the rate of movement depends on the continuity of pores and their diameter. In practice the rise from the water table is rarely more than 2 cm for coarse sands, typically 15 cm in finer textured soils, but it can be substantially greater in silty soils and in chalk. The upward movement of water in these very fine pores is very slow. Capillary rise is used to aid the watering of plants grown in containers (see capillary benches). Several 'self-watering' containers also depend on capillary rise from a water store in their base (see aggregate culture p397).

Drying of a wet soil

Soil water is lost from the soil surface by evaporation and from the rooting zone by plant transpiration.

Evaporation

The rate of water loss from the soil by **evaporation** depends on the drying capacity of the atmosphere just above the ground and the water content in the surface layers. The evaporation rate is directly related to the **net radiation** (see p26) from the sun which can be measured with a solarimeter (see p43). Evaporation rates increase with higher air temperatures and wind speed or lower humidity levels. As water evaporates from the surface, the water films on the soil particles become thinner. The surface tension forces in the film surface become proportionally greater as the water volume of the film decreases. This leaves water films on the particles at the surface with a high surface tension compared with those in the films on particles lower down in the soil. The increased suction gradient causes water to move slowly upwards to restore the equilibrium. Whilst the surface layers are kept moist by water moving slowly up from below, the losses by evaporation, in contrast, are quite rapid. Consequently the surface layers can become dry and the evaporation rate drops significantly after 5 to 10 'mm of water' is lost. Evaporation virtually ceases after the removal of 20 'mm of water' from the soil. Maintaining a dry layer on the soil surface helps conserve moisture in the soil below. Evaporation from the soil surface is almost eliminated by a **leaf canopy** that shades the surface, thus reducing air flow and maintaining a humid atmosphere over the soil. **Mulches** (see p335) can also reduce water loss from the soil surface.

Evapotranspiration

As a leaf canopy covers a soil the rate of water loss becomes more closely related to transpiration rates. The potential transpiration rate represents the estimated loss of water from plants grown in moist soil with a full leaf canopy. It can be calculated from weather data (see Table 19.2).

Table 19.2 **Potential transpiration rates**. The calculated water loss (mm) from a crop grown in moist soil with a full leaf canopy, over different periods of time and based on weather data collected in nine areas in the British Isles

Area	April	May	June	July	Aug	Sept	Summer	Winter	Annual
Ayr	46	81	90	83	65	38	405	70	475
Bedford	50	78	89	91	80	43	430	70	500
Cheshire	53	75	83	88	76	44	420	80	500
Channel Isles	51	86	91	99	84	46	457	103	560
Essex (NE)	50	79	98	98	83	45	450	80	530
Hertford	49	79	91	94	80	43	435	75	510
Kent (Central)	50	79	93	96	83	44	445	65	510
Northumberland	44	64	81	76	60	34	360	70	430
Dyfed	46	75	84	81	74	44	405	105	510

As roots remove water it is slowly replaced by the water film equilibrium, but rapid water uptake by plants necessitates root growth towards a water supply in order to maintain uptake rates. At any point when water loss exceeds uptake, the plant loses turgor and may wilt. This tends to happen in very drying conditions, even when the growing medium is moist. Wilting is accompanied by a reduction in carbon dioxide movement into the leaf, which in turn reduces the plant's growth rate (see photosynthesis). The plant recovers from this **temporary wilt** as the rate of water loss falls below that of the uptake, which usually occurs in the cool of the evening onwards. Continued loss of water causes the soil to reach the permanent wilting point because roots can extract no more water within the rooting zone.

> The **permanent wilting point** (PWP) is the soil's water content when a plant growing in it does not regain turgor overnight.

When the soil has reached its **permanent wilting point (PWP)** there is still water in the smallest of the soil pores, within clay particles and in combination with other soil constituents, but it is too tightly held to be removed by roots. Typical water contents of different types of soil at their permanent wilting point (PWP) are given in Table 19.1.

Available water

Roots are able to remove water held at tensions up to 15 atmospheres within the rooting zone, and gravitational water drains away. Consequently the available water for plants is the moisture in the rooting depth between field capacity and the permanent wilting point. The available water content (AWC) of different soil textures is given in Table 19.1. Fine sands have very high available water reserves because they hold large quantities of water at FC and there is very little water left in the soil at PWP (see sands p304). Clays have lower available water reserves because a large proportion of the water they hold is held too tightly for roots to extract (see clay p305).

> **Available water** (to the plant) is the water held in the soil between field capacity and the permanent wilting point.

Roots remove the water from films at field capacity very easily. Even so, plants can wilt temporarily and any restriction of rooting makes wilting more likely. Water uptake is also reduced by high soluble salt concentrations (see osmosis) and by the effect of some pests and diseases (see vascular wilt diseases). As the soil dries out, the water films become thinner and the water is more difficult for the roots to extract. After about half the available water content has been removed, temporary wilting becomes significantly more frequent. Irrigating before available water falls to this point helps maintain growth rates. Plants grown under glass are often irrigated more frequently to keep the growing medium near to field capacity. This ensures maximum growth rates since the roots have access to 'easy' water, i.e. water removed by low suction force.

Soil consistency

The number of days each year that are available for soil cultivation depends on the weather, but more specifically on soil consistency (sometimes referred to as the workability of the soil). It also influences the timing and effect of cultivations on the soil.

> **Soil consistency** describes the effect of water on those physical properties of the soil.

It is assessed in the field by prodding and handling the soil. A very wet soil can lose its structure and flow like a thick **fluid**. In this state it has no **load-bearing strength** to support machinery. As the soil dries out it becomes **sticky**, then plastic. When **plastic** the soil is readily moulded. In general, the soil is difficult to work in this condition because it still tends to stick to surfaces, has insufficient load-bearing strength, is readily compacted and is easily smeared by cultivating equipment. As the soil dries further it becomes **friable**. At this stage it is in the ideal state for cultivation because it has adequate load-bearing strength, but the soil aggregates readily crumble. If the soil dries out further to a **harsh** (or hard) **consistency** the load-bearing strength improves considerably, but whilst coarse sands and loams still readily crumble in this condition, soils with high clay, silt or fine sand content form hard resistant clods. The **friable range** can be extended by adding organic matter (see humus). At a time when bulky organic matter is more difficult to obtain it is important to note that a fall in soil humus content narrows the friable range. This allows less latitude in the timing of cultivations and increases the chances of cultivations being undertaken when they damage the soil structure.

> **Timeliness** is the cultivation of the soil when it is at the right consistency.

Whereas many sands and silts can be cultivated at field capacity, clays and clay loams do not become friable until they have dried out to well below field capacity, i.e. heavier soils need more time for evaporation to remove water through the soil surface.

Drainage

As gravitational (excess) water leaves the macropores, the air that takes its place ensures that the root zone is replenished with 'fresh air'. Horticultural soils should return to at least 10 per cent air capacity in the top half metre within one day of being saturated (see porosity). Some soils, notably those over chalk or gravel, are naturally free draining.

> **Drainage** is the removal of gravitational (excess) water from the root zone.

However, many have underlying materials which are impermeable or only slowly permeable to water, and in such cases **artificial drainage**, sometimes referred to as field drainage or under drainage, is put in to carry away the gravitational water (see Figure 19.3). This helps the soil to restore air content rapidly without reducing the available water content. **Well-drained soils** are those that are rarely saturated within the upper 90 cm except during or immediately after heavy rain. Uniform brown, red, or yellow colours indicate an **aerobic** soil, i.e. a soil in which oxygen is available. **Poorly-drained** soils are saturated within the upper 60 cm for at least half the year and are predominantly grey which is typical of **anaerobic** soil conditions. Between these extremes, **imperfectly drained** soils are those that are saturated in the top 60 cm for several months each year. These soils tend to have less bright colours than well-drained soil; grey and ochre colours are usually seen at 450 mm giving a characteristic rusty mottled appearance (see Figure 17.7).

Symptoms of poor drainage

Symptoms of poor drainage include:

- grey or mottled soil colours;
- restricted rooting;
- reduced working days for cultivation;
- weed problems;
- pest and disease problems;

- excess fertilizer requirements;
- topsoils water-logged for long periods in warm conditions have a smell of bad eggs (see sulphur cycle p325).

Soil pits dug in appropriate places reveal the extent of the drainage problem and help pinpoint the cause, which is the basis of finding the solution. The level of water that develops in the pit indicates the current water table. Further indications of poor drainage are the presence of high organic matter levels (see organic soils) and small black nodules of manganese dioxide.

Soil colours show the history of water-logging in the soil. Whereas free drainage is indicated by uniform red, brown or yellow soil throughout the subsoil, the iron oxide which gives soil these colours in the presence of oxygen is reduced to grey or blue forms in **anaerobic** conditions, i.e. when no oxygen is present. Zones of soil that are saturated for prolonged periods have a dull grey appearance, referred to as **gleying** (see Figure 17.7). Reliance on colour alone as an indication of drainage conditions is not recommended, because it persists for a long time after efficient drainage is established (see also soil types p301–3).

Structural damage, whether caused by water (see stability), machinery (see cultivation), or by accumulations of iron (see natural pans), is an obstruction to water flow in the soil profile. Pans or platy structures near the surface can be broken with cultivating equipment on arable land or spiking on grassland; but subsoiling is used to burst those deeper in the soil. If water cannot soak away from well-structured rooting zones, artificial drainage is required.

Artificial drainage

The low permeability of many subsoils, which create a perched water table, is the major reason for artificial drainage in horticultural soils. Clay, clay loam and silty clays, when wetted, become almost impermeable as the clay swells and the cracks close; clay is 'puddled' to form a liner for ponds. This 'top water problem' is dealt with by putting in pipes to intercept the trapped gravitational water. Straight lines of pipes are placed at an even **gradient** from the highest point to the outfall in a ditch or main drain (see Figure 19.4). The pipes are laid below cultivation depth in a series of parallel lines across the slope to the headland of the area to be drained. Where a valley or the lower areas lie within the area to be drained, a herringbone pattern is used.

Silt traps should be placed at regular intervals to help to service the system at points where there is a change of gradient or direction (see Figure 19.4). The spacing between the lines of pipes depends on the permeability of the soil, a maximum of 5 metre intervals being necessary in clay subsoils. Soil permeability and the land use dictate the **depth of the drains** which is normally more than 60 cm. Drains should be set deeply in cultivated land where heavy equipment and deep cultivation will not disturb the pipes. Shallow drains can be used where rapid drainage is a high priority and the pipes are not likely to be crushed by heavy vehicles or severed by cultivating equipment, e.g. gardens and sports grounds. Pipe drainage is usually combined with secondary treatments, such as mole drainage or subsoiling, to achieve effective drainage at reasonable cost. Deep subsoiling improves soil permeability and the pipes carry the water away. Installation costs can be reduced because pipes can be laid further apart. Similarly mole drainage over and at right angles to the pipes enables them to be spaced 50–100 metres apart.

The pipes are made of clay or plastic. The **diameter** of the pipe depends on the gradient available and the amount of water to be carried when wet conditions prevail. Tiles (clayware) are usually 300 mm pipes either 75 mm or 100 mm in diameter (see Figure 19.1). These lead into a ditch or a larger main drain. The tiles are butted tightly together to allow entry of water, but not soil particles. It is recommended that they are covered with permeable fill, usually stones or clinker, to improve water movement into the drains. Plastic pipes consist of very long lengths of pipe perforated by many small holes and usually covered with a rough felt to keep out soil particles.

An outlet into a ditch is very vulnerable to damage and so it should consist of a strong, long pipe set flush in a concrete or brick headwall so that it is neither dislodged by erosion in the ditch nor by people using it as a foothold. Outlet pipes should be glazed to prevent frost damage. Vermin traps should be fitted to prevent pipes being blocked by nests or dead animals (see Figure 19.4).

Mole drainage is very much cheaper than pipe drainage. A mole plough draws a 75 mm 'bullet', followed by a 100 mm plug, through the soil at a depth of 500–750 mm from a ditch up the slope of a field or across a pipe drain system with permeable backfill (see Figure 19.4). The soil should be plastic at the working depth so that a tunnel to carry water is created. The soil above should be drier so that some cracks are produced as the implement is drawn slowly along. These cracks improve the soil structure and conduct water to the mole drain. Sandy and stony areas are unsuitable because tunnels are not properly formed or collapse as water flows. Tunnels drawn in clay soils can remain useful for 10–15 years, but in wetter areas their useful life may be nearer 5 or even as little as 2 years.

Sandslitting is used on sports grounds to remove water from the surface as quickly as possible. It involves cutting narrow trenches at frequent intervals in the soil and infilling with carefully graded sand that conducts water from surface to a free draining zone under the playing surface.

(a) Simple Interceptor Drainage System:

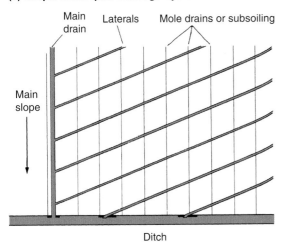

(b) Outfalls

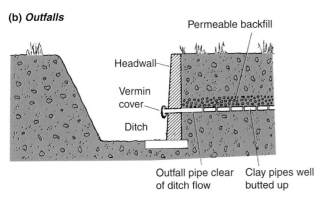

(c) Silt trap/Inspection Chamber

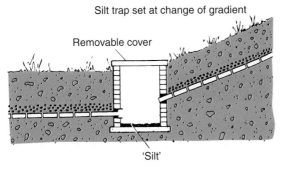

Figure 19.4 Drainage

French drains can be placed around impermeable surfaces, such as concrete hard standings and patios, to intercept the run-off.

Maintenance of drainage systems

Artificial drainage is very expensive to install and must be serviced to ensure that the investment is not wasted. **Ditches** need regular attention because they are open to the elements (see Figure 19.5). Weed growth should be controlled; rubbish cleared out and collapsed banks repaired, because obstructions lead to silting up or undercutting of the bank. The design of the ditches depends on the soil type and should be maintained when being repaired. The batter (the slope of the sides) on sandy soils has to be less steep than on clays.

Drain outlets are a particularly vulnerable part of the drainage system, especially if not set into a headwall. They should be marked with a stake (holly trees were traditionally used in some areas) and inspected regularly after the soil returns to field capacity. Blockages should be cleared with rods and vermin traps refitted where appropriate.

Silt traps need to be cleaned out regularly to prevent accumulated soil being carried into the pipes. Wet patches in the field indicate where a blockage has occurred. The pipe should be exposed and the cause of the obstruction removed. Silted-up pipes can be rodded, broken sections replaced, or dislodged pipes realigned.

At all times it should be remembered that the drains only carry away water that reaches the pipes. Every effort must be made to maintain good soil permeability and to avoid compaction problems. Subsoilers (see p315) should be used to remedy subsoil structural problems. Once drainage has been installed the soil dries more quickly, leading to better soil structure because cracks appear more extensively and for longer periods. Deeper layers of the soil are dried out as roots explore the improved root environment, which adds to the improving cycle.

Groundwater ('bottom water') problems occur where the water table is too high and drains at the desired depth are of no use, because there is nowhere low enough to discharge the water. An artificially low water table can be created by pumping water out of the ditches up into a network of waterways. A line of windmills that provided the required energy was a familiar sight in lowland areas such as the Fens, where such land reclamation was undertaken. Goundwater problems are a common feature of many gardens where attempts to introduce drainage systems are thwarted by there being nowhere suitable (or legal) to discharge the water.

Figure 19.5 Ditch. This is open to the elements and easily choked if not regularly maintained

Irrigation

Irrigation is used to prevent plant growth being limited by water shortage.

Irrigation should be seen as a husbandry aid in addition to otherwise sound practice. It is assumed in the following that water is being added to a well-drained soil. The need for irrigation depends mainly on available water in the rooting zone and the effect of water stress on the plant's stage of growth. The very large quantities of water required for commercial production are illustrated clearly in the estimates for growing in protected culture, where all the requirements have to be delivered to the crop by irrigation. In the British Isles, the daily consumption of water from a full cover crop, such as tomatoes or cucumbers, is about 20 000 litres per hectare in March rising to double that in June. This amounts to about 9000 cubic metres per year (approximately 750 000 gallons per acre). A more exact estimate can be obtained by measuring the light levels outside the greenhouse; 2200 litres per hectare are required for each megajoule per square metre (see p43). This can be compared with the 6000 cubic metres of rainfall that could be collected, on average, from the roof of a hectare of glass in the south-east of England (see p338). To take advantage of this contribution there would need to be substantial storage facilities and the water quality issues would need to be addressed.

Response periods are the growth stages when the use of irrigation during periods of rainfall deficiency is likely to be worthwhile. In general all plants benefit from moist seedbeds and eliminating water stress maximizes vegetative growth. Initiation of flowering and fruiting is favoured by drier conditions. The response periods of a range of plants grown in the UK is given in Table 19.3.

Irrigation plans

In general, water should not be added to outdoor soils until moisture levels fall to 50 per cent of available water content in the rooting zone. Outdoors 25 mm of water is the minimum that should be added at any one time in order to reduce the frequency of irrigation, to reduce water loss by evaporation and to prevent the development of shallow rooting. On most soils the amount of water added should be such as to return the soil to field capacity. Addition of water to clays and clay loams should be minimized so as not to reduce the vital drying and wetting cycles, and if they have to be irrigated they should not be returned to field capacity in case rain follows (see ponding). Irrigation should never result in fertilizers being leached from the rooting depth unless it is the specific objective, as in flooding of greenhouse soils (see conductivity).

Most recommendations are given in a simplified form taking the above points into account. The recommended plan is usually expressed in terms of how much water to apply, at a given soil moisture deficit, for a named crop on a soil of stated available water content. Thus for an outdoor grown summer lettuce crop grown on soils of a medium available water content, 25 mm of water should be added when a 25 mm soil moisture deficit occurs. This would require the application of 250 000 litres per hectare or 25 litres per square metre. Further examples are given in Table 19.3.

Soil moisture deficit (SMD) is the amount of water required to return the growing medium to field capacity.

Soil moisture deficit (SMD) can be calculated by keeping a soil water balance sheet. The account is conveniently started after rain returns the soil to field capacity, i.e. when SMD is zero.

Table 19.3 Irrigation guide

| Plants | Growth stages at which to irrigate | | Irrigation plan (mm of water at mm SMD) | | |
| | | | A | B | C |
	Response periods	Time of year when they occur	low* AWC	medium AWC	high** AWC
Beans, runner	Early flowering onwards	June to August	25 at 25+	50 at 50+	50 at 75
Brussels sprouts	When lower buttons 15–18 mm diameter	Aug to October	40 at 40	40 at 40	40 at 40
Carrots	Throughout life	June to Sept	25 at 25	40 at 50	
Cauliflowers	Throughout life	April to June	25 at 25	25 at 25	
Flower; perennials	Throughout life	April to Sept	25 at 25	50 at 75	50 at 75
Lettuce summer	Throughout life	April to Aug	25 at 25	25 at 25	25 at 50
Nursery stock trees and shrubs	a. to establish newly planted stock	April to June	25 at 25	25 at 50	25 at 50+
	b. established stock	May to July	25 at 25	25 at 50	25 at 50+
	c. to aid early lifting	September	25 at 25+	25 at 50+	25 at 50+
Potatoes, first early	After tuberization reaches 10 mm dia	May to June	25 at 25	25 at 50	25 at 50
Maincrop and second earlies	From time tubers reach marble stage onwards	June to August	25 at 25	25 at 50	25 at 50
Rhubarb	When pulling has stopped	May to Sept	40 at 50+	40 at 50+	50 at 75+
Strawberries		Sept to Oct	50 at 50	50 at 75	50 at 75
Top fruit Apples Pears		July to Sept July to Aug	When SMD is more than 50 mm apply 50 mm of water to suffice for two weeks. Then continue irrigation to make the total water supply (rain + irrig) equal to 50 mm/fortnight for the remainder of July, 40 mm/fortnight in Aug and 25 mm/fortnight in Sept.		

*less than 40 mm available water per 300 mm soil, e.g. gravels, coarse sands.

**more than 65 mm available water per 300 mm soil, e.g. silts, peats.

In Britain it is assumed that, unless it has been a dry winter, the soil is at field capacity until the end of March. From the first day of April a day-by-day check can be made of water gains and losses. A worked example of a weekly water balance sheet is given in Table 19.4; a daily water balance sheet may be more appropriate in some situations.

Rainfall varies greatly from year to year from one locality to the next and so it should be determined on site (see rain-gauge) or obtained from a local weather station. Water loss for each month does not vary very much over the years and so potential transpiration rates based on past records can be used in the calculation. There are potential transpiration

Table 19.4 **A weekly water balance sheet** for established nursery stock grown on sandy loam (AWC 55 mm per 300 mm) in Essex. The irrigation plan is to apply 25 mm water if a 50 mm SMD is reached (see Table 19.3). Water loss estimated from Table 19.2.

| Week beginning | Water loss (mm) (A) | Water gains (mm) | | SMD at end of week (mm) (D) |
		rainfall (B)	irrigation (C)	
				D + A − (B + C) = new D
March 31				0
April 7	11	10	0	0 + 11 − (10 + 0) = 1
14	11	8	0	1 + 11 − (8 + 0) = 4
21	12	16	0	4 + 12 − (16 + 0) = 0
28	12	5	0	0 + 12 − (5 + 0) = 7
May 4	16	28	0	7 + 16 − (28 + 0) = 0*
11	17	10	0	0 + 17 − (10 + 0) = 7
18	18	14	0	7 + 18 − (14 + 0) = 11
25	18	4	0	11 + 18 − (4 + 0) = 25
June 1	18	10	0	25 + 18 − (10 + 0) = 33
8	23	14	0	33 + 23 − (14 + 0) = 42
15	23	18	0	42 + 23 − (18 + 0) = 47
22	23	20	0	47 + 23 − (20 + 0) = 50**
29	23	10	25	50 + 23 − (10 + 25) = 38
July 6	22	8	0	38 + 22 − (8 + 0) = 52**
13	22	18	25	52 + 22 − (18 + 25) = 31
20	22	24	0	31 + 22 − (24 + 0) = 29
27	22	5	0	29 + 22 − (5 + 0) = 46

*Soil Moisture Deficit (SMD) cannot be less than zero because water above FC drains away.

**Irrigation might have been delayed if prolonged heavy rain forecast.

rate figures available for all localities having weather stations. Examples are given in Table 19.2. These figures can be used when calculating water loss, but until there is 20 per cent leaf canopy a maximum SMD of 20 mm is not exceeded because in the early stages water loss is predominantly from the soil surface by evaporation (see p340).

In protected cropping all the water that plants require has to be supplied by the grower, who must therefore have complete control over irrigation. With experience the grower can determine water requirements by examining plants, soil, root balls or by tapping pots. A **tensiometer** can be used to indicate the soil water tension but, while this is useful to indicate when to water, it does not show directly how much water is needed. **Evaporimeters** distributed through the planting can give the water requirement by showing how much water has been evaporated. A **solarimeter** measures the total radiation received from the sun and the readings obtained can be used to calculate water losses, often expressed in litres per square metre for convenience.

Methods of applying water

These should be carefully related to plant requirements, climate and soil type. On a small scale, **watering cans** or **hoses** fitted with trigger lances can be used, but care should be taken to avoid damaging the structure of the growing medium. Water can be sprayed from fixed or mobile equipment, but it is essential that the rate of application is related to soil infiltration rate The droplet size in the spray should not be large enough to damage the surface structure (see tilth). Indoors, **spray lines** can be fitted with nozzles to control the direction and quantity of water. Overhead lines can lead to very high humidity levels and wet foliage, predisposing some plants to disease (see grey mould). Consequently, it should be restricted to watering low level crops, e.g. lettuce, deliberately increasing humidity ('damping down' or 'spraying over'), or winter flooding (see conductivity). **Trickle** lines deliver water very slowly to the soil, leaving plant foliage and the soil surface dry, which ensures a drier atmosphere and reduced water loss. However, care is needed because there is very little sideways spread of water into coarse sand, loose soil, or a growing medium that has completely dried out. **Drip** irrigation is a variation on the trickle method, but the water is applied through pegged down thin, flexible 'spaghetti' tubes to exactly where it is needed, e.g. in each pot or at the base of each plant.

Simple **flooded benches** are sometimes used to water pot plants. The shallow tray of the benched area is filled with water which the pots absorb, after which the excess water is drained off ('ebb and flow'). This tends to produce a high humidity around the plants and **capillary benches** have come to be preferred. The pots stand on, and the contents make contact with, a level 50 mm bed of sand kept saturated at the base by an automatic water supply. Water lost by evaporation at the surface and from the plant is replaced by **capillary rise**. The sand must be fine enough to lift the water but coarse enough to ensure that the flow rate is sufficient. **Capillary matting** made of fibre woven to a thickness and pore size to hold, distribute and/or lift water has many uses in watering containerized crops both indoors and outside. Containers with built-in water reserves and easy watering systems utilize capillarity to keep the rooting zone moist. Sub-irrigated sand beds are used for standing out container plants in nursery production and are less wasteful of water than the more usual overhead spray lines.

Water quality

Water used in horticulture is taken from different sources and has different dissolved impurities. Soft water has very few impurities, whereas **hard water** contains large quantities of calcium and/or magnesium salts which raise the pH of the growing medium, especially where impurities accumulate (see liming). Even small quantities of **micro-elements,** such as boron or zinc, have to be allowed for when making up nutrient solutions which are to be re-circulated (see hydroponics). Water taken from boreholes in coastal areas can have high concentrations of salt that can lead to **salt concentration problems**. The quantity of dissolved salt in water can be measured by its conductivity; the higher the salt concentration the greater its electrical conductivity. Providing the levels of useful salts are not too high, the water can be used, but the additional nutrient levels (fertilizers) must be suitably adjusted (see conductivity).

In the re-circulation systems that are again becoming more prevalent in protected culture, the salts not used by plants can become concentrated in the water. These dissolved salts can interfere with the uptake of useful salts such as potassium, making it difficult to create a balanced feed within the safe conductivity limits and to reduce the plant growth rates as they become too concentrated. Water drawn from rivers, lakes or even on-site reservoirs may contain algal, bacterial or fungal pollution, which can lead to blocked irrigation lines or plant disease (see hygienic growing).

Rainwater is increasingly being used as a major source of water. It is usually of high quality, i.e. low conductivity, but there can be contamination related to the location or the method of collection or storage, e.g. high levels of zinc when collected through galvanized gullies. Good quality rainwater can be used to dilute otherwise unsuitable water to bring it into use. Alternatively, poor quality water can be treated using reverse osmosis; water under pressure is forced through a membrane which holds back most of the dissolved salts. Alternatively, deionization can be used, this involves passing the water over resins to remove the unwanted salts. In both cases, an environmentally sound method for disposal of the concentrated solution produced remains a problem. High energy distillation and electrodialysis methods are generally too expensive for cleaning water for growing. To avoid disease problems, water supplies can be sterilized. On a commercial scale this is usually done by heat sterilization. Ultra-violet light or ozone treatments are usually more expensive and the use of hydrogen peroxide tends to be less effective.

Water conservation

The need to manage water efficiently is a major concern in the use of scarce resources. Responsible action is increasingly supported by legislation and the higher price of using water. Clearly the major factors that determine the level of water use are related to the choice of plant species to be grown and the reasons for growing. **The selection of drought-tolerant rather than water intensive plantings is fundamental**. Some growing systems are inherently less water intensive, but in most of them there are many ways in which water use can be reduced if certain principles are kept in mind and acted upon appropriately:

- **Whenever possible, the use of artificial water should be avoided**.
- **Use recycled water**. The recycling of water and the capture of rainwater are important considerations in the choice of water source.

- **Minimize evaporation of water**. This is best achieved by not spraying water into the air and by minimizing the time when the soil surface is moist. When water does have to be applied overhead, this should be undertaken in cool periods.
- **Increase the water reservoir of the soil**. The application of water can be reduced by increasing the growing medium's water holding capacity; most soils can be improved by the addition of suitable organic matter.
- **Encourage plant root systems**. Plants should be encouraged to establish as quickly as possible but, after the initial watering-in, infrequent applications will encourage the plant to put down deep roots by searching for water. Most importantly, soil pans should be eliminated and good soil structure maintained to increase the rooting depth.
- **Minimize water lost through drainage**. Where application is partially controlled, the correct relationship between water applied and water holding (see water-holding capacity) helps to prevent leaching, which leads to nutrient loss (see nitrogen), as well as wasting water. Thus returning an outdoor soil to less than field capacity helps avoid losses by drainage or run-off in the event of unexpected rainfall.

Water is lost more rapidly from a moist than from a dry soil surface. After just 10 mm of water has been lost from the surface, the rate of evaporation falls significantly. Infrequent application thus helps, but even more effective is the delivery of water to specific spots next to plants (see trickle lines) or from below through pipes to the rooting zone. Avoid bringing moist soil to the surface. If hoeing is undertaken it should be confined to the very top layers; this also reduces the risk of root damage. Losses from the surface can be reduced considerably by plant cover and almost eliminated by the use of mulches. Loss of water from the plants themselves is reduced when they are grouped together rather than spaced out.

Unless maximum growth rates are the main consideration, **reduced application** saves water, money and staff time without detriment to most plantings. In production horticulture, the introduction of sophisticated moisture-sensing equipment and computer controls has enabled water to be delivered more precisely when and where it is needed. This has led to considerable reduction in water use.

Nutrient loss and run-off from **overhead watering** used in container nursery stock production can be minimized by matching application to rainfall, growing medium, container size, plant species, stage of growth and time of year. Nozzles should be maintained to ensure even water application. Loss from sub-irrigated capillary sand beds tends to be lower. Recirculation (closed) systems should be considered in new developments. The quantity of water required for **flooding** soils in protected culture (e.g. to remove excess nutrients when a crop sensitive to high salt levels, such as lettuce, is to be grown after a tolerant one, such as tomatoes), can be reduced by discontinuing the liquid feeding of the previous crop as soon as possible.

In non-recirculating (open) **hydroponics systems** excessive water waste should be avoided by using flow meters to measure the quantity of run-off and comparing it with standard figures for the growing system used. A run-off of over 30 per cent is usually considered to be excessive and the amount and frequency of nutrient applications delivered by the nozzles or drippers should be reviewed. Closed systems (see NFT p394) recirculate the nutrient solution, but this is not always practical. Where they are used, the system must not be emptied illegally into watercourses or soakaways. It is recommended that the volume in the system be run down before discharge and the waste nutrient solution be sprayed on to crops during the growing season. Permission to empty into public sewers might be granted, but it is usually subject to a charge depending on volume and contamination level.

Check your learning

1. Define field capacity and available water.

2. State four symptoms of poor drainage.

3. State what is meant by drainage.

4. Explain what is meant by soil moisture deficit.

5. Explain what is meant by irrigation response periods.

Further reading

Baily, R. (1990). *Irrigated Crops and Their Management*. Farming Press.

Brown, L.V. (2008). *Applied Principles of Horticultural Science*. 3rd edn. Butterworth Heinemann.

Castle, D.A., McCunnell, J., and Tring, I.M. (1984). *Field Drainage: Principles and Practice*. Batsford Academic.

Farr, E. and Henderson, W.C. (1986). *Land Drainage*. Longman Handbooks in Agriculture.

Hope, F. (1990). *Turf Culture: A Manual for the Groundsman*. Cassell.

Ingram, D.S. *et al.* (eds) (2002). *Science and the Garden*. Blackwell Science Ltd.

McIntyre, K. and Jakobsen, B. (1998). *Drainage for Sportsturf and Horticulture*. Horticultural Engineering Consultancy.

MAFF Advisory Booklet: No. 2067 (1979). *Irrigation Guide*.

MAFF Advisory Booklet: No. 2140 (1980). *Watering Equipment for Glasshouse Crops*.

MAFF Advisory Booklet: No. 776 (1981). *Water Quality for Crop Irrigation*.

MAFF Advisory Reference Book No. 138 (1976). *Irrigation*.

MAFF Technical Bulletin No. 34 (1970). *Climate and Drainage*.

MAFF Advisory Leaflets: Nos 721–740 (various dates). *Getting Down to Drainage*.

Chapter 20   Soil pH

Summary

This chapter includes the following topics:

- **The importance of soil pH**
- **The pH scale**
- **Acidity and alkalinity**
- **Nutrient availability**
- **Plant selection**
- **Changing pH**

with additional information on:

- **Measuring pH**
- **Lime requirement**
- **Types of lime**

Figure 20.1 **Hydrangeas**. Blue flowers are produced by plants in a growing medium at a low pH whereas pink occurs in those nearer neutral; with less determinate colours in between

The importance of soil pH

> pH is the negative log of the hydrogen ion concentration of a solution, e.g. soil water.

The pH scale is a means of expressing the degree of acidity or alkalinity of the soil. Temperate area soils are usually between pH 4 and 8; the vast majority are between 5.5 and 7.5.

The ideal growing condition for most plants is a soil of about pH 6.5, which is slightly acid. At this level most of the plant nutrients are available for uptake by the roots. Alkaline conditions are usually created by the presence of large quantities of calcium ('lime'), which interferes with the uptake and utilization of several plant nutrients.

Acidity, alkalinity and the pH scale

Pure water is neutral, i.e. neither acid nor alkaline. It is made up of two hydrogen atoms and one oxygen atom, expressed as the familiar formula H_2O. Most of the water is made up of water molecules in which the atoms stay together, but a tiny proportion dissociates, i.e. they form ions (see p129). Equal numbers of positive ions, cations, and negative ions, anions, are formed. In the case of water, an equal number of hydrogen cations, H^+, and hydroxyl anions, OH^-, exist within the clusters of water molecules (H_2O). When, as in water, the concentration of hydrogen ions is the same as that of hydroxyl ions, there is neutrality.

Many compounds form ions as they dissolve and mix with water to produce a solution. If the concentration of hydrogen ions is increased, the concentration of hydroxyl ions decreases and **acidity** increases. Likewise, the addition of hydroxyl ions and a decrease in hydrogen ions leads to increased **alkalinity**. All acids release hydrogen ions when dissolved in water. Whereas a **strong acid** (such as hydrochloric acid) fully dissociates when dissolved (i.e. every molecule splits into ions), only a part of a **weak acid** (such as carbonic or acetic acid) breaks up into ions. Bases, such as caustic soda, dissolve in water and increase the concentration of hydroxyl ions.

The **pH scale** expresses the amount of acidity or alkalinity in terms of hydrogen ion concentration. pH values less than 7 are acid and the lower the figure the greater the acidity. Values greater than 7 indicate increasing alkalinity and pH 7 is neutral. In order to present the scale simply, negative logarithms are used: **'pH' is the negative logarithm**, the mathematical symbol for which is 'p', **of the hydrogen ion concentration**, abbreviated as 'H'. It is important to note that, as the scale is logarithmic, a one-unit change represents a ten-fold increase or decrease in hydrogen ion concentration and a change of two units represents a hundred-fold change. Thus a solution of pH 3 is ten times more acid than one of pH 4, one hundred times more acid than pH 5, and a thousand times more than pH 6.

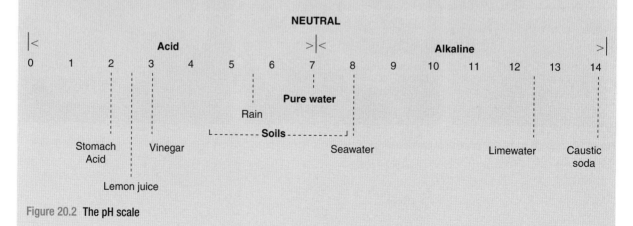

Figure 20.2 **The pH scale**

Measuring pH

pH can be measured using a pH meter or by the use of indicators; both are used for testing soils. Soils sent to laboratories for analysis would normally be tested with pH meters. Colour indicator methods are most commonly used by growers and gardeners who do their own testing.

Colour indicator method

Required:

- Clean test-tubes; ideally ones that have stoppers that can be removed top and bottom for easy cleaning;
- BDH soil pH indicator;
- BDH pH colour chart (kept in an envelope to prevent fading);
- Distilled water;
- Barium sulphate;
- Clean spatula or similar.

The soil to be tested should be a **representative sample** of the area being analysed? (see p379). Before the pH test is started, its texture should be identified (see p307).

- The soil, preferably dried and sieved, is now added to the test-tube; for 'loams' this should be to a depth of 2 cm (see below for heavier and lighter soils).
- Add to this the same depth of barium sulphate (2 cm).
- Shake together.
- Half fill the test-tube with distilled water (some specially designed tubes have a mark to fill to).
- Add soil pH indicator; for most soils only a few drops are required (more can be added later).
- Place bung in top of tube and shake.
- Leave to stand.
- Find the soil pH by matching the **hue** of the colour with one of the colours on the chart. This should be done by having the light coming over the shoulder whilst holding the tube against the white background alongside the colour square on the chart.
- If the colour is too concentrated to see the hue add a little distilled water; a few more drops of indicator if too weak.

This procedure is suitable for most soils, but it should be noted that the barium sulphate is added to help clear the water in order to see the colour clearly (by flocculating the clay so that it sinks more quickly to the bottom). If the soil texture indicates a very light

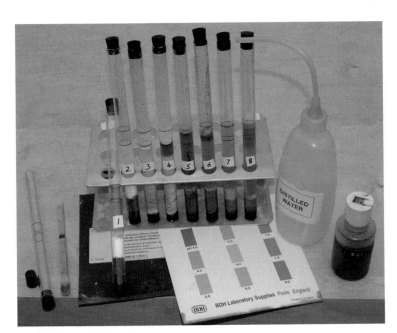

Figure 20.3 Soil pH testing. Equipment required includes test tubes, spatula, barium sulphate, pH indicator, colour chart. 1 Barium sulphate added to soil. 2 Distilled water added. 3 Shake. 4 Add indicator and leave to stand. 5,6,7,8 Check colour against the chart to find soil pH

soil (low clay) or a very heavy one (high clay) then the proportions should be altered as follows:

	Soil:Barium sulphate
'Loams'	2 cm:2 cm
Light soils	3 cm:1 cm
Heavy soils	1 cm:3 cm

Changes in soil pH

The balance of hydrogen ions and basic ions determines soil acidity. A clay particle with abundant hydrogen ions acts as a weak acid, whereas if fully charged with bases (such as calcium, Ca) it has a neutral or alkaline reaction. In practice, soil pH is usually regulated by the presence of calcium cations; **soils become more acid as calcium is leached from the soil** faster than it is replaced. This is the tendency in temperate areas where rainfall (carbonic acid see p297) exceeds evaporation over the year.

Soil pH falls and becomes more acid as 'lime' is lost from the soil.

Hydrogen ions take over the soil's cation exchange sites (see p305) and the pH falls. Soils with large reserves of calcium (containing pieces of chalk or limestone) do not become acid because they are kept base-saturated. In contrast, calcium ions are readily leached from free-draining sands in high rainfall areas and these soils tend to go acid rapidly (see podsols p302). In addition to the carbonic acid in rainfall, there are several other sources of acid that affect the soil:

- **Acid rain** (polluted rain and snow) is directly harmful to vegetation, but also contributes to the fall in soil pH.
- **Organic acids** derived from the microbial breakdown of organic matter, e.g. humic acids, also lead to an increase in soil acidity.
- **Fertilizers.** The bacterial nitrification of **ammonia** to nitrate yields acid hydrogen ions. Consequently fertilizers containing ammonium salts prevent calcium from attaching to soil colloids and cause calcium loss in the drainage water. Other fertilizers have much less effect.
- **Crop removal.** Calcium and magnesium are plant nutrients and the soil's lime reserves are therefore gradually reduced by the removal of plants.

In climates where the evaporation exceeds rainfall over the year, the dissolved salts are brought to the surface. As the water is lost from the soil by evaporation, the dissolved salts accumulate on the surface. These are usually basic (alkaline) in action so the soil pH rises. An extreme example of this is the salt (sodic) desert, e.g. Utah Salt Flats.

Buffering capacity

Buffering capacity is the ability of water to maintain a stable pH. Pure water has no buffering capacity; the addition of minute quantities of acid or alkali has an immediate effect on its pH. In the laboratory buffer tablets can be added to water to enable the solution to be maintained at a specified pH which would resist change despite the addition of some acid or alkali. This is useful for standardizing a pH meter, usually setting the instrument at precisely pH 7 and pH 4 for work on soil pH.

The buffering capacity of soil water reduces the effect of acidity coming from rainfall or from pollution, e.g. acid rain. Chalky or limestone soils, for instance, are very alkaline and can neutralize acids more effectively than acid peat soils. The cation exchange capacity of clays reduces the effect because the hydrogen ions exchange with calcium ions on the clay's colloid surface. Since the number of hydrogen ions being released or absorbed is small compared with the clay's reserve, the pH changes very little. High humus soils similarly have the advantage of a high buffering capacity. A related buffer effect is seen when acids, such as the carbonic acid of rain, are incorporated into soils with 'free' lime present; the acid dissolves some of the carbonate with no accompanying change in pH.

Effects of soil pH

Nutrient availability

The influence of soil pH on plant **nutrient availability** is demonstrated in Figure 20.4. It can be seen that for mineral soils the most favourable level for ensuring the availability of all nutrients is pH 6.5, while peat soils should be limed to pH 5.8 in order to maximize overall nutrient availability. In acid conditions some nutrients, such as manganese, and other soil minerals, such as aluminium, may become toxic.

Other effects on soil

Beneficial **soil organisms** (see p320) are affected by soil acidity and liming. A few soil-borne disease-causing organisms tend to occur more frequently on lime deficient soils (see clubroot), whereas others are more prevalent in well-limed soils. Calcium sometimes improves soil structure and soil stability. It is probable that this is mainly because it encourages **root activity** and creates conditions favourable for decomposition of organic matter, yielding **humus** (see p326). Free lime in clay soils sometimes, but not always, leads to better crumb formation on drying and shrinking.

Plant selection

Plant tolerance to soil pH and calcium levels varies considerably, but all plants are adversely affected when the soil becomes too acid. At very low pH some elements, such as aluminium, become soluble at levels that are toxic to plants. Table 20.1 shows the point below which the growth of common horticultural plants is significantly reduced. Although aluminium is not an essential nutrient for plants it is required to produce

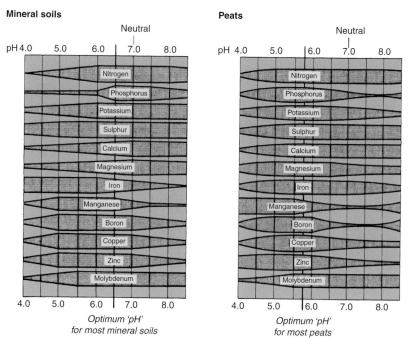

Mineral soils   **Peats**

Figure 20.4 **Effect of soil pH on nutrient availability**. The availability of a given amount of nutrient is indicated by the width of the band. The growing media should be kept at a pH at which all essential nutrients are available. For most plants the optimum pH is 6.5 in mineral soils and 5.8 in peats

Table 20.1 **Soil acidity and plant tolerance**

pH below which plant growth may be restricted on mineral soils:			
Celery	6.3	Rose	5.6
Daffodil	6.1	Raspberry	5.5
Bean	6.0	Cabbage	5.4
Lettuce	6.1	Strawberry	5.1
Carnation	6.0	Tomato	5.1
Chrysanthemum	5.7	Apple	5.0
Carrots	5.7	Potato	4.9
Hydrangea (pink)	5.9	Hydrangea (blue)	4.1

a blue flower rather than a pink one in hydrangeas (see Figure 20.1). Commercially, growers do not just use a compost with a low pH, but also add aluminium sulphate to get the blue colouring.

In the case of **calcifuges** the highest point before growth is affected by the presence of calcium should be noted, e.g. for *Rhododendrons* and some *Ericas* this is at pH 5.5 in mineral soils. Such plants are unable to metabolize many of the nutrients when there is more calcium present; typically they show signs of lime induced chlorosis (see p372).

In contrast, **calcicoles** are well adapted to utilizing soil nutrients in the presence of calcium, but are unable to survive in acid conditions where they shown signs of aluminium toxicity (dead tissues) and phosphate deficiency (stunting and blue or reddish stem and leaves).

Calcifuge or 'lime-hating' plants do not tolerate the level of calcium in soils normally found at pH 5.5. Consequently, they must be grown in more acid conditions.

Calcicoles, or 'lime-loving' plants, have evolved a different metabolism and are tolerant of high soil pH.

Changing soil pH

Raising soil pH

Soil pH can be raised by the addition of lime. Lime is most commonly applied as ground chalk, ground limestone or slaked lime. When lime is added to an acid soil it neutralizes the soluble acids and the calcium cations replace the exchangeable hydrogen on the soil colloid surface (see cation exchange). Eventually hydrogen ions are completely replaced by bases and **base saturation** is achieved, producing a soil of pH 7 or more. However, care should be taken not to overlime a soil because of its effect on the availability of plant nutrients.

Lime requirement is expressed as the amount of calcium carbonate in tonnes per hectare required to raise the pH of the top 150 mm of soil to the desired pH.

The **lime requirement** of soil can be estimated from knowledge of the required increase in pH and the soil texture (see buffering capacity p358). A pH of 6.5 is recommended for temperate plants on mineral soils; pH 5.8 on peats. The amount of a liming material needed to meet the lime requirement will depend on the neutralizing value of the lime chosen and its fineness.

Liming materials

Neutralizing Value (NV) is determined in the laboratory by comparing a materials ability to neutralize soil acidity with that of the standard, pure calcium oxide.

Liming materials can be compared by considering their ability to neutralize soil acidity, fineness, and cost to deliver and spread. The neutralizing value (NV) of a lime indicates its power to overcome acidity.

A neutralizing value of 50 signifies that 100 kg of that material has the same effect on soil acidity as 50 kg of calcium oxide. The **fineness** of the lime is important because it indicates the rate at which it affects the soil acidity (see surface area p304). It is expressed, where relevant, in terms of the percentage of the sample that will pass through a 100 mesh sieve. Liming materials commonly used in horticulture are listed below with some of their properties. The relationship between the different forms of calcium is shown in Figure 20.5.

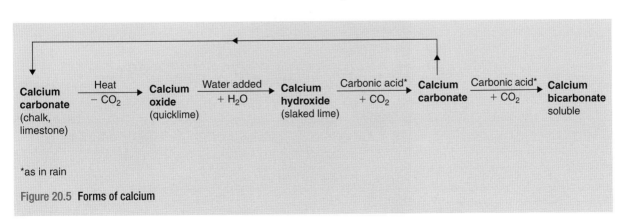

*as in rain

Figure 20.5 Forms of calcium

Calcium carbonate is the most common liming material. Natural soft **chalk** (or **limestone**) that is high in calcium carbonate is quarried and ground (NV = 48). It is a cheap liming material, easy to store and safe to handle. A sample in which 40 per cent will pass through a 100 mesh sieve can be used at the standard rate to meet the lime requirement. Coarser samples although cheaper to produce, easier to spread and longer lasting in the soil, require heavier dressings. **Shell sands**, mainly calcium carbonate, have neutralizing values from 25 to 45, i.e. whilst the purest samples can be used at nearly the same rate as chalk, up to twice as much of a poorer sample is required to have the same effect.

Calcium oxide (also known as quicklime, burnt lime, cob lime or caustic lime) is produced when chalk or limestone are very strongly heated in a lime kiln. Calcium oxide has a higher calcium content than calcium carbonate and, consequently, a higher neutralizing value. Pure calcium oxide is used as the standard to express neutralizing value (100) and the impure forms have lower values (usually 85–90). If used instead of ground limestone, only half the quantity needs to be applied.

In contact with moisture, lumps of calcium oxide slake, i.e. react spontaneously with water to produce a fine white powder, calcium hydroxide, with release of considerable heat. This was an effective way of obtaining a fine lime from the quarried material before there was heavy rolling machinery to grind the coarse lumps. The lime kilns that were used are still a common sight, especially in small ports round the coast (see Figure 20.6) Although rarely used now, calcium oxide has to be used with care because it is a fire risk, 'burns' flesh and scorches plant tissue.

Calcium hydroxide, hydrated or slaked lime, is derived from calcium oxide by the addition of water. The fine white powder formed is popular in horticulture. It has a higher neutralizing value than calcium carbonate and its fineness ensures a rapid effect on the growing medium. Once exposed to the atmosphere it reacts with carbon dioxide to form calcium carbonate.

It should be noted that all forms of processed lime quickly revert to calcium carbonate when added to the soil. Calcium carbonate, which is

Figure 20.6 Lime kilns. Used to 'burn' (heat) chalk or limestone (calcium carbonate) to produce quicklime (calcium oxide)

insoluble in pure water, gradually dissolves in the weak carbonic acid of the soil solution around the roots (see Figure 20.5).

Magnesian limestone, also known as Dolomitic limestone, is especially useful in the preparation of composts because it both neutralizes acidity and introduces magnesium as a nutrient. Magnesium limestone has a slightly higher neutralizing value (50–55) than calcium limestone, but tends to act more slowly.

Liming materials also provide the essential nutrients calcium and, when present, magnesium to the soil. Bicarbonate is formed from the carbonate in carbonic acid, e.g. rainwater or soil water, around respiring roots to provide a soluble form that can be taken up by plants (see Figure 20.5).

Lime application

Unless very coarse grades are used, lime raises the soil pH over a one- to two-year period, although the full effect may take as long as four years; thereafter pH falls again. Consequently lime application should be planned in the planting programme. It is normally worked into the top 15 cm of soil. If deeper incorporation is required, the quantity used should be increased proportionally. The lime should be evenly spread and regular moderate dressings are preferable to large infrequent applications. Very large applications needed in land restoration work should be divided for application over several years.

Care should be taken that the surface layers of the soil do not become too acid even when the lower topsoil has sufficient lime. Top layers are the first to become depleted with consequent effect on plant establishment. This tendency has to be carefully looked for in turf management as this can lead to the formation of 'thatch' (see Figure 18.6). Applications of organic manures or ammonium fertilizers should be delayed until lime has been incorporated. If mixed they react to release **ammonia** which can be wasteful and sometimes harmful.

Decreasing soil pH

Soil pH can be lowered by the addition of acids or sulphur to reduce the base saturation of the mineral soil. Some acid industrial by-products can be used, but the most usual method is to apply agricultural **sulphur**, which is converted to sulphuric acid by soil micro-organisms. The sulphur requirement depends on the pH change required and the soil's buffering capacity. The application of large quantities of **organic matter** gradually makes soils more acid. Acid fertilizers such as ammonium sulphate reduce soil pH over a period of years in outdoor soils and can be used in liquid feeding to offset the tendency of hard water to raise pH levels in composts. In some circumstances it has been appropriate to grow plants in a raised bed of **acid peat** or to work large quantities of peat into the topsoil; an approach that is not sympathetic to avoiding the destruction of peat wetlands.

Check your learning

1. State what is meant by pH.

2. State the range of pH that is suitable for most garden plants.

3. Describe how a crop of beans can be grown successfully on an allotment which has been tested and the soil found to have a pH of 5.5.

4. Describe the effect of 'overliming' an allotment soil.

5. Explain how plant selection is affected by the pH of the garden soil.

Further reading

Archer, J. (1988). *Crop Nutrition and Fertilizer Uses*. 2nd edn. Farming Press.

Brown, L.V. (2008). *Applied Principles of Horticultural Science*. 3rd edn. Butterworth-Heinemann.

Cresser, M.S. *et al.* (1993). *Soil Chemistry and Its Applications*. CUP.

Hay, R.K.M. (1981). *Chemistry for Agriculture and Ecology*. Blackwell Scientific Publications.

Ingram, D.S. *et al.* (eds). (2002). *Science and the Garden*. Blackwell Science Ltd.

Pratt, M. (2005). *Practical Science for Gardeners*. Timber Press.

Simpson, K. (1986). *Fertilizers and Manures*. Longman Handbooks in Agriculture.

Chapter 21   Plant nutrition

Summary

This chapter includes the following topics:

- Major nutrients
- The nitrogen cycle
- Minor nutrients
- Sources of nutrients in soil
- Organic and inorganic fertilizers
- Types of fertilizers
- Methods of applying fertilizers
- Green manures

with additional information on the following:

- Fertilizer programmes
- Soil conductivity

Figure 21.1 Range of fertilizers to supply plant nutrients

Major nutrients

Nitrogen

> **Nitrogen** is needed by plants to form **proteins** and is associated with **leafy growth**.

Nitrogen is taken up by plants as the nitrate and, to a lesser extent, the ammonium ion. Nitrates and ammonium ions are utilized in the plant to form protein. Plants use large quantities of nitrogen; it is associated with vegetative growth. Consequently large dressings of nitrogen are given to leafy crops, whereas fruit, flower or root crops require limited nitrogen balanced by other nutrients to prevent undesirable characteristics occurring.

The nitrogen cycle

Although plants live in an atmosphere largely made up of nitrogen they cannot utilize gaseous nitrogen. They are able to take up soluble nitrogen from the soil water as nitrates and ammonium ions. Both are derived from proteins by a chain of bacterial reactions as shown in Figure 21.2.

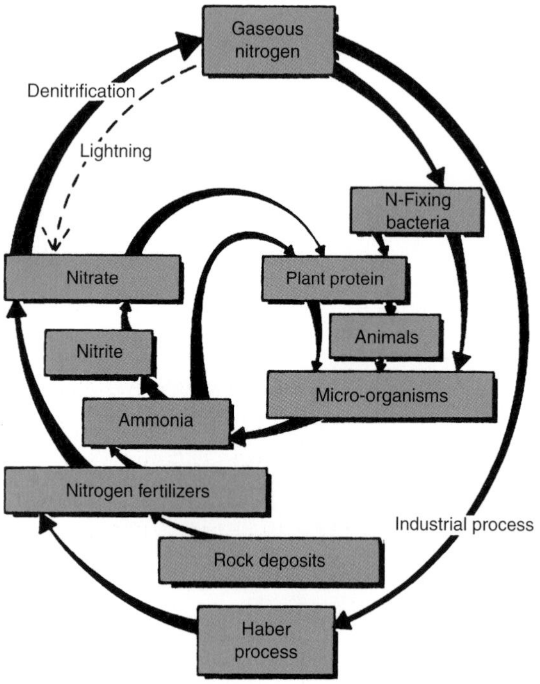

Ammonifying bacteria convert the proteins they attack to ammonia. **Ammonia** from the breakdown of protein in organic matter or from **inorganic nitrogen** fertilizers is converted to **nitrates** by **nitrifying bacteria**. This is accomplished in two stages. Ammonia is first converted to nitrites by *Nitrosomonas spp*. **Nitrites** are toxic to plants in small quantities, but they are normally converted to nitrates by *Nitrobacter spp*. before they reach harmful levels. Ammonifying and nitrifying bacteria thrive in aerobic conditions. Where there is no oxygen, anaerobic organisms dominate. Many anaerobic bacteria utilize nitrates and in doing so convert them to **gaseous nitrogen**. This **denitrification** represents an important loss of nitrate from the soil, which is at its most serious in well-fertilized, warm and waterlogged land.

Figure 21.2 **Nitrogen cycle**. The recycling of the element nitrogen by organisms is illustrated. Note the importance of nitrates that can be taken up and used by plants to manufacture protein. Micro-organisms also have this ability but animals require nitrogen supplies in protein form. Gaseous nitrogen only becomes available to organisms after being captured by nitrogen-fixing organisms or via nitrogen fertilizers manufactured by man. In aerobic soil conditions, bacteria convert ammonia to nitrates (nitrification) whereas in anaerobic conditions nitrates are reduced to nitrogen gases (denitrification)

Nitrogen fixation

Although plants cannot utilize gaseous nitrogen, it can be converted to plant nutrients by some micro-organisms. *Azotobacter* are free-living bacteria that obtain their nitrogen requirements from the air. As they die and decompose, the nitrogen trapped as protein is converted to ammonia and then to nitrates by other soil bacteria. *Rhizobia spp*. which live in root nodules on some **legumes** (see Figure 21.3) also trap nitrogen to the benefit of the host plant. Finally, nitrogen gas can be converted to ammonia industrially in the **Haber process**, which is the basis of the artificial nitrogen fertilizer industry.

Excess nitrogen produces soft, lush growth making the plant vulnerable to pest attack and more likely to be damaged by cold. Very large

Figure 21.3 Rhizobium nodules on legume

quantities of nitrogen are undesirable since they can harm the plant by producing high salt concentrations at the roots (see **conductivity**) and are lost by leaching. Large quantities are usually applied as a split dressing, e.g. some in base dressing and the rest in one or more top dressings.

Nitrates are mobile in the soil, which makes them vulnerable to leaching. In the British Isles it is assumed that all nitrates are removed by the winter rains so that virtually none are present until the soils warm up and nitrification begins or artificial nitrogen is applied (see nitrogen cycle). Nitrates leached through the root zone may find their way into the groundwater that is the basis of the water supply in some areas. Nitrification also leads to the **loss of bases**; for every 1 kg N in the ammonia form that is oxidized to nitrate and leached, up to 7 kg of calcium carbonate or its equivalent is lost. Nitrogen is also lost from the root zone by denitrification, especially in warm, waterlogged soil conditions. When in contact with calcareous material, ammonium fertilizers are readily converted to ammonia gas which is lost to the soil unless it dissolves in surrounding water. For this reason urea or ammonia-based fertilizers should not be applied to such soils as a top dressing or used in contact with lime. **Nitrogen fertilizers** used in horticulture and their nutrient content are given in Table 21.2.

Ammonium nitrate is now commonly used in horticulture. In pure form it rapidly absorbs moisture to become wet; on drying it 'cakes' and can be a fire risk. Pure ammonium nitrate can be safely handled in polythene sacks and as prills. **Ammonium sulphate** has a highly acid reaction in the growing medium. **Urea** has a very high nitrogen content and in contact with water it quickly releases ammonia. Its use as a solid fertilizer is limited, but it is utilized in liquid fertilizer or foliar sprays. The addition of a sulphur coating to urea not only creates a controlled release action, but also a fertilizer with an acid reaction. Other manufactured organic fertilizers include urea formaldehydes (nitroform, ureaform, etc.) which release nitrogen as they are decomposed by micro-organisms, isobutylidene urea (IBDU) which is slightly soluble in water and releases urea and crotonylidene (CDU, e.g. Crotodur). The latter breaks down very slowly and evenly, which makes it ideal for applying to turf.

Natural organic sources of nitrogen, including dried blood, hoof and horn and shoddy, amongst others, are generally considered to provide slow release nitrogen, but in warm greenhouse conditions decomposition is quite rapid.

Phosphorus

Phosphorus is taken up by plants in the form of the phosphate anion $H_2PO_4^{3-}$. Phosphorus is mobile in the plant and is constantly being recycled from the older parts to the newer growing areas. In practice this means that, although seeds have rich stores of phosphorus, phosphate is needed in the seedbed to help establishment. Older plants have a very low phosphate requirement compared with quick growing plants harvested young. **Most soils contain very large quantities of phosphorus, but only a small proportion is available to plants.** The concentration of available phosphate ions in the soil water and on soil colloids is at its highest between pH 6 and 7. Phosphorus is released from soil organic matter by micro-organisms (see mineralization), but most of it and any other soluble phosphorus, including that from fertilizers, is quickly converted to insoluble forms by a process known as **phosphate fixation**. Insoluble aluminium, iron and manganese phosphates are formed at low pH and insoluble calcium phosphate at high pH. The carbonic acid in the vicinity of respiring roots and organisms in the rhizosphere, such as mycorrhizae (see p322), facilitate phosphorus uptake. The low solubility of phosphorus in the soil makes it virtually immobile, with the result that roots have to explore for it. Soils should be cultivated to allow roots to explore effectively; compacted or waterlogged areas deny plants phosphorus supplies. Phosphate added to the soil should be placed near developing roots (see band placement) in order to reduce phosphorus fixation and ensure that it is quickly found. If applied to the surface, phosphate fertilizers should be cultivated into the root zone.

Unlike soils, most artificial growing media have no reserves of phosphorus and when added in soluble form it remains mobile and subject to leaching. Incorporating phosphorus in liquid feeds in hard water is complicated by the precipitation of insoluble calcium phosphates that lead to blocked nozzles. Slow release phosphates are often selected in these situations to reduce losses and to eliminate the need for phosphorus in the liquid feeds.

Phosphorus nutrition used to be based on organic sources such as bones, but now phosphate fertilizers are mainly derived from rock phosphate ore (see Table 21.2). **Slow-acting forms**, such as rock phosphate, bone meal and basic slag, can be analysed in terms of their 'citric soluble' phosphate content, this being a good guide to their usefulness in the first season. Such materials should be finely ground to enhance their effectiveness. These phosphates are applied mainly to grassland, tree plantings and in the preparation of herbaceous borders, to act as long-term reserves, particularly on phosphate deficient soils. Magnesium ammonium phosphate (MagAmp, Enmag), calcium metaphosphate and potassium metaphosphate contain other nutrients, but are slow release phosphates for use in soilless growing media. Treating rock phosphates with acids produces **water-soluble phosphates**. Superphosphate, derived from rock phosphate by treating with sulphuric acid, is composed of a water-soluble phosphate and calcium sulphate (gypsum), whereas triple superphosphate, derived from a phosphoric acid treatment, is a more concentrated source of phosphorus with fewer impurities. Both superphosphate and triple superphosphate are widely used in horticulture and are available in granular or powder form. Whilst they have a neutral effect on soil pH they tend to reduce the pH of composts. High-grade monoammonium phosphate is used as a phosphorus source in liquid feeds because it is low in iron and aluminium impurities that lead to blockage in pipes and nozzles.

Figure 21.4 Potassium deficiency

Potassium

Potassium is taken up by the roots as the potassium cation and is distributed throughout the plant in inorganic form where it plays an important role in plant metabolism. For balanced growth the nitrogen to potassium ratio should be 1:1 for most crops, but 2:3 for roots and legumes. Leafy crops take up large amounts of potassium, especially when given large amounts of nitrogen. Where potassium supplies are abundant some plants, especially grasses, take up 'luxury' levels, i.e. more than needed for their growth requirements. Consequently, if large proportions of the plant are taken off the land, e.g. as grass clippings, there is a rapid depletion of potassium reserves. Potassium forms part of clay minerals and is released by chemical weathering. The potassium in soil organic matter is very rapidly recycled and exchangeable potassium **cations** (see p371), held on the soil colloids and in the soil solutions, are readily available to plant roots. Potassium is easily leached from sands low in organic matter and from most soilless growing media (see Figure 21.4).

Potassium and magnesium ions mutually interfere with uptake of each other. This **ion antagonism** is avoided when the correct ratio between 3:1 and 4:1 available potassium to magnesium is present in the growing medium. Availability of potassium is also reduced by the presence of calcium (see induced deficiency).

The main potassium fertilizers used in horticulture are detailed in Table 21.2. Although cheaper and widely used in agriculture, potassium chloride causes scorch in trees and can lead to salt concentration problems because the chloride ion accumulates as the potassium is taken up. Commercial potassium sulphate can be used in base dressings for composts, but only the more expensive refined grades should be used in liquid feeding. More usually potassium nitrate is used to add both potassium and nitrate to liquid feeds, but it is hygroscopic. Most potassium compounds are very soluble so that the range of slow release formulations is limited to resin-coated compounds.

Magnesium

Magnesium is an essential plant nutrient in leaves and roots and is taken up as a cation. There are large reserves in most soils, especially clays, and those soils receiving large dressings of farmyard manure. Deficiencies (see Figure 21.5) are only likely on intensively cropped sandy soils if little organic manure is used. Magnesium ion uptake is also interfered with by large quantities of potassium or calcium ions because of **ion antagonism**. Chalky and over-limed soils are less likely to yield adequate magnesium for plants. Magnesium fertilizers include magnesian limestones containing a mixture of magnesium and calcium carbonate that raise soil pH (see liming). Magnesium sulphate, as kieserite, provides magnesium ions without affecting pH levels and in a purer form, Epsom salts, is used for liquid feeding and foliar sprays (see Table 21.2).

Figure 21.5 Magnesium deficiency

Calcium

Calcium is an essential plant nutrient taken up by the plant as calcium cations. Generally a satisfactory pH level of a growing medium indicates suitable calcium levels (see liming). Gypsum (calcium sulphate) can be used where it is desirable to increase calcium levels in the soil without affecting soil pH. Deficiencies are infrequent and usually caused by lime being omitted from composts. Inadequate calcium in fruits is a more complex problem involving the distribution of calcium within the plant (see page 258). Calcium nitrate or chloride solutions can be applied to apples to ensure adequate levels for safe storage (see plant tissue analysis).

Sulphur

Sulphur taken up as sulphate ions is a nutrient required in large quantities for satisfactory plant growth. It is not normally added specifically as a fertilizer because the soil reserves are replenished by re-circulated organic matter and a steady supply from winds off the sea in the form of dimethyl sulphide (DMS) that gives the distinctive smell of the seaside. Air pollution has added considerably to the supply reaching the land. Several fertilizers used to add other nutrients are in sulphate form, e.g. ammonium sulphate, superphosphate and potassium sulphate, and as such supply sulphur as well (see Table 21.2). However, as air pollution is reduced and fewer sulphate fertilizers are used, it is becoming necessary for growers in some parts of the world to take positive steps to include sulphur in their fertilizer programme.

Minor nutrients

Minor nutrients, also known as trace elements or micro-elements, are present in plants in very small quantities, but are just as essential for healthy growth as major elements. However, they can be toxic to plants if too abundant. This means that rectifying deficiencies with soluble salts has to be undertaken carefully.

Deficiencies

Simple deficiencies are those in which too little of the nutrient is present in the growing medium. Most soils have adequate reserves of trace elements, so simple deficiencies in them are uncommon, especially if replenished with bulky organic matter. Sandy soils tend to have low reserves and so too have several organic soils from which trace elements have been leached. In horticulture simple deficiencies of trace elements are mainly associated with growing in soil-less composts which require careful supplementation.

Induced deficiencies are those in which sufficient nutrients are present, but other factors, such as soil pH or **ion antagonism**, interfere with plant

nutrient availability. On mineral soils boron, copper, zinc, iron and manganese become less available in alkaline soils, whereas molybdenum availability is reduced severely in soils with pH levels below 5.5, as shown in Figure 20.4. Trace element problems are aggravated in dry soils or where waterlogging, root pathogens or poor soil structure reduces root activity.

Iron deficiency is induced by the presence of large quantities of calcium and this **'lime induced' chlorosis** (yellowing) occurs on over-limed soils and calcareous soils. The natural flora of chalk and limestone areas is calcicoles. Other plants grown in such conditions usually have a typically yellow appearance. Deficiencies can also be induced by high levels of copper, manganese, zinc and phosphorus. Top fruit and soft fruit are particularly susceptible, as well as crops grown in complete nutrient solutions. The problem is overcome by using iron chelates.

Boron deficiency tends to occur when pH is above 6.8. It is readily leached from peat. Crops grown in peat are particularly susceptible when pH levels rise (Figure 20.4). Boron can be applied to soils before seed sowing in the form of borax or 'Solubor'.

Manganese deficiency is more frequent on organic and sandy soils of high pH. Plant uptake can be reduced by high potassium, iron, copper and zinc levels. Manganese availability is greatly increased at low pH and can reach toxic levels which most commonly occur after steam sterilization of acid, manganese-rich soils. High phosphorus levels can be used to reduce the uptake of manganese in these circumstances.

Copper deficiency usually occurs on peat and sands, notably reclaimed heath-land, and in thin organic soils over chalk. High rates of nitrogen can accentuate the problem. Soils can be treated with copper sulphate or plants can be sprayed with copper oxychloride.

Zinc deficiencies are not common and are usually associated with high pH.

Molybdenum deficiency occurs in most soil types at a low pH (see Figure 20.4). Availability becomes much reduced below pH 5.5, especially in the presence of high manganese levels. Cauliflowers are particularly susceptible and soils are limed to solve the problem. Sodium or ammonium molybdate can be added to growing media or liquid feeds where molybdenum supplies are inadequate.

Basic chemistry

Some knowledge of chemistry can give insights into biology and thus horticulture. Chemistry deals with the reactions that occur when different chemical substances are brought into contact with each other.

Elements. All substances are made up of elements. There are 92 stable elements, such as oxygen and hydrogen. Each element occurs in nature as atoms.

Atoms. The first step in chemical understanding involves the atom and its structure. All substances are made of atoms. Atoms are about one ten millionth of a millimetre in size. Each one may be considered to have a central nucleus (containing protons and neutrons),

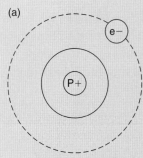

Hydrogen atom is reactive

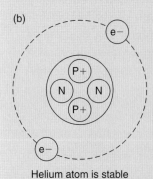

Helium atom is stable

P = proton
N = neutron
e = electron

Figure 21.6 Atoms.
(a) hydrogen (b) helium

and one or more outer shells (or orbitals) of circulating electrons (see Figure 21.6). Each proton and each neutron contributes a unit of 1 to the atomic weight of the element. In the carbon atom, there are 6 protons and 6 neutrons giving carbon an atomic weight of 12. The proton has a positive charge whilst the neutron has no charge. The circulating electrons can be considered to have no measurable weight, but have a negative charge equal but opposite to the proton. In each element, the number of electrons equals the number of protons. In the carbon atom, there are 6 protons and 6 electrons.

At its simplest, the chemistry of an element may be considered in terms of the group of electrons' 'quest for stability'. The first (inner) orbital is described as stable when it has either zero or two electrons. Outer orbitals are considered stable when they have zero or eight electrons. Four examples are used to illustrate this principle.

Hydrogen is the simplest of the elements. It has one proton and one electron in its one (inner) orbital. Its atomic weight is 1. For stability, it can pair its electron with another hydrogen atom's electron. In this way the two atoms have a stable orbital of two electrons. **Helium** is an inert (non-reacting) gas. It has two protons and two electrons and therefore does not react with other substances (see Figure 21.4).

Ionic compounds. Ionic compounds, such as salts, are made up of charged atoms (ions). In the case of common salt (sodium chloride), the **sodium** and **chlorine** atoms react together (see Figure 21.7). The metallic sodium atom gives away its negative electron, thus becoming a positively charged ion called a **cation**. At the same time, the chlorine atom receives the negative electron from the sodium, thus becoming a negatively charged ion called an **anion**. Ionic substances, such as sodium chloride and many fertilizers, allow an electric current to pass through them when they are dissolved in water. Horticulturists are able to assess the strength of dissolved fertilizers in soils and composts by measuring this current (conductivity see p380).

Compound ions. Some anions occur in a compound form. Carbonate (CO_3^{2-}), Nitrate (NO_3^-), Nitrite (NO_2^-), phosphate (PO_4^{3-}) and sulphate (SO_4^{2-}) are some examples. One cation, ammonium (NH_4^+) is commonly found in fertilizers. In contrast to ionic compounds, the element carbon **shares** electrons with the element it reacts with to produce molecules, many of which are very large (see **carbon chemistry** p111).

Oxygen is a gas like hydrogen but differs from it in several ways. It is much heavier, having 8 protons and 8 neutrons (and 8 electrons) and an atomic weight of 16. Its inner electron orbital is stable, with 2 electrons, leaving six electrons in the second orbital. Oxygen therefore needs to receive two electrons to become stable. It can be seen that oxygen's combining power is twice that of hydrogen. Oxygen is thus said to have a **valency** (combining power) of two while hydrogen's valency is one. When oxygen and hydrogen react together, it becomes clear from the description above that the substance produced by the reaction is H_2O, or **water**, since the two hydrogens are needed to fill oxygen's orbital.

Horticultural plants require 15 elements for their growth. Table 21.1 lists the atomic weights and valency for each of the elements.

Molecular weight. Examples of simple molecules are given in the 'common compounds' section of Table 21.1. The molecular weight of a substance can be calculated by adding the individual atomic weights of the elements within it. For example, the molecular weight of ammonium nitrate (NH_4NO_3) is $14 + 1 + 1 + 1 + 1 + 14 + 16 + 16 + 16 = 80$. Note that there are two nitrogen atoms in the compound, contributing 28 parts out of the total of 80 i.e. 35 per cent. Fertilizers of this type are not quite as pure and typically contain 33 per cent nitrogen (see p325).

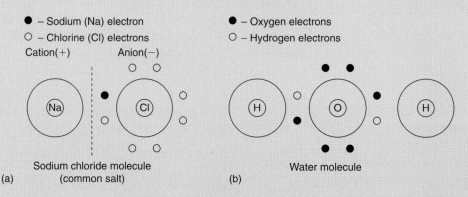

● – Sodium (Na) electron
○ – Chlorine (Cl) electrons
Cation(+)   Anion(−)

● – Oxygen electrons
○ – Hydrogen electrons

Sodium chloride molecule
(common salt)
(a)

Water molecule
(b)

Figure 21.7 (a) Ionic bonding; sodium chloride (common salt). Sodium loses an electron and becomes a sodium cation; chlorine gains an electron and becomes an anion; (b) covalent bonding, e.g. water; one oxygen and two hydrogen atoms share electrons in their outer orbits (shells) to gain stability.

Table 21.1 Chemical information on horticulturally useful elements

Element	Symbol	Atomic weight	Valency	Common compound
Hydrogen	H	1	1	Water (H_2O)
Boron	B	11	3	Borax
Carbon	C	12	4	Carbon dioxide (CO_2)
Nitrogen	N	14	2,4	Ammonium nitrate
Oxygen	O	16	2	Oxides
Magnesium	Mg	24	2	Kieserite
Phosphorus	P	31	3	Superphosphate
Sulphur	S	32	2	Ammonium sulphate
Potassium	K	40	1	Potassium nitrate
Calcium	Ca	41	2	Lime
Manganese	Mn	55	2	Manganese sulphate
Iron	Fe	56	2,3	Ferrous sulphate
Copper	Cu	64	1,2	Copper sulphate
Zinc	Zn	65	2	Zinc sulphate
Molybdenum	Mo	96	2–5	Ammonium molybdate

Sources of nutrients

Nutrients are supplied naturally from the decomposition of organic matter and are released as clay weathers in the soil. In horticulture, additional supplies are made by the use of organic and inorganic fertilizers, bulky organic matter and through green manuring.

Fertilizers

Fertilizers are concentrated sources of plant nutrients that are added to growing media.

Straight fertilizers are those that supply only one of the major nutrients: nitrogen, phosphorus, potassium or magnesium (see Table 21.2). The amount of nutrient in the fertilizer is expressed as a percentage. Nitrogen fertilizers are described in terms of percentage of the element nitrogen in the fertilizer, i.e. per cent N. Phosphate fertilizers have been described in terms of the equivalent amount of phosphoric oxide, i.e. per cent P_2O_5, or increasingly as percentage phosphorus, per cent P. Likewise potash fertilizers, i.e. per cent K_2O, or percentage potassium, per cent K. Magnesium fertilizers are described in terms of per cent of Mg. The percentage figures clearly show the quantities of nutrient in each 100 kg of fertilizer.

Compound fertilizers are those that supply two or more of the nutrients nitrogen, phosphorus and potassium. The nutrient content expressed as for straight fertilizers is, by convention, written on the bag in the order nitrogen, phosphorus and potassium. For example, 20–10–10 denotes 20 per cent N, 10 per cent P_2O_5 and 10 per cent K_2O.

Table 21.2 Nutrient analysis of fertilizers

	N %	P$_2$O$_5$ (P) %	K$_2$O (K) %	Mg %	Ca %	S %	Na %
Ammonium nitrate	33–35						
Ammonium sulphate	20–21					24	
Bone meal	3	20 (9)					
Calcium nitrate	15.5				20		
Calcium sulphate					23	18	
Chilean potassium nitrate	15		10 (8)				20
Dried blood	12–14						
Hoof and horn	12–14						
Kieserite				15		21	
Meat and bone meal	5–10	18 (8)					
Monoamm phosphate	12	37 (15)					
Phosphoric acid		54 (24)					
Potassium chloride			59 (49)				
Potassium nitrate	14		46 (38)				
Potassium sulphate			50 (42)			17	
Shoddy (wool waste)	2–15						
Superphosphate		18–20 (8–9)			20	12–14	
Triple superphosphate		47 (20)			14		
Urea	46						

Fertilizer regulations require that further details of trace elements, pesticide content and phosphorus solubility should appear where applicable on the invoice. Fertilizers and manures are available in many different forms. Generally the term **organic** implies that the fertilizer is derived from living organisms, whereas **inorganic** fertilizers are those derived from non-living material. However, in the context of organic growing it is necessary to look at specific requirements of the regulations (Table 21.3).

Application methods

Fertilizers are applied in several different ways.

- **Base dressings** are those that are incorporated in the growing medium. Combine drilling with seeds and fertilizer running into the same drill can achieve this. In horticulture, however, **band placement** of fertilizers is far more common, involving equipment that drills the seeds in rows and places a band of fertilizer in parallel a few centimetres below and to one side. The risk of retarded germination

Table 21.3 Sources of nutrients for use in organic growing

Farmyard and poultry manure	Rock potash
Slurry or urine	Sulphate of potash*
Composts from spent mushroom and vermiculture substrates	Limestone
Composts from organic household refuse	Chalk
Composts from plant residues	Magnesium rock
Processed animal products from slaughterhouses and fish industries	Calcareous magnesium rock
Organic by-products of foodstuffs and textile industries	Epsom salt (magnesium sulphate)
Seaweeds and seaweed products	Gypsum (Calcium sulphate)
Sawdust, bark and wood waste	Trace elements (boron, copper, iron, manganese, molybdenum, zinc)*
Wood ash	Sulphur*
Natural phosphate rock	Stone meal
Calcinated aluminium phosphate rock	Clay (bentonite, perlite)
Basic slag	

*Need recognized by control body.

or scorch of young plants due to high soluble fertilizers placed near seeds is thus avoided (see salt concentration). There is much less risk if fertilizer is **surface broadcast**, i.e. scattered on prepared soil surface, or broadcast on the surface to be cultivated-in during the final stages of seedbed preparation.

- **Top dressings** are fertilizers added to the soil surface but not incorporated. Such fertilizers must be soluble and not fixed by soil because the nutrient is carried to the roots by soil water. Nitrogen is the material most frequently applied by this method mainly because the large applications to crops require a base dressing and one or more top dressings to minimize the risk of scorch and loss by leaching.
- **Liquid feeding** is the application of fertilizer diluted in water to the root zone; fertigation if incorporated in irrigation system.
- **Foliar feeding** is the application of a liquid fertilizer in suitably diluted form to be taken up through leaves. This technique is usually restricted to the application of trace elements.

Formulations

Quick-acting fertilizers contain nutrients in a form which plant roots can take up, and dissolve as soon as they come in contact with water, e.g. ammonium nitrate and potassium chloride. Many are obtainable as powders or crystals. These are difficult to spread or place evenly, but several are formulated this way to help in the preparation of liquid feeds. For this purpose they must be readily soluble and free of impurities that might lead to blockages in the feed lines. Some of the less soluble fertilizers, as well as lime, are spread in a finely divided form for maximum effect on soil. One of the major problems with many of the fertilizers is their **hygroscopic** nature, i.e. they pick up water from the atmosphere and create storage and distribution problems. Powdered forms, in particular, go sticky as they take up water then 'cake' (form hard lumps) as they dry. Fertilizers formulated as **granules** are more satisfactory for accurate placement or broadcasting. They flow better, can be metered, and are thrown more accurately. **Prilled** fertilizers represent an improvement on granules because of their uniform spherical shape.

Slow release fertilizers are those in which a large proportion of the nutrient is released slowly. Several of these fertilizers, such as rock phosphate, are insoluble or only slightly soluble and the nutrients are released only after many months, even years. Micro-organisms break down organic products and the rate at which nutrients become available depends on their activity (see bacteria). Some slow release artificial fertilizers, such as those based on urea formaldehyde, dissolve slowly in the soil solution whilst others are formulated in such a way that the soluble fertilizer they contain diffuses slowly through a resin coat, e.g. Osmocote, or sulphur coating. Some of these slow-release fertilizers have been formulated in such a way as to release nutrients at a rate that matches a plant's uptake and as such are sometimes referred to as **controlled-release** fertilizer. **Frits**, made from fine glass powders containing nutrient elements, are used either to release soluble materials slowly or to overcome the trace element problem caused by the narrow limits between deficiency and toxicity (see trace elements). Frits ease the difficulties experienced in mixing tiny amounts evenly through large volumes of compost. **Ion-exchange resins** release their nutrients by exchange with cations in the surrounding water. These resins help to overcome the problems of high salt concentration and leaching of nutrients from growing media based on inert materials (see aggregate culture).

Some plant nutrients are formed as **chelates** to maintain availability in extreme conditions where the mineral salt is 'locked up'. There are many different chelating or sequestrating agents selected, for each element to be protected and for each unavailability problem. Iron is chelated with EDDHA to form the product Chel 138 or Sequestrine 138 which releases the element in all soils including those with a high pH (see iron deficiency). EDTA, effective where there are high levels of copper, zinc and manganese, is used to chelate iron to be applied in foliar sprays.

Bulky organic matter

Compost, straw, farmyard manure, bark and peat are important in horticulture as a means of maintaining organic matter and humus levels (see p329–30). However, they tend to be low in nutrients and some, such as straw and bark, lead to locking up of nutrients (see C:N ratio p325). They can be evaluated on the basis of their effect on the physical properties of soil and their small, nutrient content.

Green manures

Green manuring is the practice of growing plants primarily to develop and maintain soil structure and fertility (see p330). The plants used are typically agricultural crops that cover the ground quickly and yield a large amount of leaf to incorporate. The seeds are normally broadcast sown in the autumn when there are no other overwintering plants. The plants are then dug or ploughed in when the land is needed again.

Fertilizer programmes

The fertilizer applications required to produce the desired plant growth vary according to the type of plant, climate, season of growth, other sources of nutrients applied and the nutrient status of the soil. General advice is available in many publications including those of the national advisory services and horticultural industries. Examples are given in Table 21.4.

Growing medium analysis

The nutrient status of growing media varies greatly between the different materials and within the same materials as time passes. The nutrient levels change because they are being lost by plant uptake, leaching and fixation and gained by the weathering of clay, mineralization of organic matter, and the addition of lime and fertilizers.

There are many visual symptoms which indicate a deficiency of one or more essential nutrients (see minerals), but unfortunately by the time they appear the plant has probably already suffered a check in growth or change in the desired type of growth. The concentration of minerals in the plant and the nature of growth are linked so that **plant tissue analysis**, usually on selected leaves, can provide useful information, particularly in the diagnosis of some nutrient deficiencies; e.g. it is used to identify the calcium levels in apples in order to check their storage qualities. However, nutrient supply is usually assessed by analysis of the growing medium. There is general agreement about the methodology for **analyzing soils**. However, there needs to be some care where analysis of other growing media is undertaken, because there are considerable differences between the methods particularly with regard to dilution of the nutrient extractant.

A representative sample of the growing medium is taken and its nutrient status determined. This involves extracting the **available nutrients** and measuring the quantities present. The **pH level** is determined and, where appropriate, the **lime requirement**. The **conductivity** of growing media from protected culture is also measured. The **nitrogen status** of soils is usually determined from previous cropping because, outdoors, nitrates are washed out over winter and their release from organic matter reserves is very variable. In protected planting, nitrate and ammonia levels are usually determined. Results are often given in the form of an index number in order to simplify their presentation and interpretation. The ADAS soil analysis index is based on a ten point scale from 0 (indicating levels which correspond

Table 21.4 Examples of fertilizer requirements

Carrots

N, P or K index		0	1	2	3	4	over 4
		(kg/ha)					
Carrots, early bunching	N	60	25	Nil	–	–	–
	P_2O_5	400	300	250	150	125	Nil
	K_2O	200	125	100	Nil	Nil	Nil
Carrots, (maincrop)	fen peats N	Nil	Nil	Nil	–	–	–
	other soils N	60	25	Nil	–	–	–
	all soils P_2O_5	300	250	200	125	60	Nil
	K_2O	200	125	100	Nil	Nil	Nil

Carrots on sandy soils respond to salt: 150 kg/ha Na (400 kg/ha salt) should be applied and potash reduced by 60 kg/ha K_2O. Salt must be worked deeply into the soil before drilling or be ploughed in.

Dessert apples: mature trees

	Summer rainfall	Cultivated or overall herbicide	Grass/herbicide strip	Grass
		(kg/ha per year)		
Nitrogen (N)	more than 350 mm	30	40	90
	less than 350 mm	40	60	120

P, K or Mg index	0	1	2	3	over 3
		(kg/ha per year)			
P_2O_5	80	40	20	20	Nil
K_2O	220	150	80	Nil	–
Mg	60	40	30	Nil	–

For the first 3 years, fertilizer is not required by *young trees* grown in herbicide strips, provided that deficiencies of phosphate, potash and magnesium are corrected before planting by thorough incorporation of fertilizer.

Lettuce: base dressings for border soils under glass

Nitrate P, K or Mg index	Ammonium nitrate	Triple superphosphate	Sulphate of potash	Kieserite
		g/m^2		
0	30	150	160	110
1	15	140	110	80
2	Nil	130	50	30
3	Nil	110	Nil	Nil
4	Nil	80	Nil	Nil
5	Nil	45	Nil	Nil
Over 5	Nil	Nil	Nil	Nil

Increase the nitrogen application by 50 per cent for summer grown crops. Lettuce is sensitive to low soil pH and the optimum pH is the range 6.5–7.0. It is also sensitive to salinity, and growth may be retarded on mineral soils when the soil conductivity index is greater than 2.

to probable failure of plants if nutrient is not supplied) to 9 (indicating excessively high levels of nutrient present); most outdoor soils normally give levels of 1 to 3.

Fertilizer recommendations

The results of the **growing medium analysis** are interpreted with the appropriate **nutrient requirement tables** to determine the actual amount of fertilizer to apply. These tables usually have growing medium nutrient status indices to aid interpretation and results are normally given in kg of nutrient per hectare or grams of nutrient per square metre (Table 21.4). In some cases the amount of named fertilizer required is stated; if another fertilizer is to be used to supply the nutrient the quantity needed must be calculated using the nutrient content figures (Table 21.2). It is important that throughout the fertilizer planning process the same units are used, i.e. per cent P_2O_5 *or* P per cent; K_2O *or* per cent K. Conversion figures are:

$$\%P_2O_5 = \%P \quad \times 2.29$$
$$\%P = \%P_2O_5 \times 0.44$$
$$\%K_2O = \%K \quad \times 1.20$$
$$\%K = \%K_2O \times 0.83$$

Sampling

Normally only a small proportion of the whole growing medium is submitted for analysis and therefore it must be a **representative sample** of the whole. This is not easy because of the variability of growing media, particularly soils. It is recommended that each sample submitted for testing should be taken from an area no greater than 4 hectares (see Figure 21.8). The material sampled must itself be uniform and so only areas with the same characteristics and past history should be put in the same sample. Irrespective of the area involved, from small plot to 4 hectare field, at least 25 sub-samples should be taken by walking a zigzag path, avoiding the atypical areas such as headlands, wet spots, old paths, hedge lines, old manure heaps, etc. The same amount of soil should be taken from each layer to a depth of 150 mm. This is most easily achieved with a soil auger or tubular corer.

Peat bags should be sampled with a cheese-type corer by taking a core at an angle through the planting hole on the opposite side of the plant to the drip nozzle from each of 30 bags chosen from an area up to a maximum of 0.5 hectares. Discard the top 20 mm of each core and if necessary take more than 30 cores to make up a one litre sample for analysis. Samples should be submitted to analytical laboratories in clean containers capable of completely retaining the contents. They

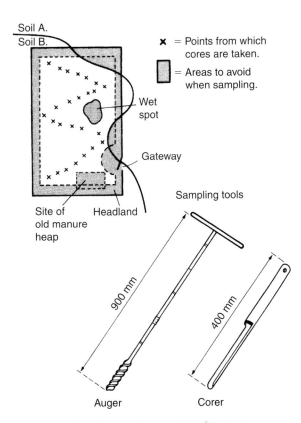

Figure 21.8 Sampling growing media. Suitable tools to remove small quantities of growing media are corers or augers which have the advantage of removing equal quantities from the top and bottom of the sampled zone. The material to be sampled must be clearly identified, then 25 cores should be removed in a zigzag that avoids anything abnormal

should be accompanied by name and address of supplier, the date of sampling and any useful background information. All samples must be clearly identified. Further details of sampling methods in greenhouses or orchards, bags, pots, straw bales, water, etc., are obtainable from the advisory services used. Remember, the result of the analysis can be no better than the extent to which the sample is representative of the whole.

Soil conductivity

The **soil solution** is normally a weaker solution than the plant cell contents. In these circumstances plants readily take up water through their roots by osmosis. As more salt, such as soluble fertilizer, is added to the soil solution, **salt concentrations** are increased and less water, on balance, is taken up by roots. When salt concentrations are balanced as much water passes out of the roots as into them. When salt concentrations are greater in the soil the roots are plasmolyzed. The root hairs, then the roots, are 'scorched', i.e. irreversibly damaged, and the plant dries up.

Symptoms of high salt concentration above ground are related to the water stress created. Plants wilt more often and go brown at the leaf margin. Prolonged exposure to these conditions produces hard, brittle plants, often with a blue tinge. Eventually severe cases become desiccated.

Salt concentration levels are measured indirectly using the fact that the solution becomes a better conductor of electricity as salt concentration is increased. The conductivity of soil solution is measured with a **conductivity meter** (see ionic compounds, p373).

Salt concentration problems are most common where fertilizer salts accumulate, as in climates with no rainfall period to leach the soil and in protected culture. Periods when rainfall exceeds evaporation, as in the British Isles during winter, ensure that salts are washed out of the ground. Any plant can be damaged by applications of excess fertilizer. Some plants, such as tomatoes and celery, are more tolerant than others, but seedlings are very sensitive. Young roots can be scorched by the close proximity of fertilizer granules in the seedbed (see band placement).

In **protected culture** large quantities of fertilizer are used and residues can accumulate, particularly if application is not well adjusted to plant use. Sensitive plants, such as lettuce, are particularly at risk when following heavily-fed, more tolerant plants, such as tomatoes or celery. Salt concentration levels should be carefully monitored and feeding adjusted accordingly, applying water alone if necessary. Soils can be **flooded** with water between plantings to leach excess salts. Large quantities of water are needed, but should be applied so that the soil surface is not damaged. Every effort should be made to minimize the effect on the environment and quantities of water needed to flush out the excess salts by reducing the nutrient levels as the crop comes to an end.

Check your learning

1. Describe the main effect that nitrogen has on plant growth.

2. Describe how gaseous nitrogen may become used by plants.

3. Explain why phosphates are needed in seedling compost.

4. Describe the symptoms of potash deficiency in plants.

5. Describe the advantages of using organic fertilizers.

6. Distinguish between base and top dressings.

7. Explain what is meant by controlled release fertilizers and give examples.

8. Explain why green manures are used.

Further reading

ADAS/ARC. Robinson, J.B.D. (ed.) (1982). *The Diagnosis of Mineral Disorders in Plants*. Vol. 1. Introduction. HMSO.

ADAS/ARC. Robinson, J.B.D. (ed.) (1987). *The Diagnosis of Mineral Disorders in Plants*. Vol. 2 Vegetable crops. HMSO.

ADAS/ARC. Winsor, G. and Adams, P. (eds) (1987). *The Diagnosis of Mineral Disorders in Plants*. Vol. 3 Glasshouse Crops. HMSO.

Archer, J. (1988). *Crop Nutrition and Fertilizer Uses*. 2nd edn. Farming Press.

Brown, L.V. (2004). *Applied Principles of Horticultural Science*. 3rd edn. Butterworth-Heinemann.

Cresser, M.S. *et al.* (1993). *Soil Chemistry and Its Applications*. CUP.

DEFRA (2003). *Fertilizer Recommendations*. HMSO.

Hay, R.K.M. (1981). *Chemistry for Agriculture and Ecology*. Blackwell Scientific Publications.

Haylin, J.L. *et al.* (2004). *Soil Fertility and Fertilizers: An Introduction to Nutrient Management*. Prentice Hall.

Ingram, D.S. *et al.* (eds) (2002). *Science and the Garden*. Blackwell Science Ltd.

Marschner, H. (1995). *Mineral Nutrition in Higher Plants*. 2nd edn. Academic Press.

Postgate, J. (1998). *Nitrogen Fixation*. 3rd edn. CUP.

Roorda von Eysinga, J.P.N.L. and Smilde, K.W. (1980). *Nutritional Disorders in Chrysanthemums*. Centre for Agricultural Publishing and Documentation, Wageningen.

Roorda von Eysinga, J.P.N.L. and Smilde, K.W. (1981). *Nutritional Disorders in Glasshouse Tomatoes, Cucumbers and Lettuce*. Centre for Agricultural Publishing and Documentation, Wageningen.

Simpson, K. (1986). *Fertilizers and Manures*. Longman Handbooks in Agriculture.

Chapter 22   Alternatives to growing in the soil

Summary

This chapter includes the following topics:

- **Alternative growing media**
- **Air-filled porosity, stability and water holding**
- **Compost ingredients**
- **Peat alternatives**
- **Compost mixes**

with additional information on the following:

- **Compost mixing**
- **Plant containers**
- **Hydroculture**

Figure 22.1 **Hanging baskets** are both popular for the garden, but are also used in public areas by Local Authorities and businesses wishing to make a good impression

Alternative growing media

In addition to open ground or greenhouse borders, plants may be grown in pots, troughs, bags and other containers where restricted rooting makes more critical demands on the growing medium for air, water and nutrients. Soil is an inappropriate material to use in containers as it tends to collapse when kept wet; try watering a pot full of soil and note that it is not long before the container is only half full of soil. Soil is replaced in this situation by alternative growing media generally called **composts**. These materials are also called plant substrates, plant growing media, or just 'mixes' or 'media'.

Compost ingredients need to ensure adequate air space after wetting, with a stability to withstand prolonged watering without collapse. The need for the material to have good water-holding capacity depends on the irrigation system to be used. The nutrient content of the soil alternative needs to be allowed for and it is often advantageous to use one that has none as they can be added more precisely. The material should also be 'partially sterile' (free from pest and diseases) and free from toxics. Increasingly, in intensive production, the preferred alternative to growing in the soil is to use hydroponics (see p394). The weaknesses of soil for sportsground construction leads to its replacement with alternatives, e.g. graded sand on golf greens (see p397–8).

Air-filled porosity (AFP)

Air-filled porosity is the proportion of the volume of a growing medium which contains air, after it has been saturated and allowed to drain.

The importance of supplying water to plants in a restricted root volume is usually understood, but the difficulties associated with achieving it whilst maintaining adequate **air-filled porosity (AFP)** are less well appreciated. Roots require oxygen to maintain growth and activity. As temperatures rise the plant requires more, but the amount of oxygen that is dissolved in water decreases. Even in cool conditions, the oxygen that can be extracted from the water provides only a fraction of the roots requirements. So, unless the plants have special modifications to transport oxygen through their tissues, as in aquatic plants, there has to be good gaseous movement through the growing medium. Many large interconnected pores allow rapid entry of oxygen (see soil structure). Creating successful physical conditions depends on the use of components that provide a high proportion of macropores.

It is generally considered that 10–15 per cent AFP is needed for a wide range of plants. Azaleas and epiphytic orchids require 20 per cent or more, whereas others, including chrysanthemums, lilies and poinsettia, tolerate 5–10 per cent AFP and carnations, conifers, geraniums, ivies and roses can be grown at levels as low as 2 per cent.

Ensuring that a growing medium in a container has adequate air-filled porosity is made difficult because water does not readily leave the container unless it is in good contact through its holes with similar-sized pore spaces as when placed on sand or capillary matting. However, when standing out on gravel or wire the water will cling to the particles in the container (see surface tension p338). This can be tested by fully watering a pot of compost, holding it until it has finished dripping then touching the compost through a hole; normally a stream of water will run down your finger. Furthermore, unless stood out on appropriate material, the lower layers of the compost remain saturated (i.e. no air) irrespective of the height or width of the container. This makes it particularly difficult to get good aeration in shallow trays, modules and blocks. (You may understand this better if you fully wet a washing sponge and leave it to drain. After water has left the sponge under the influence of gravity, the lower layers remain saturated.)

Stability

Very large quantities of water have to be applied to composts over the course of a season, so the materials chosen must have very good **stability**. Fine sand and silt soils collapse too quickly and reduce the size of the pore spaces. Even clay crumbs, unless reinforced with high humus content, collapse quite quickly. The sizes of the components used must be selected carefully to ensure that they create macropores, but also so that the gaps between the larger particles are not subsequently filled in by smaller particles ('fines'). This is most easily achieved by using closely graded coarse particles. The reverse is achieved when combining many different-sized particles, as one would in mixing concrete, where the object is to minimize the air spaces as shown in Figure 22.2.

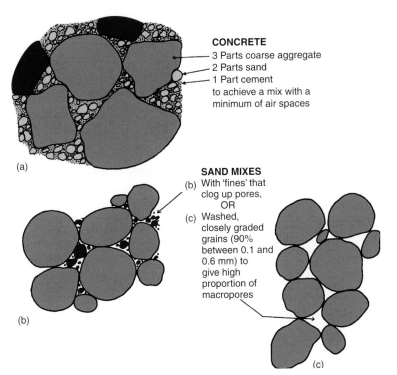

CONCRETE
3 Parts coarse aggregate
2 Parts sand
1 Part cement
to achieve a mix with a minimum of air spaces

(a)

SAND MIXES
(b) With 'fines' that clog up pores, OR
(c) Washed, closely graded grains (90% between 0.1 and 0.6 mm) to give high proportion of macropores

(b)

(c)

Figure 22.2 **Pore spaces** in (a) concrete mix (b) and (c) sand mixes

Water-holding capacity

The water-holding capacity of compost ingredients varies enormously. Peat is significantly better than most others. However, the importance of this depends on how the plants in the compost are to be irrigated. It is a major consideration if the plants are in small hanging baskets watered by hand and there are benefits in using absorbent polymers (see p390) that improve water-holding more than peat alone. Peat presents a problem if it dries out, because it does not rewet easily unless treated with wetters (see p390). On the other hand, if there is to be a constant supply of water through one of the many self-watering systems (see p350), this water-holding capacity is far less significant and the emphasis should be on choosing material that is stable and provides the right air-filled porosity.

Compost ingredients

Over the years growers have added a wide variety of materials, such as leaf mould, pine needles, spent hops, old mortar, crushed bricks, composted animal and plant residues, peat, sand and grit, to selected soils to produce a compost with suitable physical properties. To supplement the nutrient released from the materials in the compost, if any, various slow release organic manures or small dressings of powdered soluble inorganic fertilizers have been added to the mixtures to provide the necessary nutrition.

The correct physical and nutritional conditions are vital to successful growing in a restricted rooting volume. Significant developments occurred as a result of the work done in the 1930s at the John Innes Institute, where the importance of 'sterile' (pest and disease free), stable and uniform ingredients was demonstrated. The range of composts that resulted from this work established the methods of achieving uniform production and reliable results with a single potting mixture suitable for a wide range of plant species.

Loam composts

Loam composts, typified by John Innes composts, are based on loam sterilized to eliminate the soil-borne fungi (see damping off) and insects that largely caused the unreliable results from traditional composts. There is a risk of ammonia toxicity developing after sterilization of soil with pH greater than 6.5 or very high in organic matter (see nitrogen cycle). Induced nutrient deficiencies are possible in soils with a pH greater than 6.5 or less than 5.5. Furthermore, loam should have sufficient clay and organic matter present to give good structural stability (the original specification identifies 'turfy clay loam'). Peat and sand are added to further improve the physical conditions: the peat gives a high water-holding capacity and the coarse sand ensures free drainage and therefore good aeration. There are two main John Innes composts, one for seed sowing and cuttings, the other for potting.

John Innes seed compost consists of 2 parts loam, 1 part peat and 1 part sand by volume. Well-drained turfy clay loam low in nutrients and with a pH between 5.8 and 6.5; undecomposed peat graded 3 mm to 10 mm with a pH between 3.5 and 5.0; and lime-free sand graded 1–3 mm should be used. 1200 g of superphosphate and 600 g of calcium carbonate (lime) are added to each cubic metre of compost.

John Innes potting (JIP) composts consist of 7 parts by volume turfy clay loam, 3 parts peat and 2 parts sand. To allow for the changing nutritional requirements of a growing plant, the nutrient level is adjusted by adding appropriate quantities of JI base fertilizer which consists of 2 parts by volume hoof and horn, 2 parts superphosphate and 1 part potassium sulphate. To prepare JIP 1, 3 kg JI Base fertilizer and 600 g of calcium carbonate are added to one cubic metre of compost. To prepare JIP 2 and JIP 3, double and treble fertilizer levels respectively are used.

Whilst the standard JI composts are suitable for a wide range of species, some modification is required for some specialized plants. For example, calcifuge plants such as *Ericas* and *Rhododendrons* should be grown in a JI(S) mix in which sulphur is used instead of calcium carbonate. All loam-based composts should be made up from components of known characteristics and according to the specification given. Such composts are well proven and are relatively easy to manage because of the water-absorbing and nutrient-retention properties of the clay present.

These composts are commonly used by amateurs, for valuable specimens, and for tall plants where pot stability is important; but loam-based composts have been superseded in horticulture generally by cheaper alternatives. The main disadvantage of loam-based composts has always been the difficulty of obtaining suitable quality loam ('turfy clay loam'), as well as the high costs associated with steam sterilizing. Furthermore, the loam must be stored dry before use and the composts are heavy and difficult to handle in large quantities. Many of the loam-based composts currently produced have relatively low loam content and consequently exhibit few of its advantages.

Loamless or soilless composts

Loamless composts introduced the advantages of a uniform growing medium, but with components that are lighter, cleaner to handle, cheaper to prepare and which do not need to be sterilized (unless being used more than once). Many have low nutrient levels which enable growers to manipulate plant growth more precisely through nutrition, but the control of nutrients is more critical, as many components have a low **buffering capacity**. **Peat** has until recently been the basis of most loamless composts. It is used alone or in combination with materials, such as sand, to produce the required rooting environment. Peats are derived from partially decomposed plants and their characteristics depend on the plant species and the conditions in which they are formed (see Chapter 18). Peats vary and respond differently to herbicides, growth regulators and lime. All peats have a high cation exchange capacity, which gives them some buffering capacity. The less decomposed sphagnum peats have a desirable open structure for making composts and all peats have high water-holding capacities.

Alternatives to peat

Whilst peat remains a popular choice as a compost ingredient, great efforts are being made to find alternatives in order to preserve the wetland habitats where peat is harvested. A list of some of the materials used is given in Table 22.1. Much progress has been made by using suitably processed bark or coconut fibre in composts. Along with several other organic sources they are waste-based and recycling them helps in conserving resources. All such alternatives must be free of toxics and pathogens. Several inorganic materials, such as sand and grit, have always been used in composts, but there is now a wider choice available. Most of the inorganic alternatives are made from non-renewable

Table 22.1 Alternatives to peat

Organic materials	Inorganic materials
Pine	Expanded aggregates
Coir	Extracted minerals
Garden compost	Hydroponics
Heather/bracken	Perlite
Leaf-mould	Polystyrene
Lignite	Rockwool
Recycled landfill	Dredgings/warp
Refuse-driven humus	Vermiculite
Seaweed	Topsoil
Sewage sludge	
Spent hops and grains	
Spent mushroom compost	
Straw	
Vermicomposts	
Wood chips	
Woodwastes	
Wood fibre	

resources (sand, loam, pumice) or consume energy in their manufacture (plastic foams, polystyrene) or both (vermiculite, perlite, rockwool).

Where possible an environmentally friendly alternative is used, but there is considerable debate about the relative merits of some of those being used because of the associated energy use in their manufacture or transport. However, peat is being replaced successfully by different substitutes, many only available locally, according to the needs of the various sectors of the industry.

Sand, grit or gravel is used in composts, frequently in combination with peat. They have no effect on the nutrient properties of composts except by diluting other materials. They are used to change physical properties. As sand or gravel is added to lightweight materials the density of the compost can be increased, which is important for ballast when tall plants are grown in plastic pots. Sand is also used as an inert medium in aggregate culture. Sand should be introduced with caution because it tends to reduce the air-filled porosity (AFP) of the final mix. It is important that the sands used should have low lime levels; otherwise they may induce a high pH and associated mineral deficiencies (see trace elements).

Pulverized bark has been used as a mulch and soil conditioner for many years. More recently it has been tried in compost mixtures as a replacement for peat. There are many different types of bark and they have different properties. Its problems include the presence of toxics, overcome by composting, and a tendency to 'lock-up' nitrogen (see carbon to nitrogen ratio), which can be offset by extra nitrogen in the feed. When composted with sewage sludge, a material suitable as a plant-growing medium is produced. It is increasingly being incorporated into growing mixes in the attempt to reduce the use of peat.

However, the great variation of barks, especially when they are from a mixed source, makes it difficult to incorporate into growing mixes. Much of the conifer bark tends to be stringy. Consequently the main role of bark is in mulching. The import of bark is strictly controlled by the Forestry Commission to prevent the introduction of pests and diseases. Wood-fibres based on stabilized shredded wood are being used to increase the air-filled porosity of mixes, but they tend to be dusty and not easily dispersed in compost mixes. Sawdust and off-cuts from the chipboard industry are also being tested for use in growing, but there are problems associated with their stability and fungal growth in the freshly stored material.

Coconut wastes such as **coir** (the dust particles) are proving to be useful in growing mixes. The material has good water-holding capacity, rewetting and air-filled porosity characteristics. It has a pH between 5 and 6, which makes it suitable for a wide range of plants, but it cannot replace peat directly in mixes for calcifuges. It has a carbon:nitrogen ratio of 80:1 which means that allowance has to be made for its tendency to 'lock-up' nitrogen (see p325).

Perlite is a mineral that is crushed and then expanded by heat to produce a white, lightweight aggregate (see Figure 22.3). The granules are porous and the rough surface holds considerably more water than gravel or polystyrene balls. It tends to be used to improve aeration of growing media generally and for the rewetting of peat. It is devoid of nutrients and has no cation exchange capacity. Graded samples may be used in aggregate culture, but it tends to be used to add to mixes to improve the uptake of water.

Vermiculite is a mica-like mineral expanded to twenty times its original size by rapid conversion to steam of its water content. The finished product is available in several grades, all of which produce growing media with good aeration and water-holding properties (see Figure 22.3). There is a tendency for the honeycomb structure to break down and go 'soggy'. Consequently, for long-term planting, it tends to be used in mixtures with the more stable peat or perlite. Some vermiculites are alkaline, but the slightly acid samples are preferred in horticulture. Vermiculite has a high cation exchange capacity, which makes it particularly useful for propagation mixes. Most samples contain some available potassium and magnesium.

Rockwool is an insulation material derived from a granite-like rock crushed, melted, and spun into threads. The resulting slabs of lightweight, spongy, absorbent, inert and sterile rockwool provide ideal rooting conditions with high water-holding capacity and good aeration. Shredded rockwool can be used in compost mixes (see Figure 22.3). Its pH is high but is easily reduced by watering with a slightly acid nutrient solution. It is frequently used in tomato and cucumber production and film-wrapped cubes are available for plant raising and pot plants. It is necessary to use a complete nutrient feed (see aggregate culture). It has some buffering capacity, but this is very low on a volume basis. The main problem areas lie in calcium and phosphorus supply and the control of pH and salt concentration. Some rockwool has been formulated with clay to overcome some of these problems. This increases its cation exchange properties, making it very suitable for interior landscaping. Rockwool is also available in water-absorbent and water-repelling forms. Mixtures of these enable formulators to achieve the right balance between air-filled porosity, water-holding and capillary lift. Rockwool is available as granules that provide a flexible alternative for those who produce their own mixes. However, it is most usually supplied as wrapped slabs,

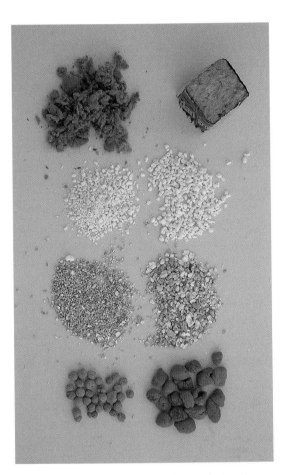

Figure 22.3 **Growing media**. Top to bottom: rockwool, perlite, vermiculite, expanded clay aggregates

cubes, propagation blocks and plugs that are modularized to create a complete growing system.

Pumice is a porous volcanic rock that is prepared for use as a growing medium by crushing, washing (to remove salt and 'fines') and grading. It is most commonly used to grow long-term crops such as carnations in troughs or polysacks.

Expanded polystyrene balls or flakes provide a very lightweight inert material, which can be added to soils or composts as a physical conditioner. It is non-porous and so reduces the water-holding capacity of the growing medium while increasing its aeration, thus making it less liable to waterlogging when over-watered. This has made it an attractive option for winter propagation mixes. However, it is less popular than it might be because it is easily blown away and sticks to most surfaces.

Plastic foams of several different types are becoming popular for propagation because of their open porous structure. They are available as flakes and balls for addition to composts or as cubes into which the cuttings can be pushed.

Chopped straw has been used with some success. Generally the main types available, wheat and barley, break down too easily and a practicable method of stabilizing them has not yet been found. Stable, friable material has been derived from bean and oil seed rape straws, although care is needed in mixes because of the high potassium levels.

Lignite is very variable soft brown coal formed from compressed vegetation; often found at the base of the larger peat bogs. The dusts, 'fines', have been used as carriers for fertilizers and the more granular material can be used to replace grit in mixes, often bringing an improved water retention.

Absorbent polymers have the ability to hold vast quantities of water that is available to plants. However, this is considerably reduced in practice because water absorption falls as the salt concentration of the water increases and the release patterns appear to be very similar to that of some compost ingredients, such as sphagnum moss peats.

Wetters, or non-phytotoxic detergents, are included in mixes to enable water to wet dry composts. They reduce the surface tension of the water, which improves its penetration of the pores. This speeds up the wetting process and maximizes the water-holding capacity of the materials used. Wetters should be selected with care because the different types need to be matched with the peat in the mix and above all must not be harmful to the plants.

Compost mixes

Materials alone or in combination are prepared and mixed to achieve a rooting environment that is free from pests and disease organisms and has adequate air-filled porosity, easily available water, and suitable bulk density for the plant to be grown. While lightweight mixes are usually advantageous, 'heavier' composts are sometimes formulated to give pot stability for taller specimens. This should not be achieved by compressing the lightweight compost, but by incorporating denser materials such as sand. Quick-growing plants are normally the aim and loosely filling containers with the correct compost formulation, consolidated with a presser board and settling it with applications of water obtain this. Firming with a rammer reduces the total pore space whilst increasing the amount of compost and nutrients in the container. The reduction in air-filled porosity and available water with an increase in soluble salt concentration leads to slower growing, harder plants (see conductivity).

The addition of nutrients must take into account not only the plant requirements, but also the nutrient characteristics of the ingredients used. Most loamless composts require trace element supplements and many, including those based on peat, need the addition of all major nutrients and lime. The Glasshouse Crops Research Institute developed **general purpose potting composts** based on a peat/sand mix (see Table 22.2). They contain different combinations of nutrients and consequently their storage life differs. One of the range of composts has a slow release phosphate, removing the need for this element in a liquid feed (see phosphorus). The **GCRI seed compost** contains equal parts by volume of sphagnum peat and fine, lime-free sand. To each cubic metre of seed compost is added 0.75 kg of superphosphate, 0.4 kg potassium nitrate and 3.0 kg calcium carbonate. Variations on these mixtures are

Table 22.2 GCRI composts

Constituents	Seed composts		Potting composts		
			Urea formaldehyde types*		
			Winter use	Summer use	High P type**
Peat:sand (per cent by volume)	50:50	75:25	75:25	75:25	75:25
Base dressings (kg/m³)					
Ammonium nitrate	Nil	0.4	Nil	Nil	0.2
Urea formaldehyde	Nil	Nil	0.5	1.0	Nil
Magnesium ammonium phosphate	Nil	Nil	Nil	Nil	1.5
Potassium nitrate	0.4	0.75	0.75	0.75	0.4
Superphosphate	0.75	1.5	1.5	1.5	Nil
Ground chalk	3.0	2.25	2.25	2.25	2.25
Ground magnesian limestone	Nil	2.25	2.25	2.25	2.25
Fritted trace elements (WM225)	Nil	0.4	0.4	0.4	0.4

*Composts containing urea formaldehyde should not be stored longer than seven days.

**For longer term crops where there is a risk of phosphorus deficiency and liquid feeding with phosphate is not desired, use commercial magnesium ammonium phosphate. This also contains 11 per cent K_2O.

formulated with alternatives to peat, taking into account their different properties particularly with regard to their particle size, water-holding capacity and final air-filled porosity.

Compost mixing

It is most important when making up the desired compost formulation to achieve a uniform product and, commercially, it must be undertaken with a minimum labour input. The ingredients of the compost must be as near as possible to the specification for the chosen formulation. Materials must not be too moist when mixing because it then becomes impossible to achieve an even distribution of nutrients. There are several designs of **compost mixer**. Continuous mixers are usually employed by specialist compost mixing firms and require careful supervision to ensure a satisfactory product. Batch mixers of the 'concrete mixer' design are produced for a wide range of capacities to cover most nursery needs. Many of the bigger mixers have attachments which aid filling. Emptying equipment is often linked to automatic tray or pot-filling machines.

Ingredients used in loamless composts or growing modules do not normally require partial sterilization unless they are being reused, but **sterilizing equipment** is certainly needed to prepare loams for inclusion in loam-based composts. Where steam is used it is injected through perforated pipes on a base plate and rises through the material being sterilized. In contrast a steam–air mixture injected from the top under an air-proof covering is forced downwards to escape through a permeable base.

Storage of prepared composts should be avoided if possible and should not exceed three weeks if slow release fertilizers are incorporated. If nitrogen sources in the compost are mineralized, **ammonium ions** are produced followed by a steady increase in **nitrates** (see p366). These changes lead to a rise in compost pH followed by a fall. As nitrates increase, the salt concentration rises towards harmful levels (see conductivity). Peat-based composts can become infested during storage by sciarid flies.

Plant containers

The characteristics of the container affect the root environment, as does the standing-out area. There is an enormous range of containers used to meet the many different requirements of growing plants (Figure 12.2). **Clay pots** are porous and water is lost from the walls by evaporation. Consequently clay pots dry out quicker than plastic pots, especially in the winter and, although air does not enter through the walls, this can help improve air-filled porosity. The higher evaporation rate also keeps the clay pots slightly cooler, which can be beneficial in hot conditions. Likewise the contents of white plastic pots can be as much as 4°C lower than in other colours. Pots of white or light green plastic can transmit sufficient light to adversely affect root growth and encourage algal growth.

Biodegradable containers such as those made from paper have become popular because they can be planted directly. Some materials decompose more rapidly than others and there can be a temporary 'lockup' of nitrogen, but most peat containers are now manufactured with added available nitrogen. It is essential that these containers are soaked and surrounding soil is kept moist after planting or the roots fail to escape from the dry wall.

The air to water characteristics of the mixture in the container depend not only on the nature of the contents, but also on the characteristics of the base on which the container stands. If containers are stood out on wire mesh or on stones, relatively little water leaves so the oxygen content remains poor. **It is also important to retain contact between the compost and the standing out material through adequate holes in the base, whether to help drainage or to ensure the uptake of water if irrigated from below.**

Blocks

Blocks are made of a suitable compressed growing medium into which the seed is sown with no container or simply a net of polypropylene. Aeration tends to be poorer than in pots, but the high surface area helps make this a successful means of growing some vegetables on a large scale. One type of block comes in the form of a dry compressed disc that expands quickly on soaking ready to receive the seed in the shallow depression in the top surface. This technique has been replaced in large measure by the use of rockwool blocks, particularly when these are to be grown on in rockwool modules (see Fig. 22.3).

Modules

Increasingly, traditional seedbed, bare-rooted or block transplant techniques have been replaced by raising a wide variety of plants in modules. A module is made by adding a loose growing medium mix to a tray of cells. The cells are variously wedge or pyramid-shaped, so designed to enable a highly mechanized transplanting process to be used. Fine, free-flowing mixes of peat, polystyrene or bark are used to fill the cells, which have large drainage holes and no rim to hold free water. Roots in the wedge-shaped cells are 'air-pruned' as they reach the edge of the cell, which encourages secondary root development. 'Plugs' are mini-modules in which each transplant develops in less than $10\,cm^3$ of growing medium and are used for bedding plants, as well as vegetable production. The rate of establishment is largely determined by the water stress experienced by the transplant. Irrigation of the module or plug is found to be more successful than applying water to the surrounding growing medium.

Figure 22.4 **A hydroponics system**

Hydroponics

Hydroponics (water culture) involves the growing of plants in water. The term often includes the growing of plants in solid rooting medium watered with a complete nutrient solution, which is more accurately called '**aggregate culture**'. Plants can be grown in nutrient solutions with no solid material so long as the roots receive oxygen and suitable anchorage and support is provided. The advantages of hydroponics, compared with soil, in temperate areas includes accurate control of the nutrition of the plant and hence better growth and yield. There is a constant supply of available water to the roots. Evaporation is greatly reduced and loss of water and nutrients through drainage is minimal in recirculating systems. There can be a reduction in labour and growing medium costs and a quicker 'turn round' time between crops in protected culture. The disadvantages include the high initial costs of construction and the controls of the more elaborate automated systems. Active roots require a constant supply of oxygen, but oxygen only moves slowly through water. This can be resolved by pumping air through the water that the plants are grown in, but it is usually achieved on a large scale by growing in thin films of water as created in the nutrient film technique (NFT) or a variation on the very much older aggregate culture methods.

Nutrient film technique (NFT)

This is a method of growing plants in a shallow stream of nutrient solution continuously circulated along plastic troughs or gullies. The method is commercially possible because of the development of relatively cheap non-phytotoxic plastics to form the troughs, pipes and tanks (see Figure 22.5). There is no solid rooting medium and a mat of roots develops in the nutrient solution and in the moist atmosphere above it. Nutrient solution is lifted by a pump to feed the gullies directly or via a header tank. The ideal flow rate through the gullies appears to be 4 litres per minute. The gullies have a flat bottom, often lined with capillary matting to ensure a thin film throughout the trough. They are commonly made of disposable black or white polythene set on a graded soil or on adjustable trays. There must be an even slope with a minimum gradient of 1 in 100; areas of deeper liquid stagnate and adversely affect root growth (see aerobic conditions).

The nutrient solution can be prepared on site from basic ingredients or proprietary mixes. It is essential that allowance be made for the local water quality, particularly with regard to the micro-elements, such as boron or zinc, which can become concentrated to toxic levels in the circulating solution. The nutrient level is monitored with a conductivity

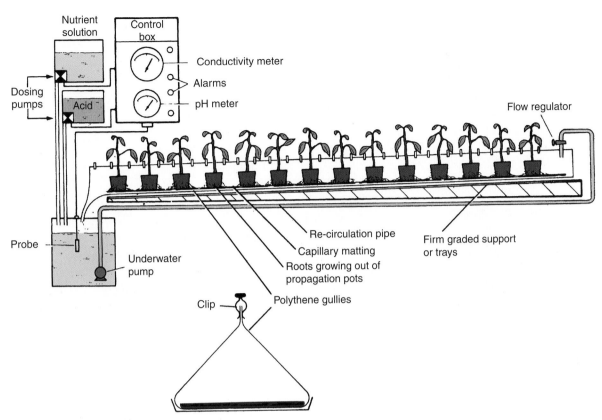

Figure 22.5 **Nutrient film technique layout**. The nutrient solution is pumped up to the top of the gullies. The solution passes down the gullies in a thin film and is returned to the catchment tank. The pH and nutrient levels in the catchment tank are monitored and adjusted as appropriate.

meter and by careful observation of the plants. Maintenance of pH between 6 and 6.5 is also very important. Nutrient and pH control is achieved using, as appropriate, a nutrient mix, nitric acid or phosphoric acid to lower pH and, where water supplies are too acid, potassium hydroxide to raise pH. Great care and safety precautions are necessary when handling the concentrated acids during preparation.

The commercial NFT installations have automatic control equipment in which conductivity and pH meters are linked to dosage pumps. The high and low level points also trigger visual or audible alarms in case of dosage pump failure. Dependence on the equipment may necessitate the grower installing failsafe devices, a second lift pump and a standby generator. A variation on this method is to grow crops such as lettuce in gullies on suitably graded glass house floors (see Figure 22.6).

Aggregate culture

In aggregate culture the nutrient solution is broken up into water films by an essentially inert solid medium, such as coarse sand or gravel. More commonly today materials such as **rockwool, perlite, polyurethane foam, duraplast foam** or **expanded clay aggregates** (see Figure 22.7) are used. These are in the form of polythene wrapped slabs or 'bolsters' of granules sitting on a polythene covered floor graded across the row. Polyurethane slabs are often placed underneath them to help create even slopes and insulate them from the cooler soil below. These growing

Figure 22.6 (a) **NFT lettuce crop** with close up (b) showing gullies and nutrient solution delivery

Figure 22.7 **Tomato crop in rockwool growing system**

containers, on which the plants sit, are drip fed with a complete nutrient solution at the top with the surplus running out through slits near the bottom on the opposite side. When this method was first developed the NFT systems were copied, i.e. the water was recirculated, but it was soon found to be difficult where the quality of water was poor and there was a risk of a build-up of water-borne pathogens and trace elements. It was found that the surplus nutrient solution was most easily managed by allowing it to run to waste into the soil. However, this **open system** presents environmental problems and increasingly a **closed system** has had to be adopted. It is now becoming more usual to run the waste to a storage sump via collection gullies or pipes. Some of this can be used to irrigate outdoor crops if nearby. To recirculate the water it is necessary to have equipment to remove the excess salts or accept a gradual deterioration of the nutrient solution and then flush it out to a sump when it becomes unacceptable. Sources of infection such as *Pythium* are minimized by isolation from soil and using clean water; the risks of recirculating pathogens is addressed by using one of the four main methods of sterilization (see water quality).

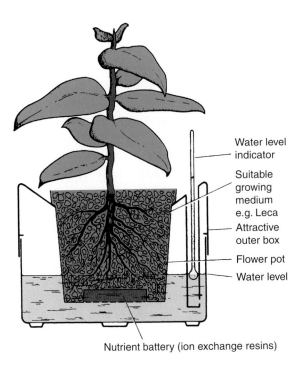

Water level indicator

Suitable growing medium e.g. Leca

Attractive outer box

Flower pot

Water level

Nutrient battery (ion exchange resins)

Figure 22.8 Plant pots with water reserves. Plants grown in a variety of growing media can be fitted with reservoirs that supply water by capillarity. A water level indicator is frequently incorporated and in some systems the nutrients are supplied from ion exchange resins. While this system can be used for any pot size it is particularly attractive in large displays

Rockwool slabs are a very successful way of growing which lend themselves to a modular system. It is widely used for a range of commercial crops, such as tomatoes, cucumbers, peppers, melons, lettuce, carnations, roses, orchids and strawberries, in protected culture. It is not biodegradable so the vast quantity of rockwool now utilized has produced a serious disposal problem. The slabs can be used successfully several times, if sterilized on each occasion, but eventually they lose their structure. Tearing them up and incorporating them in composts or soils can deal with a limited amount, but far more can now be recycled in the production of new slabs.

Several types of **expanded clay aggregates** used in the building industry, such as Leca or Hortag, have been used particularly in interior landscaping (see Figure 22.3). Smooth but porous granules 4–8 mm in diameter, giving a capillary rise of about 100 mm, are used to create an ideal rooting environment with a dry surface which makes it an attractive method of displaying house plants (see Figure 22.8). All forms of aggregate culture require feeding with all essential minerals. Trace element deficiencies occur less frequently when clay aggregates are used. Ion exchange resins are an ideal fertilizer formulation in these circumstances because the nutrients are released slowly, remove harmful chlorides and fluorides from irrigation water, and aid pH control.

Sports surfaces

The specifications for sports playing surfaces are such that turf has increasingly given way to artificial alternatives, typified by the trend toward playing 'lawn' tennis on 'clay' courts. This is partly attributable to maintenance requirements, but at the higher levels of sport it is because the users or the management expect play to continue with a minimum of interference by rainfall. The usual problem is that the soil in which the turf grows does not retain its structure under the pounding it receives from players and machinery, especially when it is in the wet plastic state. Turf is still preferred by many, but to achieve the high standards required it has to be grown in a much modified soil (see also sand slitting) or, increasingly, in an alternative such as sand. The most extreme approach is to grow the turf in pure sand isolated from the soil, sometimes within a plastic membrane. The high cost of these methods is such that it is only used to create small areas such as golf greens.

Normally the existing topsoil is removed from the site and the subsoil is compacted to form a firm base and graded to carry water away to drains. Drainage pipes are laid, above which is placed a drainage layer usually

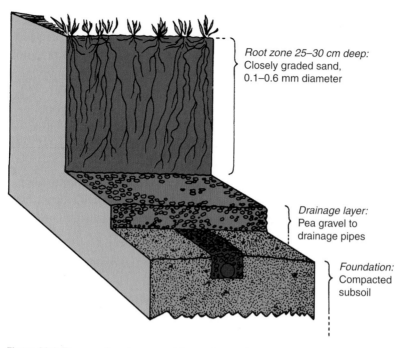

Root zone 25–30 cm deep:
Closely graded sand,
0.1–0.6 mm diameter

Drainage layer:
Pea gravel to
drainage pipes

Foundation:
Compacted
subsoil

Figure 22.9 **Pure sand root zone** used for sportsground surfaces

consisting of washed, pea-sized gravel, as shown in Figure 22.9. Because it is considerably coarser than the sand placed on it, this layer prevents the downward percolation of water (see water films) and creates a perched water table. This helps to give the root zone a large reserve of available water whilst ensuring that gravitational water, following heavy rain or excess irrigation, is removed very rapidly.

A 25–30 cm root zone of free-draining sand is placed uniformly over the drainage layer, evenly consolidated. Allowance has to be made for continued settling over the first year. It is essential that the sand used has a suitable particle size distribution, ideally 80–95 per cent of the particles being between 0.1 and 0.6 mm diameter. A minimum of 'fines' is essential to avoid clogging up of the pores in the root zone (see Figure 22.2). Sometimes a small amount of organic matter is worked into the top 5 cm to help establish the grass, although success is probably as easily achieved with no more than regular light irrigation and liquid feeding.

Some very sophisticated all-sand systems, such as the **cell system**, are constructed so that the root zone is sub-divided into bays with vertical plastic plates and supplied with drains that can be closed so that the water in each of them can be controlled. Tensiometers are used to activate valves that allow water back into the drainage pipes to sub-irrigate the turf.

Check your learning

1. State the main disadvantages of growing in soils.

2. Describe what is required of a material to be used in a compost.

3. State the advantages and the disadvantages of loam based composts.

4. State the advantages of loamless compost.

5. Explain why alternatives are being sought to replace peat in growing plants.

6. State the advantages of hydroculture growing systems.

Further reading

Bragg, N. (1998). *Grower Handbook 1 – Growing Media*. Grower Books.

Bunt, A.C. (1988). *Media and Mixes for Container Grown Plants*. Unwin Hyman.

Cooper, A. (1979). *The ABC of NFT*. Grower Books.

Handreck, K.A. and Black, N.D. (2002). *Growing Media for Ornamentals and Turf*. Revised 3rd edn. New South Wales University Press.

McIntyre, K. and Jakobsen, B. (1998). *Drainage for Sportsturf and Horticulture*. Horticultural Engineering Consultancy.

Mason, J. (1990). *Commercial Hydroponics*. Kangaroo Press.

Molyneux, C.J. (1988). *Practical Guide to NFT*. Nutriculture Ltd.

Pryce, S. (1991). *The Peat Alternatives Manual*. Friends of the Earth.

Smith, D. (1998). *Grower Manual 2 Growing in Rockwool*. Grower Books.

Index